RILEM State-of-the-Art Reports

RILEM STATE-OF-THE-ART REPORTS
Volume 32

RILEM, The International Union of Laboratories and Experts in Construction Materials, Systems and Structures, founded in 1947, is a non-governmental scientific association whose goal is to contribute to progress in the construction sciences, techniques and industries, essentially by means of the communication it fosters between research and practice. RILEM's focus is on construction materials and their use in building and civil engineering structures, covering all phases of the building process from manufacture to use and recycling of materials. More information on RILEM and its previous publications can be found on www.RILEM.net.

The RILEM State-of-the-Art Reports (STAR) are produced by the Technical Committees. They represent one of the most important outputs that RILEM generates – high level scientific and engineering reports that provide cutting edge knowledge in a given field. The work of the TCs is one of RILEM's key functions.

Members of a TC are experts in their field and give their time freely to share their expertise. As a result, the broader scientific community benefits greatly from RILEM's activities.

RILEM's stated objective is to disseminate this information as widely as possible to the scientific community. RILEM therefore considers the STAR reports of its TCs as of highest importance, and encourages their publication whenever possible.

The information in this and similar reports is mostly pre-normative in the sense that it provides the underlying scientific fundamentals on which standards and codes of practice are based. Without such a solid scientific basis, construction practice will be less than efficient or economical.

It is RILEM's hope that this information will be of wide use to the scientific community.

Indexed in SCOPUS, Google Scholar and SpringerLink.

More information about this series at http://www.springer.com/series/8780

Denys Breysse · Jean-Paul Balayssac
Editors

Non-Destructive In Situ Strength Assessment of Concrete

Practical Application of the RILEM TC 249-ISC Recommendations

Editors
Denys Breysse
University Bordeaux
I2M-UMR CNRS 5295
Talence, France

Jean-Paul Balayssac
LMDC
Université de Toulouse, INSA/UPS
Génie Civil
Toulouse, France

ISSN 2213-204X   ISSN 2213-2031   (electronic)
RILEM State-of-the-Art Reports
ISBN 978-3-030-64902-9   ISBN 978-3-030-64900-5   (eBook)
https://doi.org/10.1007/978-3-030-64900-5

This Springer imprint is published by the registered company Springer Nature Switzerland AG
The registered company address is: Gewerbestrasse 11, 6330 Cham, Switzerland

RILEM Members

Preface

This book results from the activities of RILEM TC 249 ISC "In situ Strength Assessment of Concrete" (2012–2020). The authors would like to acknowledge the support of RILEM during this period. They also want to underline the quality of the exchanges between all members of the committee (active and correspondent members) which have contributed to the achievement of this book. This is why this book does not appear as just a series of independent chapters. Each chapter is written by several active authors and is the result of a collective work. Finally, both chairman and deputy chair thank all the authors for their efficient contribution to the iterative reviewing process.

Talence, France
Toulouse, France

Denys Breysse
Jean-Paul Balayssac

RILEM Publications

The following list is presenting the global offer of RILEM Publications, sorted by series. Each publication is available in printed version and/or in online version.

RILEM Proceedings (PRO)

PRO 1: Durability of High Performance Concrete (ISBN: 2-912143-03-9; e-ISBN: 2-351580-12-5; e-ISBN: 2351580125); *Ed. H. Sommer*

PRO 2: Chloride Penetration into Concrete (ISBN: 2-912143-00-04; e-ISBN: 2912143454); *Eds. L.-O. Nilsson and J.-P. Ollivier*

PRO 3: Evaluation and Strengthening of Existing Masonry Structures (ISBN: 2-912143-02-0; e-ISBN: 2351580141); *Eds. L. Binda and C. Modena*

PRO 4: Concrete: From Material to Structure (ISBN: 2-912143-04-7; e-ISBN: 2351580206); *Eds. J.-P. Bournazel and Y. Malier*

PRO 5: The Role of Admixtures in High Performance Concrete (ISBN: 2-912143-05-5; e-ISBN: 2351580214); *Eds. J. G. Cabrera and R. Rivera-Villarreal*

PRO 6: High Performance Fiber Reinforced Cement Composites—HPFRCC 3 (ISBN: 2-912143-06-3; e-ISBN: 2351580222); *Eds. H. W. Reinhardt and A. E. Naaman*

PRO 7: 1st International RILEM Symposium on Self-Compacting Concrete (ISBN: 2-912143-09-8; e-ISBN: 2912143721); *Eds. Å. Skarendahl and Ö. Petersson*

PRO 8: International RILEM Symposium on Timber Engineering (ISBN: 2-912143-10-1; e-ISBN: 2351580230); *Ed. L. Boström*

PRO 9: 2nd International RILEM Symposium on Adhesion between Polymers and Concrete ISAP '99 (ISBN: 2-912143-11-X; e-ISBN: 2351580249); *Eds. Y. Ohama and M. Puterman*

PRO 10: 3rd International RILEM Symposium on Durability of Building and Construction Sealants (ISBN: 2-912143-13-6; e-ISBN: 2351580257); *Ed. A. T. Wolf*

PRO 11: 4th International RILEM Conference on Reflective Cracking in Pavements (ISBN: 2-912143-14-4; e-ISBN: 2351580265); *Eds. A. O. Abd El Halim, D. A. Taylor and El H. H. Mohamed*

PRO 12: International RILEM Workshop on Historic Mortars: Characteristics and Tests (ISBN: 2-912143-15-2; e-ISBN: 2351580273); *Eds. P. Bartos, C. Groot and J. J. Hughes*

PRO 13: 2nd International RILEM Symposium on Hydration and Setting (ISBN: 2-912143-16-0; e-ISBN: 2351580281); *Ed. A. Nonat*

PRO 14: Integrated Life-Cycle Design of Materials and Structures—ILCDES 2000 (ISBN: 951-758-408-3; e-ISBN: 235158029X); (ISSN: 0356-9403); *Ed. S. Sarja*

PRO 15: Fifth RILEM Symposium on Fibre-Reinforced Concretes (FRC)—BEFIB'2000 (ISBN: 2-912143-18-7; e-ISBN: 291214373X); *Eds. P. Rossi and G. Chanvillard*

PRO 16: Life Prediction and Management of Concrete Structures (ISBN: 2-912143-19-5; e-ISBN: 2351580303); *Ed. D. Naus*

PRO 17: Shrinkage of Concrete—Shrinkage 2000 (ISBN: 2-912143-20-9; e-ISBN: 2351580311); *Eds. V. Baroghel-Bouny and P.-C. Aïtcin*

PRO 18: Measurement and Interpretation of the On-Site Corrosion Rate (ISBN: 2-912143-21-7; e-ISBN: 235158032X); *Eds. C. Andrade, C. Alonso, J. Fullea, J. Polimon and J. Rodriguez*

PRO 19: Testing and Modelling the Chloride Ingress into Concrete (ISBN: 2-912143-22-5; e-ISBN: 2351580338); *Eds. C. Andrade and J. Kropp*

PRO 20: 1st International RILEM Workshop on Microbial Impacts on Building Materials (CD 02) (e-ISBN 978-2-35158-013-4); *Ed. M. Ribas Silva*

PRO 21: International RILEM Symposium on Connections between Steel and Concrete (ISBN: 2-912143-25-X; e-ISBN: 2351580346); *Ed. R. Eligehausen*

PRO 22: International RILEM Symposium on Joints in Timber Structures (ISBN: 2-912143-28-4; e-ISBN: 2351580354); *Eds. S. Aicher and H.-W. Reinhardt*

PRO 23: International RILEM Conference on Early Age Cracking in Cementitious Systems (ISBN: 2-912143-29-2; e-ISBN: 2351580362); *Eds. K. Kovler and A. Bentur*

PRO 24: 2nd International RILEM Workshop on Frost Resistance of Concrete (ISBN: 2-912143-30-6; e-ISBN: 2351580370); *Eds. M. J. Setzer, R. Auberg and H.-J. Keck*

PRO 25: International RILEM Workshop on Frost Damage in Concrete (ISBN: 2-912143-31-4; e-ISBN: 2351580389); *Eds. D. J. Janssen, M. J. Setzer and M. B. Snyder*

PRO 26: International RILEM Workshop on On-Site Control and Evaluation of Masonry Structures (ISBN: 2-912143-34-9; e-ISBN: 2351580141); *Eds. L. Binda and R. C. de Vekey*

PRO 27: International RILEM Symposium on Building Joint Sealants (CD03; e-ISBN: 235158015X); *Ed. A. T. Wolf*

PRO 28: 6th International RILEM Symposium on Performance Testing and Evaluation of Bituminous Materials—PTEBM'03 (ISBN: 2-912143-35-7; e-ISBN: 978-2-912143-77-8); *Ed. M. N. Partl*

PRO 29: 2nd International RILEM Workshop on Life Prediction and Ageing Management of Concrete Structures (ISBN: 2-912143-36-5; e-ISBN: 2912143780); *Ed. D. J. Naus*

PRO 30: 4th International RILEM Workshop on High Performance Fiber Reinforced Cement Composites—HPFRCC 4 (ISBN: 2-912143-37-3; e-ISBN: 2912143799); *Eds. A. E. Naaman and H. W. Reinhardt*

PRO 31: International RILEM Workshop on Test and Design Methods for Steel Fibre Reinforced Concrete: Background and Experiences (ISBN: 2-912143-38-1; e-ISBN: 2351580168); *Eds. B. Schnütgen and L. Vandewalle*

PRO 32: International Conference on Advances in Concrete and Structures 2 vol. (ISBN (set): 2-912143-41-1; e-ISBN: 2351580176); *Eds. Ying-shu Yuan, Surendra P. Shah and Heng-lin Lü*

PRO 33: 3rd International Symposium on Self-Compacting Concrete (ISBN: 2-912143-42-X; e-ISBN: 2912143713); *Eds. Ó. Wallevik and I. Níelsson*

PRO 34: International RILEM Conference on Microbial Impact on Building Materials (ISBN: 2-912143-43-8; e-ISBN: 2351580184); *Ed. M. Ribas Silva*

PRO 35: International RILEM TC 186-ISA on Internal Sulfate Attack and Delayed Ettringite Formation (ISBN: 2-912143-44-6; e-ISBN: 2912143802); *Eds. K. Scrivener and J. Skalny*

PRO 36: International RILEM Symposium on Concrete Science and Engineering —A Tribute to Arnon Bentur (ISBN: 2-912143-46-2; e-ISBN: 2912143586); *Eds. K. Kovler, J. Marchand, S. Mindess and J. Weiss*

PRO 37: 5th International RILEM Conference on Cracking in Pavements—Mitigation, Risk Assessment and Prevention (ISBN: 2-912143-47-0; e-ISBN: 2912143764); *Eds. C. Petit, I. Al-Qadi and A. Millien*

PRO 38: 3rd International RILEM Workshop on Testing and Modelling the Chloride Ingress into Concrete (ISBN: 2-912143-48-9; e-ISBN: 2912143578); *Eds. C. Andrade and J. Kropp*

PRO 39: 6th International RILEM Symposium on Fibre-Reinforced Concretes—BEFIB 2004 (ISBN: 2-912143-51-9; e-ISBN: 2912143748); *Eds. M. Di Prisco, R. Felicetti and G. A. Plizzari*

PRO 40: International RILEM Conference on the Use of Recycled Materials in Buildings and Structures (ISBN: 2-912143-52-7; e-ISBN: 2912143756); *Eds. E. Vázquez, Ch. F. Hendriks and G. M. T. Janssen*

PRO 41: RILEM International Symposium on Environment-Conscious Materials and Systems for Sustainable Development (ISBN: 2-912143-55-1; e-ISBN: 2912143640); *Eds. N. Kashino and Y. Ohama*

PRO 42: SCC'2005—China: 1st International Symposium on Design, Performance and Use of Self-Consolidating Concrete (ISBN: 2-912143-61-6; e-ISBN: 2912143624); *Eds. Zhiwu Yu, Caijun Shi, Kamal Henri Khayat and Youjun Xie*

PRO 43: International RILEM Workshop on Bonded Concrete Overlays (e-ISBN: 2-912143-83-7); *Eds. J. L. Granju and J. Silfwerbrand*

PRO 44: 2nd International RILEM Workshop on Microbial Impacts on Building Materials (CD11) (e-ISBN: 2-912143-84-5); *Ed. M. Ribas Silva*

PRO 45: 2nd International Symposium on Nanotechnology in Construction, Bilbao (ISBN: 2-912143-87-X; e-ISBN: 2912143888); *Eds. Peter J. M. Bartos, Yolanda de Miguel and Antonio Porro*

PRO 46: Concrete Life'06—International RILEM-JCI Seminar on Concrete Durability and Service Life Planning: Curing, Crack Control, Performance in Harsh Environments (ISBN: 2-912143-89-6; e-ISBN: 291214390X); *Ed. K. Kovler*

PRO 47: International RILEM Workshop on Performance Based Evaluation and Indicators for Concrete Durability (ISBN: 978-2-912143-95-2; e-ISBN: 9782912143969); *Eds. V. Baroghel-Bouny, C. Andrade, R. Torrent and K. Scrivener*

PRO 48: 1st International RILEM Symposium on Advances in Concrete through Science and Engineering (e-ISBN: 2-912143-92-6); *Eds. J. Weiss, K. Kovler, J. Marchand, and S. Mindess*

PRO 49: International RILEM Workshop on High Performance Fiber Reinforced Cementitious Composites in Structural Applications (ISBN: 2-912143-93-4; e-ISBN: 2912143942); *Eds. G. Fischer and V. C. Li*

PRO 50: 1st International RILEM Symposium on Textile Reinforced Concrete (ISBN: 2-912143-97-7; e-ISBN: 2351580087); *Eds. Josef Hegger, Wolfgang Brameshuber and Norbert Will*

PRO 51: 2nd International Symposium on Advances in Concrete through Science and Engineering (ISBN: 2-35158-003-6; e-ISBN: 2-35158-002-8); *Eds. J. Marchand, B. Bissonnette, R. Gagné, M. Jolin and F. Paradis*

PRO 52: Volume Changes of Hardening Concrete: Testing and Mitigation (ISBN: 2-35158-004-4; e-ISBN: 2-35158-005-2); *Eds. O. M. Jensen, P. Lura and K. Kovler*

PRO 53: High Performance Fiber Reinforced Cement Composites—HPFRCC5 (ISBN: 978-2-35158-046-2; e-ISBN: 978-2-35158-089-9); *Eds. H. W. Reinhardt and A. E. Naaman*

PRO 54: 5th International RILEM Symposium on Self-Compacting Concrete (ISBN: 978-2-35158-047-9; e-ISBN: 978-2-35158-088-2); *Eds. G. De Schutter and V. Boel*

PRO 55: International RILEM Symposium Photocatalysis, Environment and Construction Materials (ISBN: 978-2-35158-056-1; e-ISBN: 978-2-35158-057-8); *Eds. P. Baglioni and L. Cassar*

PRO 56: International RILEM Workshop on Integral Service Life Modelling of Concrete Structures (ISBN 978-2-35158-058-5; e-ISBN: 978-2-35158-090-5); *Eds. R. M. Ferreira, J. Gulikers and C. Andrade*

PRO 57: RILEM Workshop on Performance of cement-based materials in aggressive aqueous environments (e-ISBN: 978-2-35158-059-2); *Ed. N. De Belie*

PRO 58: International RILEM Symposium on Concrete Modelling—CONMOD'08 (ISBN: 978-2-35158-060-8; e-ISBN: 978-2-35158-076-9); *Eds. E. Schlangen and G. De Schutter*

PRO 59: International RILEM Conference on On Site Assessment of Concrete, Masonry and Timber Structures—SACoMaTiS 2008 (ISBN set: 978-2-35158-061-5; e-ISBN: 978-2-35158-075-2); *Eds. L. Binda, M. di Prisco and R. Felicetti*

PRO 60: Seventh RILEM International Symposium on Fibre Reinforced Concrete: Design and Applications—BEFIB 2008 (ISBN: 978-2-35158-064-6; e-ISBN: 978-2-35158-086-8); *Ed. R. Gettu*

PRO 61: 1st International Conference on Microstructure Related Durability of Cementitious Composites 2 vol., (ISBN: 978-2-35158-065-3; e-ISBN: 978-2-35158-084-4); *Eds. W. Sun, K. van Breugel, C. Miao, G. Ye and H. Chen*

PRO 62: NSF/ RILEM Workshop: In-situ Evaluation of Historic Wood and Masonry Structures (e-ISBN: 978-2-35158-068-4); *Eds. B. Kasal, R. Anthony and M. Drdácký*

PRO 63: Concrete in Aggressive Aqueous Environments: Performance, Testing and Modelling, 2 vol., (ISBN: 978-2-35158-071-4; e-ISBN: 978-2-35158-082-0); *Eds. M. G. Alexander and A. Bertron*

PRO 64: Long Term Performance of Cementitious Barriers and Reinforced Concrete in Nuclear Power Plants and Waste Management—NUCPERF 2009 (ISBN: 978-2-35158-072-1; e-ISBN: 978-2-35158-087-5); *Eds. V. L'Hostis, R. Gens and C. Gallé*

PRO 65: Design Performance and Use of Self-consolidating Concrete—SCC'2009 (ISBN: 978-2-35158-073-8; e-ISBN: 978-2-35158-093-6); *Eds. C. Shi, Z. Yu, K. H. Khayat and P. Yan*

PRO 66: 2nd International RILEM Workshop on Concrete Durability and Service Life Planning—ConcreteLife'09 (ISBN: 978-2-35158-074-5; ISBN: 978-2-35158-074-5); *Ed. K. Kovler*

PRO 67: Repairs Mortars for Historic Masonry (e-ISBN: 978-2-35158-083-7); *Ed. C. Groot*

PRO 68: Proceedings of the 3rd International RILEM Symposium on 'Rheology of Cement Suspensions such as Fresh Concrete (ISBN 978-2-35158-091-2; e-ISBN: 978-2-35158-092-9); *Eds. O. H. Wallevik, S. Kubens and S. Oesterheld*

PRO 69: 3rd International PhD Student Workshop on 'Modelling the Durability of Reinforced Concrete (ISBN: 978-2-35158-095-0); *Eds. R. M. Ferreira, J. Gulikers and C. Andrade*

PRO 70: 2nd International Conference on 'Service Life Design for Infrastructure' (ISBN set: 978-2-35158-096-7, e-ISBN: 978-2-35158-097-4); *Eds. K. van Breugel, G. Ye and Y. Yuan*

PRO 71: Advances in Civil Engineering Materials—The 50-year Teaching Anniversary of Prof. Sun Wei' (ISBN: 978-2-35158-098-1; e-ISBN: 978-2-35158-099-8); *Eds. C. Miao, G. Ye and H. Chen*

PRO 72: First International Conference on 'Advances in Chemically-Activated Materials—CAM'2010' (2010), 264 pp., ISBN: 978-2-35158-101-8; e-ISBN: 978-2-35158-115-5; *Eds. Caijun Shi and Xiaodong Shen*

PRO 73: 2nd International Conference on 'Waste Engineering and Management—ICWEM 2010' (2010), 894 pp., ISBN: 978-2-35158-102-5; e-ISBN: 978-2-35158-103-2, *Eds. J. Zh. Xiao, Y. Zhang, M. S. Cheung and R. Chu*

PRO 74: International RILEM Conference on 'Use of Superabsorsorbent Polymers and Other New Addditives in Concrete' (2010) 374 pp., ISBN: 978-2-35158-104-9; e-ISBN: 978-2-35158-105-6; *Eds. O.M. Jensen, M.T. Hasholt, and S. Laustsen*

PRO 75: International Conference on 'Material Science—2nd ICTRC—Textile Reinforced Concrete—Theme 1' (2010) 436 pp., ISBN: 978-2-35158-106-3; e-ISBN: 978-2-35158-107-0; *Ed. W. Brameshuber*

PRO 76: International Conference on 'Material Science—HetMat—Modelling of Heterogeneous Materials—Theme 2' (2010) 255 pp., ISBN: 978-2-35158-108-7; e-ISBN: 978-2-35158-109-4; *Ed. W. Brameshuber*

PRO 77: International Conference on 'Material Science—AdIPoC—Additions Improving Properties of Concrete—Theme 3' (2010) 459 pp., ISBN: 978-2-35158-110-0; e-ISBN: 978-2-35158-111-7; *Ed. W. Brameshuber*

PRO 78: 2nd Historic Mortars Conference and RILEM TC 203-RHM Final Workshop—HMC2010 (2010) 1416 pp., e-ISBN: 978-2-35158-112-4; *Eds. J. Válek, C. Groot and J. J. Hughes*

PRO 79: International RILEM Conference on Advances in Construction Materials Through Science and Engineering (2011) 213 pp., ISBN: 978-2-35158-116-2, e-ISBN: 978-2-35158-117-9; *Eds. Christopher Leung and K.T. Wan*

PRO 80: 2nd International RILEM Conference on Concrete Spalling due to Fire Exposure (2011) 453 pp., ISBN: 978-2-35158-118-6; e-ISBN: 978-2-35158-119-3; *Eds. E.A.B. Koenders and F. Dehn*

PRO 81: 2nd International RILEM Conference on Strain Hardening Cementitious Composites (SHCC2-Rio) (2011) 451 pp., ISBN: 978-2-35158-120-9; e-ISBN: 978-2-35158-121-6; *Eds. R.D. Toledo Filho, F.A. Silva, E.A.B. Koenders and E.M. R. Fairbairn*

PRO 82: 2nd International RILEM Conference on Progress of Recycling in the Built Environment (2011) 507 pp., e-ISBN: 978-2-35158-122-3; *Eds. V.M. John, E. Vazquez, S.C. Angulo and C. Ulsen*

PRO 83: 2nd International Conference on Microstructural-related Durability of Cementitious Composites (2012) 250 pp., ISBN: 978-2-35158-129-2; e-ISBN: 978-2-35158-123-0; *Eds. G. Ye, K. van Breugel, W. Sun and C. Miao*

PRO 84: CONSEC13—Seventh International Conference on Concrete under Severe Conditions—Environment and Loading (2013) 1930 pp., ISBN: 978-2-35158-124-7; e-ISBN: 978-2- 35158-134-6; *Eds. Z.J. Li, W. Sun, C.W. Miao, K. Sakai, O.E. Gjorv and N. Banthia*

PRO 85: RILEM-JCI International Workshop on Crack Control of Mass Concrete and Related issues concerning Early-Age of Concrete Structures—ConCrack 3— Control of Cracking in Concrete Structures 3 (2012) 237 pp., ISBN: 978-2-35158-125-4; e-ISBN: 978-2-35158-126-1; *Eds. F. Toutlemonde and J.-M. Torrenti*

PRO 86: International Symposium on Life Cycle Assessment and Construction (2012) 414 pp., ISBN: 978-2-35158-127-8, e-ISBN: 978-2-35158-128-5; *Eds. A. Ventura and C. de la Roche*

PRO 87: UHPFRC 2013—RILEM-fib-AFGC International Symposium on Ultra-High Performance Fibre-Reinforced Concrete (2013), ISBN: 978-2-35158-130-8, e-ISBN: 978-2-35158-131-5; *Eds. F. Toutlemonde*

PRO 88: 8th RILEM International Symposium on Fibre Reinforced Concrete (2012) 344 pp., ISBN: 978-2-35158-132-2; e-ISBN: 978-2-35158-133-9; *Eds. Joaquim A.O. Barros*

PRO 89: RILEM International workshop on performance-based specification and control of concrete durability (2014) 678 pp., ISBN: 978-2-35158-135-3; e-ISBN: 978-2-35158-136-0; *Eds. D. Bjegović, H. Beushausen and M. Serdar*

PRO 90: 7th RILEM International Conference on Self-Compacting Concrete and of the 1st RILEM International Conference on Rheology and Processing of Construction Materials (2013) 396 pp., ISBN: 978-2-35158-137-7; e-ISBN: 978-2-35158-138-4; *Eds. Nicolas Roussel and Hela Bessaies-Bey*

PRO 91: CONMOD 2014—RILEM International Symposium on Concrete Modelling (2014), ISBN: 978-2-35158-139-1; e-ISBN: 978-2-35158-140-7; *Eds. Kefei Li, Peiyu Yan and Rongwei Yang*

PRO 92: CAM 2014—2nd International Conference on advances in chemically-activated materials (2014) 392 pp., ISBN: 978-2-35158-141-4; e-ISBN: 978-2-35158-142-1; *Eds. Caijun Shi and Xiadong Shen*

PRO 93: SCC 2014—3rd International Symposium on Design, Performance and Use of Self-Consolidating Concrete (2014) 438 pp., ISBN: 978-2-35158-143-8; e-ISBN: 978-2-35158-144-5; *Eds. Caijun Shi, Zhihua Ou and Kamal H. Khayat*

PRO 94 (online version): HPFRCC-7—7th RILEM conference on High performance fiber reinforced cement composites (2015), e-ISBN: 978-2-35158-146-9; *Eds. H.W. Reinhardt, G.J. Parra-Montesinos and H. Garrecht*

PRO 95: International RILEM Conference on Application of superabsorbent polymers and other new admixtures in concrete construction (2014), ISBN: 978-2-35158-147-6; e-ISBN: 978-2-35158-148-3; *Eds. Viktor Mechtcherine and Christof Schroefl*

PRO 96 (online version): XIII DBMC: XIII International Conference on Durability of Building Materials and Components (2015), e-ISBN: 978-2-35158-149-0; *Eds. M. Quattrone and V.M. John*

PRO 97: SHCC3—3rd International RILEM Conference on Strain Hardening Cementitious Composites (2014), ISBN: 978-2-35158-150-6; e-ISBN: 978-2-35158-151-3; *Eds. E. Schlangen, M.G. Sierra Beltran, M. Lukovic and G. Ye*

PRO 98: FERRO-11—11th International Symposium on Ferrocement and 3rd ICTRC—International Conference on Textile Reinforced Concrete (2015), ISBN: 978-2-35158-152-0; e-ISBN: 978-2-35158-153-7; *Ed. W. Brameshuber*

PRO 99 (online version): ICBBM 2015—1st International Conference on Bio-Based Building Materials (2015), e-ISBN: 978-2-35158-154-4; *Eds. S. Amziane and M. Sonebi*

PRO 100: SCC16—RILEM Self-Consolidating Concrete Conference (2016), ISBN: 978-2-35158-156-8; e-ISBN: 978-2-35158-157-5; *Ed. Kamal H. Kayat*

PRO 101 (online version): III Progress of Recycling in the Built Environment (2015), e-ISBN: 978-2-35158-158-2; *Eds I. Martins, C. Ulsen and S. C. Angulo*

PRO 102 (online version): RILEM Conference on Microorganisms-Cementitious Materials Interactions (2016), e-ISBN: 978-2-35158-160-5; *Eds. Alexandra Bertron, Henk Jonkers and Virginie Wiktor*

PRO 103 (online version): ACESC'16—Advances in Civil Engineering and Sustainable Construction (2016), e-ISBN: 978-2-35158-161-2; *Eds. T.Ch. Madhavi, G. Prabhakar, Santhosh Ram and P.M. Rameshwaran*

PRO 104 (online version): SSCS'2015—Numerical Modeling—Strategies for Sustainable Concrete Structures (2015), e-ISBN: 978-2-35158-162-9

PRO 105: 1st International Conference on UHPC Materials and Structures (2016), ISBN: 978-2-35158-164-3; e-ISBN: 978-2-35158-165-0

PRO 106: AFGC-ACI-fib-RILEM International Conference on Ultra-High-Performance Fibre-Reinforced Concrete—UHPFRC 2017 (2017), ISBN: 978-2-35158-166-7; e-ISBN: 978-2-35158-167-4; *Eds. François Toutlemonde and Jacques Resplendino*

PRO 107 (online version): XIV DBMC—14th International Conference on Durability of Building Materials and Components (2017), e-ISBN: 978-2-35158-159-9; *Eds. Geert De Schutter, Nele De Belie, Arnold Janssens and Nathan Van Den Bossche*

PRO 108: MSSCE 2016—Innovation of Teaching in Materials and Structures (2016), ISBN: 978-2-35158-178-0; e-ISBN: 978-2-35158-179-7; *Ed. Per Goltermann*

PRO 109 (2 volumes): MSSCE 2016—Service Life of Cement-Based Materials and Structures (2016), ISBN Vol. 1: 978-2-35158-170-4; Vol. 2: 978-2-35158-171-4; Set Vol. 1&2: 978-2-35158-172-8; e-ISBN : 978-2-35158-173-5; *Eds. Miguel Azenha, Ivan Gabrijel, Dirk Schlicke, Terje Kanstad and Ole Mejlhede Jensen*

PRO 110: MSSCE 2016—Historical Masonry (2016), ISBN: 978-2-35158-178-0; e-ISBN: 978-2-35158-179-7; *Eds. Inge Rörig-Dalgaard and Ioannis Ioannou*

PRO 111: MSSCE 2016—Electrochemistry in Civil Engineering (2016); ISBN: 978-2-35158-176-6; e-ISBN: 978-2-35158-177-3; *Ed. Lisbeth M. Ottosen*

PRO 112: MSSCE 2016—Moisture in Materials and Structures (2016), ISBN: 978-2-35158-178-0; e-ISBN: 978-2-35158-179-7; *Eds. Kurt Kielsgaard Hansen, Carsten Rode and Lars-Olof Nilsson*

PRO 113: MSSCE 2016—Concrete with Supplementary Cementitious Materials (2016), ISBN: 978-2-35158-178-0; e-ISBN: 978-2-35158-179-7; *Eds. Ole Mejlhede Jensen, Konstantin Kovler and Nele De Belie*

PRO 114: MSSCE 2016—Frost Action in Concrete (2016), ISBN: 978-2-35158-182-7; e-ISBN: 978-2-35158-183-4; *Eds. Marianne Tange Hasholt, Katja Fridh and R. Doug Hooton*

PRO 115: MSSCE 2016—Fresh Concrete (2016), ISBN: 978-2-35158-184-1; e-ISBN: 978-2-35158-185-8; *Eds. Lars N. Thrane, Claus Pade, Oldrich Svec and Nicolas Roussel*

PRO 116: BEFIB 2016—9th RILEM International Symposium on Fiber Reinforced Concrete (2016), ISBN: 978-2-35158-187-2; e-ISBN: 978-2-35158-186-5; *Eds. N. Banthia, M. di Prisco and S. Soleimani-Dashtaki*

PRO 117: 3rd International RILEM Conference on Microstructure Related Durability of Cementitious Composites (2016), ISBN: 978-2-35158-188-9; e-ISBN: 978-2-35158-189-6; *Eds. Changwen Miao, Wei Sun, Jiaping Liu, Huisu Chen, Guang Ye and Klaas van Breugel*

PRO 118 (4 volumes): International Conference on Advances in Construction Materials and Systems (2017), ISBN Set: 978-2-35158-190-2; Vol. 1: 978-2-35158-193-3; Vol. 2: 978-2-35158-194-0; Vol. 3: ISBN:978-2-35158-195-7; Vol. 4: ISBN:978-2-35158-196-4; e-ISBN: 978-2-35158-191-9; *Ed. Manu Santhanam*

PRO 119 (online version): ICBBM 2017—Second International RILEM Conference on Bio-based Building Materials, (2017), e-ISBN: 978-2-35158-192-6; *Ed. Sofiane Amziane*

PRO 120 (2 volumes): EAC-02—2nd International RILEM/COST Conference on Early Age Cracking and Serviceability in Cement-based Materials and Structures, (2017), Vol. 1: 978-2-35158-199-5, Vol. 2: 978-2-35158-200-8, Set: 978-2-35158-197-1, e-ISBN: 978-2-35158-198-8; *Eds. Stéphanie Staquet and Dimitrios Aggelis*

PRO 121 (2 volumes): SynerCrete18: Interdisciplinary Approaches for Cementbased Materials and Structural Concrete: Synergizing Expertise and Bridging Scales of Space and Time, (2018), Set: 978-2-35158-202-2, Vol.1: 978-2-35158-211-4, Vol.2: 978-2-35158-212-1, e-ISBN: 978-2-35158-203-9; *Eds. Miguel Azenha, Dirk Schlicke, Farid Benboudjema, Agnieszka Knoppik*

PRO 122: SCC'2018 China—Fourth International Symposium on Design, Performance and Use of Self-Consolidating Concrete, (2018), ISBN: 978-2-35158-204-6, e-ISBN: 978-2-35158-205-3; *Eds. C. Shi, Z. Zhang, K. H. Khayat*

PRO 123: Final Conference of RILEM TC 253-MCI: Microorganisms-Cementitious Materials Interactions (2018), Set: 978-2-35158-207-7, Vol.1: 978-2-35158-209-1, Vol.2: 978-2-35158-210-7, e-ISBN: 978-2-35158-206-0; *Ed. Alexandra Bertron*

PRO 124 (online version): Fourth International Conference Progress of Recycling in the Built Environment (2018), e-ISBN: 978-2-35158-208-4; *Eds. Isabel M. Martins, Carina Ulsen, Yury Villagran*

PRO 125 (online version): SLD4—4th International Conference on Service Life Design for Infrastructures (2018), e-ISBN: 978-2-35158-213-8; *Eds. Guang Ye, Yong Yuan, Claudia Romero Rodriguez, Hongzhi Zhang, Branko Savija*

PRO 126: Workshop on Concrete Modelling and Material Behaviour in honor of Professor Klaas van Breugel (2018), ISBN: 978-2-35158-214-5, e-ISBN: 978-2-35158-215-2; *Ed. Guang Ye*

PRO 127 (online version): CONMOD2018—Symposium on Concrete Modelling (2018), e-ISBN: 978-2-35158-216-9; *Eds. Erik Schlangen, Geert de Schutter, Branko Savija, Hongzhi Zhang, Claudia Romero Rodriguez*

PRO 128: SMSS2019—International Conference on Sustainable Materials, Systems and Structures (2019), ISBN: 978-2-35158-217-6, e-ISBN: 978-2-35158-218-3

PRO 129: 2nd International Conference on UHPC Materials and Structures (UHPC2018-China), ISBN: 978-2-35158-219-0, e-ISBN: 978-2-35158-220-6

PRO 130: 5th Historic Mortars Conference (2019), ISBN: 978-2-35158-221-3, e-ISBN: 978-2-35158-222-0; *Eds. José Ignacio Álvarez, José María Fernández, Íñigo Navarro, Adrián Durán, Rafael Sirera*

PRO 131 (online version): 3rd International Conference on Bio-Based Building Materials (ICBBM2019), e-ISBN: 978-2-35158-229-9; *Eds. Mohammed Sonebi, Sofiane Amziane, Jonathan Page*

PRO 132: IRWRMC'18—International RILEM Workshop on Rheological Measurements of Cement-based Materials (2018), ISBN: 978-2-35158-230-5, e-ISBN: 978-2-35158-231-2; *Eds. Chafika Djelal, Yannick Vanhove*

PRO 133 (online version): CO2STO2019—International Workshop CO2 Storage in Concrete (2019), e-ISBN: 978-2-35158-232-9; *Eds. Assia Djerbi, Othman Omikrine-Metalssi, Teddy Fen-Chong*

RILEM Reports (REP)

Report 19: Considerations for Use in Managing the Aging of Nuclear Power Plant Concrete Structures (ISBN: 2-912143-07-1); *Ed. D. J. Naus*

Report 20: Engineering and Transport Properties of the Interfacial Transition Zone in Cementitious Composites (ISBN: 2-912143-08-X); *Eds. M. G. Alexander, G. Arliguie, G. Ballivy, A. Bentur and J. Marchand*

Report 21: Durability of Building Sealants (ISBN: 2-912143-12-8); *Ed. A. T. Wolf*

Report 22: Sustainable Raw Materials—Construction and Demolition Waste (ISBN: 2-912143-17-9); *Eds. C. F. Hendriks and H. S. Pietersen*

Report 23: Self-Compacting Concrete state-of-the-art report (ISBN: 2-912143-23-3); *Eds. Å. Skarendahl and Ö. Petersson*

Report 24: Workability and Rheology of Fresh Concrete: Compendium of Tests (ISBN: 2-912143-32-2); *Eds. P. J. M. Bartos, M. Sonebi and A. K. Tamimi*

Report 25: Early Age Cracking in Cementitious Systems (ISBN: 2-912143-33-0); *Ed. A. Bentur*

Report 26: Towards Sustainable Roofing (Joint Committee CIB/RILEM) (CD 07) (e-ISBN 978-2-912143-65-5); *Eds. Thomas W. Hutchinson and Keith Roberts*

Report 27: Condition Assessment of Roofs (Joint Committee CIB/RILEM) (CD 08) (e-ISBN 978-2-912143-66-2); *Ed. CIB W 83/RILEM TC166-RMS*

Report 28: Final report of RILEM TC 167-COM 'Characterisation of Old Mortars with Respect to Their Repair (ISBN: 978-2-912143-56-3); *Eds. C. Groot, G. Ashall and J. Hughes*

Report 29: Pavement Performance Prediction and Evaluation (PPPE): Interlaboratory Tests (e-ISBN: 2-912143-68-3); *Eds. M. Partl and H. Piber*

Report 30: Final Report of RILEM TC 198-URM 'Use of Recycled Materials' (ISBN: 2-912143-82-9; e-ISBN: 2-912143-69-1); *Eds. Ch. F. Hendriks, G. M. T. Janssen and E. Vázquez*

Report 31: Final Report of RILEM TC 185-ATC 'Advanced testing of cement-based materials during setting and hardening' (ISBN: 2-912143-81-0; e-ISBN: 2-912143-70-5); *Eds. H. W. Reinhardt and C. U. Grosse*

Report 32: Probabilistic Assessment of Existing Structures. A JCSS publication (ISBN 2-912143-24-1); *Ed. D. Diamantidis*

Report 33: State-of-the-Art Report of RILEM Technical Committee TC 184-IFE 'Industrial Floors' (ISBN 2-35158-006-0); *Ed. P. Seidler*

Report 34: Report of RILEM Technical Committee TC 147-FMB 'Fracture mechanics applications to anchorage and bond' Tension of Reinforced Concrete Prisms—Round Robin Analysis and Tests on Bond (e-ISBN 2-912143-91-8); *Eds. L. Elfgren and K. Noghabai*

Report 35: Final Report of RILEM Technical Committee TC 188-CSC 'Casting of Self Compacting Concrete' (ISBN 2-35158-001-X; e-ISBN: 2-912143-98-5); *Eds. Å. Skarendahl and P. Billberg*

Report 36: State-of-the-Art Report of RILEM Technical Committee TC 201-TRC 'Textile Reinforced Concrete' (ISBN 2-912143-99-3); *Ed. W. Brameshuber*

Report 37: State-of-the-Art Report of RILEM Technical Committee TC 192-ECM 'Environment-conscious construction materials and systems' (ISBN: 978-2-35158-053-0); *Eds. N. Kashino, D. Van Gemert and K. Imamoto*

Report 38: State-of-the-Art Report of RILEM Technical Committee TC 205-DSC 'Durability of Self-Compacting Concrete' (ISBN: 978-2-35158-048-6); *Eds. G. De Schutter and K. Audenaert*

Report 39: Final Report of RILEM Technical Committee TC 187-SOC 'Experimental determination of the stress-crack opening curve for concrete in tension' (ISBN 978-2-35158-049-3); *Ed. J. Planas*

Report 40: State-of-the-Art Report of RILEM Technical Committee TC 189-NEC 'Non-Destructive Evaluation of the Penetrability and Thickness of the Concrete Cover' (ISBN 978-2-35158-054-7); *Eds. R. Torrent and L. Fernández Luco*

Report 41: State-of-the-Art Report of RILEM Technical Committee TC 196-ICC 'Internal Curing of Concrete' (ISBN 978-2-35158-009-7); *Eds. K. Kovler and O. M. Jensen*

Report 42: 'Acoustic Emission and Related Non-destructive Evaluation Techniques for Crack Detection and Damage Evaluation in Concrete'—Final Report of RILEM Technical Committee 212-ACD (e-ISBN: 978-2-35158-100-1); *Ed. M. Ohtsu*

Report 45: Repair Mortars for Historic Masonry—State-of-the-Art Report of RILEM Technical Committee TC 203-RHM (e-ISBN: 978-2-35158-163-6); *Eds. Paul Maurenbrecher and Caspar Groot*

Report 46: Surface delamination of concrete industrial ffioors and other durability related aspects guide—Report of RILEM Technical Committee TC 268-SIF (e-ISBN: 978-2-35158-201-5); *Ed. Valerie Pollet*

Contents

Part I   Theory

1   In-Situ Strength Assessment of Concrete: Detailed Guidelines   3
Denys Breysse, Jean-Paul Balayssac, Maitham Alwash,
Samuele Biondi, Leonardo Chiauzzi, David Corbett, Vincent Garnier,
Arlindo Gonçalves, Michael Grantham, Oguz Gunes, Said Kenaï,
Vincenza Anna Maria Luprano, Angelo Masi, Andrzej Moczko,
Hicham Yousef Qasrawi, Xavier Romão, Zoubir Mehdi Sbartaï,
André Valente Monteiro, and Emilia Vasanelli

2   How to Identify the Recommended Number of Cores? 57
Jean-Paul Balayssac, Denys Breysse, Maitham Alwash,
Vincenza Anna Maria Luprano, Xavier Romão,
and Zoubir Mehdi Sbartaï

**3   Evaluation of Concrete Strength by Combined NDT Techniques:
Practice, Possibilities and Recommendations** 101
Zoubir Mehdi Sbartaï, Vincenza Anna Maria Luprano,
and Emilia Vasanelli

**4   Identification of Test Regions and Choice
of Conversion Models** 117
Jean-Paul Balayssac, Emilia Vasanelli,
Vincenza Anna Maria Luprano, Said Kenai, Xavier Romão,
Leonardo Chiauzzi, Angelo Masi, and Zoubir Mehdi Sbartaï

5   Identification and Processing of Outliers 161
Xavier Romão and Emilia Vasanelli

Part II Applications

6   **How Investigators Can Assess Concrete Strength with On-site
    Non-destructive Tests and Lessons to Draw from a Benchmark** . . .   183
    Denys Breysse, Jean-Paul Balayssac, Samuele Biondi,
    Adorjan Borosnyoi, Elena Candigliota, Leonardo Chiauzzi,
    Vincent Garnier, Michael Grantham, Oguz Gunes,
    Vincenza Anna Maria Luprano, Angelo Masi, Valerio Pfister,
    Zoubir Mehdi Sbartaï, and Katalin Szilagyi

7   **How Investigators Can Answer More Complex Questions
    About Assess Concrete Strength and Lessons to Draw
    from a Benchmark**   219
    Denys Breysse, Xavier Romão, Arlindo Gonçalves, Maitham Alwash,
    Jean Paul Balayssac, Samuele Biondi, Elena Candigliota,
    Leonardo Chiauzzi, David Corbett, Vincent Garnier,
    Michael Grantham, Oguz Gunes, Vincenza Anna Maria Luprano,
    Angelo Masi, Andrzej Moczko, Valerio Pfister, Katalin Szilagyi,
    André Valente Monteiro, and Emilia Vasanelli

8   **Illustration of the Proposed Methodology Based
    on Synthetic Data**   279
    Denys Breysse, Jean-Paul Balayssac, David Corbett,
    and Xavier Romão

9   **Illustration of the Proposed Methodology Based on a Real
    Case-Study**   303
    Angelo Masi, Denys Breysse, Hicham Yousef Qasrawi,
    David Corbett, Arlindo Gonçalves, Michael Grantham,
    Xavier Romão, and André Valente Monteiro

Part III Appendix

10 **Statistics**   331
    Vincent Garnier, Jean-Paul Balayssac, and Zoubir Mehdi Sbartaï

11 **Model Identification and Calibration**   341
    Zoubir Mehdi Sbartaï, Maitham Alwash, Denys Breysse,
    Arlindo Gonçalves, Michael Grantham, Xavier Romão,
    and Jean-Paul Balayssac

12 **For Those Who Want to Go Further**   359
    Xavier Romão, Denys Breysse, Jean-Paul Balayssac,
    and David Corbett

Index   379

Contributors

Maitham Alwash Department of Civil Engineering, University of Babylon, Babylon, Iraq

Jean-Paul Balayssac LMDC, Université de Toulouse, INSA/UPS Génie Civil, Toulouse, France

Samuele Biondi Engineering and Geology Department, inGeo, Gabriele d'Annunzio, University of Chieti-Pescara, Pescara, Italy

Adorjan Borosnyoi Budapest University of Technology and Economics, Presently Hilti, Budapest, Hungary

Denys Breysse University Bordeaux, I2M-UMR CNRS 5295, Talence, France

Elena Candigliota ENEA - Italian National Agency for New Technologies, Energy and Sustainable Economic Development, Bologna, Italy

Leonardo Chiauzzi School of Engineering, University of Basilicata, Potenza, Italy

David Corbett Proceq SA, Zurich, Switzerland

Vincent Garnier Laboratory of Mechanic and Acoustic, L.M.A - AMU/CNRS/ECM - UMR7031, Aix Marseille University, Marseille, France

Arlindo Gonçalves Laboratório Nacional de Engenharia Civil (LNEC), Lisboa, Portugal

Michael Grantham Consultant - Sandberg LLP, London, UK

Oguz Gunes Instanbul Technical University, Istanbul, Turkey

Said Kenai Civil Engineering Department, University of Blida, Blida, Algeria

Vincenza Anna Maria Luprano ENEA - Italian National Agency for New Technologies, Energy and Sustainable Economic Development, Department for Sustainability - Non Destructive Evaluation Laboratory, Brindisi, Italy

Angelo Masi School of Engineering, University of Basilicata, Potenza, Italy

Andrzej Moczko Wrocław University of Science and Technology, Wrocław, Poland

Valerio Pfister ENEA - Italian National Agency for New Technologies, Department for Sustainability - Non Destructive Evaluation Laboratory, Brindisi, Italy

Hicham Yousef Qasrawi Hashemite University, Zarqa, Jordan

Xavier Romão CONSTRUCT-LESE, Faculty of Engineering, University of Porto, Porto, Portugal

Zoubir Mehdi Sbartaï University Bordeaux, I2M-UMR CNRS 5295, Talence, France

Katalin Szilagyi Budapest University of Technology, Presently Concrete Consultant, Budapest, Hungary

André Valente Monteiro Laboratório Nacional de Engenharia Civil (LNEC), Lisboa, Portugal

Emilia Vasanelli Institute of Heritage Science - National Research Council (ISPC-CNR), Lecce, Italy

Part I
Theory

Chapter 1
In-Situ Strength Assessment of Concrete: Detailed Guidelines

Denys Breysse, Jean-Paul Balayssac, Maitham Alwash, Samuele Biondi,
Leonardo Chiauzzi, David Corbett, Vincent Garnier, Arlindo Gonçalves,
Michael Grantham, Oguz Gunes, Said Kenaï,
Vincenza Anna Maria Luprano, Angelo Masi, Andrzej Moczko,
Hicham Yousef Qasrawi, Xavier Romão, Zoubir Mehdi Sbartaï,
André Valente Monteiro, and Emilia Vasanelli

Abstract Guidelines describe the general process of in-situ compressive strength assessment. This process is divided into three main steps, data collection (using nondestructive testing and destructive testing), model identification and strength assessment. Three estimation quality levels (EQL) are defined depending on the

D. Breysse · Z. M. Sbartaï
University Bordeaux, I2M-UMR CNRS 5295, Talence, France

J.-P. Balayssac (✉)
LMDC, Université de Toulouse, INSA/UPS Génie Civil, Toulouse, France
e-mail: jean-paul.balayssac@insa-toulouse.fr

M. Alwash
Department of Civil Engineering, University of Babylon, Babylon, Iraq

S. Biondi
Engineering and Geology Department, inGeo, Gabriele d'Annunzio, University of Chieti-Pescara, Pescara, Italy

L. Chiauzzi · A. Masi
School of Engineering, University of Basilicata, Potenza, Italy

D. Corbett
Proceq SA, Zurich, Switzerland

V. Garnier
Laboratory of Mechanic and Acoustic, L.M.A - AMU/CNRS/ECM - UMR7031, Aix Marseille University, Marseille, France

A. Gonçalves · A. Valente Monteiro
Laboratório Nacional de Engenharia Civil (LNEC), Lisboa, Portugal

M. Grantham
Consultant - Sandberg LLP, London, UK

O. Gunes
Instanbul Technical University, Istanbul, Turkey

S. Kenaï
Civil Engineering Department, University of Blida, Blida, Algeria

© RILEM 2021
D. Breysse and J.-P. Balayssac (eds.), *Non-Destructive In Situ Strength Assessment of Concrete*, RILEM State-of-the-Art Reports 32,
https://doi.org/10.1007/978-3-030-64900-5_1

targeted accuracy of strength assessment, based on three parameters, mean value of strength and standard deviation of strength on a test region and local value of strength. All the necessary definitions (test location, test reading, test region, test result, …) are given and the different stages of data collection, i.e. planning, NDT methods, cores (dimensions, conservation, location, testing, etc) are described. The identification of the conversion model is detailed and a specific attention is paid to the assessment of test result precision (TRP). For the identification of the model parameters, two options are considered either the development of a specific model or the calibration of a prior model. A specific option is also proposed, namely the bi-objective approach. Finally, the quantification of the errors of model fitting and strength prediction is described. The global methodology is synthetized in a flowchart.

Abbreviations

a, b, c	Conversion model parameters (Sect. 1.5.4.3) (Section where it is used for the first time in the document.)
C	Calibration factor (Sect. 1.5.5.3)
C_M	Calibration factor (multiplying calibration) (Sect.1.5.5.3)
COV	Coefficient of variation (= standard deviation/average value)
COV_{rep}	Coefficient of variation of test results (for test result precision) (Sect. 1.5.5.2)
C_s	Calibration factor (shift calibration) (Sect. 1.5.5.3)
EQL	Estimation quality level (Sect. 1.1)
F	Pull-out force (Sect. 1.3.1.2)
f_c	Concrete compressive strength (Sect. 1.1)
$f_{c,i}$	Individual core strength (Sect. 1.5.5.1)
$f_{c, est}$	Estimated concrete compressive strength (Sect. 1.2.4)
$f_{c, est,cal}$	Calibrated estimated concrete compressive strength (Sect. 1.5.5.3)
$f_{c, est, o}$	Prior estimated concrete compressive strength (Sect. 1.5.5.3)

V. A. M. Luprano
ENEA - Italian National Agency for New Technologies, Energy and Sustainable Economic Development, Department for Sustainability - Non Destructive Evaluation Laboratory, Brindisi, Italy

A. Moczko
Wrocław University of Science and Technology, Wrocław, Poland

H. Y. Qasrawi
Hashemite University, Zarqa, Jordan

X. Romão
CONSTRUCT-LESE, Faculty of Engineering, University of Porto, Porto, Portugal

E. Vasanelli
Institute of Heritage Science - National Research Council (ISPC-CNR), Lecce, Italy

$f_{c, est, o, mean}$	Mean value of prior estimated concrete compressive strength (Sect. 1.5.5.3)
i	Index for test results (both core strength and NDT) (Sect. 1.5.1)
ID	Investigation domain (Sect. 1.3.3.2.1)
J	Index for model parameters (Sect. 1.5.1)
k	Index for NDT method (when several methods are used) (Sect. 1.5.4)
KL	Knowledge level (see Eurocodes) (Sect. 1.1)
M	Conversion model ($f_{c, est} = M (Tr)$) (Sect. 1.5.1)
N_c	Number of cores (Sect. 1.2.4)
NDT	Non destructive test (Sect. 1.2.4)
N_{par}	Number of parameters of the conversion model (Sect. 1.4.2)
N_{read}	Number of readings (repetitions of individual measurement) in order to derive a test result (Sect. 1.5.2.2)
N_{rep}	Number of repetitions of a test in order to derive the test repeatability (Sect. 1.5.2.2)
N_{TL}	Number of test locations (= number of NDT results) (Sect. 1.2.4)
par_j	Value of the j-th model parameter (Sect. 1.5.1)
r^2	Determination coefficient (Sect. 1.5.6.2)
R	Rebound number (test result) (Sect. 1.3.1.1)
RH	Rebound hammer (test or device) (Sect. 1.3.1.1)
RMSE	Root mean square error (Sect. 1.5.6.2)
$RMSE_{fit}$	Model fitting error (Sect. 1.5.6.2)
$RMSE_{pred}$	Model prediction error (Sect. 1.5.6.2)
sd	Standard deviation
sd_{rep}	Standard deviation of test results (for test result precision) (Sect. 1.5.2.2)
TL	Test location (Sect. 1.3.3.2.3)
TR	Test region (Sect. 1.3.3.2.5)
Tr	Test result (Sect.1.3.3.2.4)
Tr_i	Value of the i-th NDT test result (when a single NDT is used) (Sect. 1.5.1)
$Tr_{k,i}$	Value of the i-th test result of the k-th NDT method (when several NDT are combined) (Sect. 1.5.4)
TRP	Test result precision (Sect. 1.5.2.2)
true	Index for a true (or reference) value (Sect. 1.6.1.2)
U	Target precision on strength assessment (depends on EQL) (Sect. 1.1)
UPV	Ultrasonic pulse velocity (test result) (Sect. 1.3.1.1)
α	Risk level corresponding to the target precision on strength assessment (Sect. 1.1)

Preamble

The RILEM TC 249-ISC has published its Recommendations in 2019.[1] Due to the limited number of pages of Recommendations, the RILEM TC 249-ISC members

---

[1]Recommendations of RILEM TC 249-ISC on non-destructive in situ strength assessment of concrete, D. Breysse (chair) and co-authors, Materials and Structures 52(4), 71 (2019). All references

have decided to publish more detailed guidelines. The purpose of these guidelines is to provide additional information, and to make easier the development of practical applications for engineers and experts wanting to address the issue of concrete strength assessment.

The way information is organized in these guidelines does not correspond to that of the Recommendations, with a larger part devoted to explanations. Whereas the recommended concrete assessment process is organized with nine major tasks (see RILEM TC 249-ISC Recommendations, Fig. 4, Tasks T1 to T9), references to these tasks are provided at regular points in this chapter. The example given in the Recommendations has been removed, as several chapters in this book are devoted to the application of the Recommendations to different datasets.

In this chapter, three types of presentation are used:

(a) A normal font is used for most sections, which corresponds to the basic information. **Some important concepts or ideas are highlighted in bold characters**.

(b) A few sections are highlighted by using a grey highlight. They correspond to the most important information.

(c) *Some additional information is delivered in paragraphs written in italics. These paragraphs are not essential regarding the consistency of information and can be skipped if the reader wants to get only basic information.*

1.1 Scope

This document provides guidelines to assess the in-situ compressive strength of concrete f_c. Four different situations[2] requiring the evaluation of compressive strength can be listed:

– Pre-planned conformity of in-situ compressive strength e.g. for precast concrete components
– Disputes over quality of the supplied concrete or the workmanship on site
– Concrete has been declared as non-conforming
– Assessment of existing structures that are to be modified, re-designed or have been damaged

BS 6089:2010 describes a specific methodology for each of these four situations because the objectives are not exactly the same. For the first three situations, newly built structures are addressed while the last one is dedicated to existing structures.

---

to this publication are given with the same format, i.e. "RILEM TC 249-ISC Recommendations" with the number of the figure or section.

[2]Specified in BS 6089:2010.

One difficulty is to determine what existing structures are: at what age can a structure be considered an existing one? Is a damaged structure an existing structure?

The RILEM TC 249-ISC has decided to focus the guidelines on **existing reinforced concrete structures** where both age effects and reinforcing steel bars or meshes have a great influence on NDT results and on the subsequent strength definition. Age effects include possible mechanical or physical damage (cracks, concrete delamination due to reinforcement corrosion, etc.). Another specificity of existing structures, in practice, is linked to a lack of detailed information about the concrete used, and a lack of companion specimens which could be used for comparison.

> This document aims to give guidelines to assess the compressive strength of **existing concrete structures by combining cores and non-destructive measurements**. Both **mean strength at the time of investigation and its scatter** are considered.

Some elements of the guidelines will however probably be common to newly built structures, for which these guidelines may also be of some interest.

The main focus of an investigation should be:

(a) To derive representative strength estimates from the tests (including defining what is meant by representative? average, standard deviation…) at a given scale and
(b) To explain what can be done with these estimates: deriving characteristic strengths, comparing with target mandatory values, derive values for structural assessment (including their uncertainties).[3]

> These guidelines **primarily focus on the processes for estimating compressive strength (average, standard deviation) from measurements performed on site**.

As many codes are already devoted to the second point, these guidelines do not extend to the determination of characteristic strength of concrete.

The common concrete strength assessment process can be subdivided into three main stages: (a) data collection (including nondestructive testing (NDT) measurements and core strength measurements), (b) model identification, (c) model use and concrete strength estimation, as illustrated in Fig. 1.1.

---

[3]Another case of «derived» values is when one is interested in structural response, which depends on the local concrete properties.

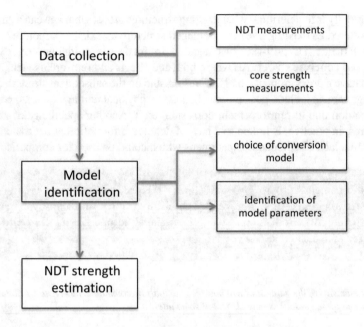

Fig. 1.1 Three main steps of the usual concrete strength assessment process

Limited information and errors at any stage influence this assessment process.[4] However, nothing is usually known about the quality of this process and the resulting precision of the strength estimates.

> **These guidelines are based on the Estimation Quality Level (EQL) concept** that corresponds to a target quality regarding the assessment.

The definition of the relevant EQL is Task 1 in the assessment process according to the RILEM TC 249-ISC Recommendations. Three EQLs are defined which correspond to:

- a target uncertainty that is supposed to be obtained on three quantities: mean strength, standard deviation of strength, local strength. Mean and standard deviation of strength are estimated over the test region.
- the process (including data collection and methods for model identification at the first two steps of Fig. 1.1) used to reach this target.

For each EQL, Table 1.1 indicates the target tolerance interval that is expected to be obtained regarding the evaluation of compressive strength. For each estimated quantity, these numbers define the uncertainty range to which it is supposed to belong,

---

[4]See Chap. 10 for the definition of uncertainty and variability in relation with these guidelines.

Table 1.1 Indicative relation between estimation quality levels (EQL) and target tolerance intervals quantifying the accuracy of strength estimates[a]

Estimated property		EQL1	EQL2	EQL3
Mean value of local strengths		±15%	±15%	±10%
Standard deviation of local strengths[b]	Relative	Not addressed	±50%	±30%
	Absolute		±4 MPa	± 2 MPa
Error on local strength[c]	Relative		20%	15%
	Absolute		6 MPa	4.5 MPa

[a]The values of the recommended number of cores that will be given in the Tables of Chap. 2 are identified for the tolerance intervals of Table 1.1
[b]The two expressions of the tolerance interval are not identical for all concretes, but an absolute error on standard deviation of strength equal to 2 MPa is close to a relative error on standard deviation of 30% for a concrete with $f_{c\ mean} = 30$ MPa and a coefficient of variation (COV) of 20%
[c]The two expressions of the tolerance interval are not identical for all concretes, but absolute errors on local strength equal to 6 MPa and 4.5 MPa are similar to relative errors on local strength of 20% and 15% of the mean strength, respectively, for a concrete with $f_{c\ mean} = 30$ MPa

with a confidence level $(1 - \alpha)$. The α value quantifies the risk level, i.e. the probability that the true (or reference) value will be outside the tolerance interval. The error on local strength (or simply root-mean-square-error (RMSE)) is always positive and the corresponding tolerance interval is one-sided. For standard deviation and error on local strength, the uncertainty range can be expressed either in relative terms or in absolute terms.

> The threeEQLs can be compared with the three different knowledge levels (KL1, KL2 and KL3) defined in Eurocode 8 Recommendations,[5] respectively corresponding to a limited, normal and full knowledge. Each KL corresponds to an appropriate confidence factor (CF) value to be adopted in the evaluation process. The CF deals with the incompleteness of knowledge that is always present in the evaluation of existing structures. To this end, the CF applies strictly to the mean strength obtained from in-situ tests and from the additional sources of information (e.g. original design specifications) to determine the design strength to be used in the calculation of the structural capacity. The recommended values of CF are 1.35, 1.2 and 1 for limited, normal and full KL, respectively. If an estimate of concrete strength is derived according the EQL approach of Table 1.1, the KL definition does not strictly apply and must be revisited.

> Even if not formalized in the Eurocodes, a knowledge level KL is related to the quality of knowledge about the estimated strength, which can be defined as the uncertainty:

- on the estimated local value of compressive strength (i.e. at a given test location),
- on the estimated mean value of compressive strength (i.e. at a given test region of the structure).
- -on the standard deviation of local values of compressive strength.

---

[5]EN 1998-3:2005, Eurocode 8: Design of structures for earthquake resistance—Part 3: Assessment and retrofitting of buildings, European Committee for Standardization (CEN), Brussels, Belgium, 2005 (see also: S. Biondi, The knowledge level in existing building assessment. 14th World Conference on Earthquake Engineering, Beijing, China, 12–17 October 2008).

Table 1.2 Merits of methods for in-situ assessment of compressive strength

Measurement result	Investigation depth	Speed of test	Ease of test	Cost	Damage to the structure	Reliability of absolute in situ strength assessment	Reliability with specific calibration
Core strength	In depth	Slow	Difficult	High	High/moderate	Reference	
Rebound number	Only at the surface	Very fast	Very easy	Very low	Minor	Low	Fair
Direct ultrasonic pulse velocity (Dir_UPV)	In depth	Moderate	Easy	Low	None	Low	Good
Indirect ultrasonicpulse velocity (Ind_UPV)	Near the surface	Fast	Easy	Low	None	Low	Fair
Pull-out force	Near the surface	Moderate	Moderate	Moderate	Moderate	Good	Good
Windsor Probe penetration depth	Near the surface	Fast	Easy	Moderate	Minor	Low	Fair
Micro-core strength	In depth	Slow	Difficult	High	Moderate	Fair	Good

One important contribution of these guidelines concerns the conversion model identification. In the context of existing buildings, a main issue is the correlation between destructive (cores) and non-destructive test results on the elements. The aim is to achieve an in-situ compressive strength estimate with an adequate precision with the lowest number of core test results, optimizing core locations while analyzing the effect of decreasing the number of cores on the precision of the assessment.

Obviously, this approach might appear to consider the failure strength on a core as the "real" strength of the element from which the core is drilled. **The strength measured on cores cannot be considered as a true value but is seen as a reference value.**

It has been widely documented that many reasons can explain the differences between core strength and "real on-site strength", among which the damage due to drilling, the modification in water content or in the state of stress and the conditions of storage of the cores (which may be different from one country to another). In some countries, concrete strength is defined from cubic specimens, which implies additional difficulties while comparing strengths measured in different situations (cores being obviously cylindrical). These issues are not discussed in these guidelines, but the reader is invited to refer to existing literature.[6]

The guidelines do not apply to situations in which no core has been taken from the existing structure, and **are limited to situations where NDT is combined with cores.**

In practice, it is allowed in some countries (e.g. Germany, China) to assess strength from NDT measurement only, without any cores. This may be useful in some very specific context when coring is impossible, for instance for safety reasons or due to regulations. In such cases, the assessment methodology usually leads to very conservative estimates.

Non-destructive tests can be combined with some semi-destructive tests not only in situ but also in the laboratory. The guidelines will consider the possibility to use also semi-destructive techniques (pull-out test for instance) for the assessment of compressive strength.

1.2 Preliminary Considerations

The purpose of the RILEM TC 249-ISC Recommendations is not to address tasks that must be carried out before the beginning of an on-site investigation. Sections 1.2.1–1.2.3 of these guidelines however recall important points to which attention must be paid.

---

[6]Conversion factors between cubes and cylinders are available in the literature, as in BS 1881–120. The Concrete Society has compared core and cube strength for a range of different mixes and types of elements. It illustrated very well the difficulties of comparing core and cube (and of course cylinder strength), see:
https://www.thenbs.com/PublicationIndex/Documents/Details?Pub=CS&DocId=267334.

1.2.1 Context and Objective of the Investigation

The problem has to be well-posed:

- What is the objective of the investigations (diagnosis, refurbishment of the structure or simply the necessity to check the structural state as a routine check-up, etc.)
- What questions have been asked by the engineers or by the structure managers?
- What are the specifications for the investigation?

  - What is the scale of the assessment (component, whole structure, part of the structure)?
  - What assessment uncertainty level is required, i.e. what is the corresponding EQL?

1.2.2 Details of the Structure to Investigate

Before investigating the structure, several questions must be addressed:

- What kind of structure is it?
- What is the age of the structure?
- What components of the structure are important with respect to the structural behavior?
- What is the exposure situation of the different structural components?
- What is the stress state of the components under consideration?
- What is the importance level of the structure (e.g. temporary structure, house, office building, heavy traffic bridge, hospital)?
- What is the structural redundancy of the structure itself?
- What is the expected service life for the structure?
- Are there any durability problems? If so, what are the assessed causes and the extent of damage?
- What useful additional information about the structure is available?

This information could have been obtained:

- during construction:

  - quality of concrete (cement type, cement content, aggregates, curing procedure, fabrication and casting conditions, compressive strength test results obtained on standard specimens...).
  - reinforcement size, type, spacing and cover.
  - any documentation, drawings and technical documents made available.

- after construction completion:

- previous investigations (if available), including their format, the ability to use the results, their reliability.
- if the structure has been used for its designed intended purpose continuously during its lifetime.
- if the structure has been rehabilitated or strengthened.
- the project specifications and/or the regulations in force at the date of the construction.
- any information about carbonation depth or possible damage.

In this regard, it has to be considered that the construction process for a concrete structure is highly dependent on the time and regional variations, i.e. it depends on the construction period, local construction methods and also on cement and aggregate availability.

1.2.3 Constraints of the Investigation

Difficulties that may arise during the test implementation stage must be identified:

- what are the constraints in terms of accessibility to the structure (e.g. components at great heights or other components that cannot be readily accessed such as columns that are founded under water)?
- which faces will be investigated (lateral, bottom or top faces)?
- test influence or test disturbance of non-structural elements (passive elements like non-structural walls, pavements, or equipment, machinery...)
- presence of spalled concrete surfaces that may fall and affect the security of the operators or the public during tests
- is coring allowed? Is there a pre-defined limit on the number of cores that can be taken or constraints regarding their size?
- what is the level of damage of the structure and what is the risk of performing the inspection?
- what is the surface condition of the components and how can it affect the use of NDT?

The list of above items is not exhaustive, and these guidelines do not offer solutions to answer all of them. Several points are covered in existing standards and guidance[7] and it is possible to refer to such documents for an explanation of the difficulties and consequences.

---

[7] See for instance:

- ASTM C823:2017, Standard practice for examination and sampling of hardened concrete in constructions, 2017.
- RILEM, STAR 207-INR, Non-destructive assessment of concrete structures: reliability and limits of single and combined techniques, D. Breysse (chair), Springer, 2012.
- IAEA, Guidebook on non-destructive testing of concrete structures, 2002.

1.2.4   Investigation as Part of a Wider Evaluation Program

The investigation strategy is constrained by the resources available, whether they are time constraints (number of hours/days devoted to the investigation) or budget constraints. Thus, the density of the investigation is a direct outcome of these constraints. It is expected that increasing the number of tests (and data) would naturally improve the precision of the assessment, but there is no simple way to quantify the precision of the assessment as a function of the investigation density.

> *If such a model was available, the correct approach would be to define the "target precision" and to define the resources to mobilize in order to reach it. Since there is no such model, it may seem natural to consider that precision tends to increase with the density of the investigation, and to replace the "target precision" with an "available amount of resources", which comes to replace "mandatory objectives" with "mandatory means".*[8]

In these guidelines, attention is paid not only to strength assessment, but also to the estimation of the precision of this assessment. These guidelines do not pretend to limit the number of investigations in relation to a given amount of resources. At the end of the investigation, if the precision of the evaluation is deemed insufficient, additional investigations remain possible.

> The assessment of concrete strength provides $f_{c, est}$ values. It is based on: (a) available **test results**[9] at test locations, (b) the use of a **conversion model** to obtain estimated strength values at the local (component) scale.

Identifying the conversion model is a part of the assessment process itself. Available information at the beginning of the assessment process is provided by a number N_{TL} of NDT measurements and a number N_c of core strengths. The NDT based assessment requires N_{TL} to be much larger than N_c, in order to deduce additional information about strength from NDT measurements at points where strength has not been assessed directly.

> Any investigation strategy must consider the following items:
> – decision about the **definition of points (i.e. number and location)** where NDT measurements are carried out (see Sect. 1.3.3),
> – decision about the **definition of points** (rules for choice, number, location) where cores are to be taken (see Sect. 1.4). These two first decision steps

---

[8]This is commonly done when, in seismic retrofitting, one defines "knowledge levels" by rough indicators about the number/density of tests for a given building surface or a given number of structural components (see note 5 and S. Biondi, The knowledge level in existing building assessment. 14th World Conference on Earthquake Engineering, Beijing, China, 12–17 October 2008).

[9]See Sect. 3.3.2 for the definitions.

interact because of constraints (cost, time...) on the whole process. An important option is the selection of core locations, which can be based on the NDT results,
- identification of the **conversion model** (see Sect. 1.5),
- **use of the conversion model** to estimate strength from NDT measurements (see Sect. 1.6),
- evaluation of the **precision of estimation**.

Regarding the conversion model, it is necessary to differentiate between (a) the **fitting error**, which measures how efficient the model is in describing the relationship between NDT results and strength at test locations where both are available and (b) the **prediction error** which measures how accurate and precise the conversion model is when it is used for estimating strength at other test locations, where only NDT measurements are available. This point is detailed at Sect. 4.3.

Additional steps in the assessment process can be the assessment of derived strength values (like average or characteristic strength, identification of weakest areas, etc.), the derivation of additional material parameters (moisture, carbonation...), and the evaluation of the precision of these additional estimates.

These guidelines apply whatever the type of NDT measurements. They apply for most common techniques (UPV, rebound hammer, pull-out) but also for less common techniques.

1.3 Planning of the Investigation

1.3.1 Methods

1.3.1.1 Description of the Methods

This section briefly describes the non-destructive methods which can be used for assessing the compressive strength of concrete (RILEM TC 249-ISC Recommendations, Task 2).

For a more detailed description, the reader is invited to refer to other guidelines or reference books which provide a more detailed description of the techniques and their application field.[10]

---

[10]See for instance:

Many methods can be proposed for assessing the compressive strength of concrete. The most common test methods are:

– core testing: compressive strength is measured in the laboratory on cylindrical cores taken from the structure,
– ultrasonic pulse velocity (UPV): this method implies the propagation of an ultrasonic wave (with a frequency generally about 50 kHz) through the material under test. Waves can be transmitted through the thickness of the component (direct method), or along the same face (indirect) or between two adjacent faces (semi-direct method) if two opposite faces are not available,
– rebound hammer (RH): this method consists of measuring the rebound number (R) of a hammer mass striking a steel plunger in contact with the concrete surface with a known force,[11]
– pull-out test (Capo-test): this method consists of measuring the load F necessary to extract a steel disc inserted into a hole drilled with an under-reaming drill bit into the concrete,[12]
– micro-core testing: the compressive strength can be measured in the laboratory on micro-cores (cores with a diameter smaller than 50 mm),
– penetration resistance (pin test or Windsor probe) whose principle consists of measuring the depth of penetration L into the concrete of a probe fired on the concrete surface with a standardized powder cartridge. The penetration depth of the probe is inversely proportional to the compressive strength.[13]

Measurement on cores is considered to provide the reference strength values, which can be used for comparison or calibration. All other methods provide a test result from which a strength value can be derived only through a conversion model.

– RILEM, STAR 207-INR, Non-destructive assessment of concrete structures: reliability and limits of single and combined techniques, D. Breysse (chair), Springer, 2012.
– J.H. Bungey, S.G. Millard, M.G. Grantham, *Testing of Concrete in Structures* (CRC Press, Boca Raton, Florida, 2018).
– C. Maierhofer, H.-W. Reinhardt, G. Dobmann, *Non-Destructive Evaluation of Reinforced Concrete Structures* (Woodhead Publishing, 2010).
– V.M. Malhotra, N.J. Carino (eds.), *Handbook of Nondestructive Testing of Concrete* (CRC Press, Boca Raton, Florida, 2003).

[11]Based on the same type of device, the rebound energy can be measured and a Q value derived.

[12]The principle of the Lok test is similar but uses a cast-in test piece when the fresh concrete is placed. It is thus not commonly used in existing structures. See for instance: A.T. Moczko, N.J. Carino, C.G. Petersen, CAPO-TEST to estimate concrete strength in bridges. ACI Mater J 113(6), 827–836 (2016).

[13]The test is standardized (ASTM C803:2010, Standard test method for penetration resistance of hardened concrete, 2010), and the concrete strength can be derived after the processing of test results.

Some authors have proposed measuring the concrete drilling resistance.[14]*The principle consists of measuring the drilling energy in relation to the depth of the hole. The energy is related to the torque measurement of the drilling machine. Even if having given promising results in the case of fire damage concrete, this technique is not considered as common practice.*

Both the number and type of chosen methods will determine the plan of investigation. Some Recommendations propose to combine two methods, the most famous being the combined use of UPV and RH measurements, which had been promoted by RILEM in the 1980s (the well-known "SonReb method" for instance[15]).

A combination of methods may be a good way to improve the precision of assessment. If combination is a perspective, the expected added-value of combination must be analyzed. Moreover, there is not a unique way of combining methods. Chap. 3 in this book provides further information about these possible ways. The combination (number and type of techniques) will depend on the structure's condition (type, age, moisture variations, level and nature of damage), on the required estimation quality level and of course on economic considerations. Its efficiency will also depend on the precision of nondestructive test results.

1.3.1.2   Selection of the Methods

Table 1.2 summarizes the merits and limitations of the proposed methods for concrete structures.[16] The core strength provides the reference value. The last two columns indicate what is the reliability of concrete strength assessment with NDT methods, either absolute (i.e. directly from NDT results) or after calibration against cores.

The choice of the methods must be also analyzed regarding:

---

[14]R. Felicetti, The drilling resistance test for the assessment of fire damaged concrete. Cem Concr Compos 28(4), 321–329 (2006).

[15]The combination of rebound hammer and velocity measurements had been initially promoted by RILEM (RILEM TC 43-CND, I. Facaoaru (chair), Draft recommendation for in situ concrete strength determination by combined non-destructive methods. Mater Struct 26, 43–49 (1993). An extensive review has been published which explains what are the main factors governing the efficiency of combination (D. Breysse, Nondestructive evaluation of concrete strength: an historical review and a new perspective by combining NDT methods. Constr Build Mater 33, 139–163, 2012).

[16]This table is based very closely on that provided by the British Standard 6089:2010, a complementary guidance document to BS EN 13791 [BS 6089:2010, Assessment of in-situ compressive strength in structures and precast concrete components - Complementary guidance to that given in BS EN 13791, British Standard, 2010] and to that published by Soutsos et al. [in RILEM, STAR 207-INR, *Non-Destructive Aassessment of Concrete Structures: Reliability and Limits of Single and Combined Techniques*, D. Breysse (chair), Springer, 2012]. Some differences may however appear depending on the context of measurements (see f.i. G. Pascale, A. Di Leo, R. Carli, V. Bonora, Evaluation of actual compressive strength of high strength concrete by NDT. 15th World Conference on NDT, Rome, Italy, 15–21 October 2000, and M.D. Machado, L.C.D. Shehata, I.A.E.M. Shehata, Correlation curves to characterize concretes used in Rio de Janeiro by means of non-destructive tests. Rev IBRACON Estrut Mater 2(2), 100–123, 2009).

- location of the position in the structure where the tests must be performed and the ease of transporting the equipment
- accessibility of the testing locations
- surface conditions of concrete (smoothness, presence of coatings, dirt, organic deposits, shells, etc.)
- damage caused by the testing. If the structure cannot be damaged by the tests (like a nuclear power plants for example), or if steel reinforcement is congested, it will be very difficult to make pull-out tests or cores.
- duration of the test.
- cost of the test.
- qualification of the operators.

1.3.2   Preliminary Recommendations Before Performing the Tests

1.3.2.1   Reinforcement

> Prior to any investigation, the position of the reinforcement must be determined. It is strongly recommended to carry out all measurements in a way that avoids any adverse effects due to the proximity of rebar.

Different systems (Covermeters, Ground Penetrating Radar or ultrasonic tomography) can be used to locate whether a suitable "window" in the steel mesh exists. A minimum distance of 20 mm to the closest rebar is recommended.[17]

RH must not be used above the steel reinforcement bars because their presence will modify the stiffness of concrete, especially if they are close to the surface. The influence of steel reinforcement bars on RH test result could be of two different and opposite levels: if the thickness of concrete cover is small, a higher stiffness will be found (higher Rebound number) due to greater steel stiffness; conversely, a lower stiffness may be found due to corrosion and consequently to concrete cover cracking and delamination. Hidden delamination of the concrete cover can also significantly reduce the measured UPV.

ForUPV this effect can also be significant, especially if the measurement mode is indirect and if the signal propagates along a bar.[18] Pull-out tests cannot be done close to the reinforcement. If a core contains a rebar, this part shall be removed before any test.

---

[17]ASTM C805 recommends: "do not conduct Rebound Hammer tests directly over reinforcing bars with cover less than 20 mm".

[18]EN 12504-2 recommends avoiding measurements close to rebar.

1.3.2.2 Highly Stressed and Damaged Sections

If cracks are present in the zone planned to be investigated, care must be taken to perform the tests far enough away from the cracks.

Indeed, if a crack is present in a core, the compressive strength can be significantly decreased compared to that of healthy concrete. If UPV is measured by a surface propagation method and if the crack is perpendicular to the wave travel, UPV is decreased.[19]

Moreover, highly stressed areas must be avoided during the tests. For instance, in a beam, if the lower part in the middle of the beam is assumed to be highly stressed, this part should be avoided for testing.

1.3.2.3 Carbonation and Exposure Effects

The effect of carbonation on the RH test can be very significant, the stiffness of a carbonated concrete being larger than that of the original concrete. Since the inner strength of the concrete is not modified by carbonation, the RH test result may lead to wrong conclusions.

When faced with carbonated concrete, several options are possible:

- *consider that areas with different levels of carbonation correspond to different concretes and may correspond to different conversion models,*
- *remove the carbonated layer at some representative locations and carry out RH measurements before and after the removal, in order to determine the effect of carbonation on the RH test result,*
- *remove the carbonated layer before performing all RH measurements, but this can be difficult and it is sometimes impossible.[20]*

The exposure conditions of the surface regarding sun, rain and wind are also very influential parameters. If a face of a beam is exposed to rain while the opposite one is sheltered, surface concrete properties can be totally different. This may greatly influence the measurements (both nondestructive and cores).

1.3.2.4 Preparation of the Surface to Investigate

The surface must be cleaned in order to eliminate possible moss, lichens and any kind of dirt. Small rough areas or surface defects can be removed before testing by using the grindstone supplied with the rebound hammer device. Coarse grain sandpaper

---

[19] A. Masi, L. Chiauzzi, An experimental study on the within-member variability of in situ concrete strength in RC building structures. Constr Build Mater 47, 951–961 (2013).

[20] This issue has been discussed into detail by J. Brozovsky, J. Zach, Influence of surface preparation method on the concrete rebound number obtained from impact hammer. 5th Pan American Conference for NDT, Cancun, Mexico, 2–6 October 2011.

can also be used. If the surface is severely damaged (high density of cracking or concrete spalling due to concrete corrosion for instance) the cover concrete can be removed and the tests can be performed on this new surface.

1.3.3 Methodology for Determining the Locations and Number of NDT Locations

1.3.3.1 General Considerations

If one assumes that the total amount of resources (either time or cost) for the investigation is fixed, a first decision concerns the balance of these resources between the part devoted on one hand to cores and direct strength measurement and, on the other hand, to semi-destructive and non-destructive tests measurements.

There is no general rule regarding the minimum number of cores (N_c). One statement is that the larger the N_c value, the better it is for the precision of the strength estimate. However, since an identical amount of resources can be devoted to one core or to several/many NDT measurements, the issue is to know how the efficiency changes when this balance between destructive and nondestructive tests changes.

> When the number of NDT measurements is being defined, it is recommended to choose their location in order to cover all areas of interest in the structure.

The most important criteria to consider are:

– one must try to cover the full range of strengths by taking measurements in all types of areas ("weak areas" as well as "good areas"). This is true at all scales, that of the structure (e.g. between floors) as well as that of a component (e.g. different elevations in a column[21]). The range of values measured at test locations will have a direct influence on the accuracy of the conversion model during later stages. This will be defined further in Sect. 1.5.5.
– if some parts of the structure or components have a stronger influence on the structural safety, more attention must be paid to these areas by taking more measurements there;
– if several NDT techniques are used and will be combined for strength estimation, one must carry out measurements with the different techniques at the same test locations as far as possible.

---

[21] The variability of concrete strength in vertical members has been documented (see for instance, H.Y. Qasrawi, Effect of the position of core on the strength of concrete of columns in existing structures. J Build Eng 25, 2019. In such a case, the investigator must choose between addressing the concrete variability or estimating the concrete strength at test locations that he considers to provide a reference for the component.

Table 1.3 definition of scales and corresponding results

Area	Measurement point	Test location TL	Test region TR
Provided data	Test reading	Test result	Distribution of test results

1.3.3.2   Definitions

This section provides some important definitions which are essential.

Investigation Domain

> The investigation domain (ID) corresponds to the structure, part of the structure or a set of components whose strength has to be assessed. The investigation domain must be also defined according the structural needs especially in relation with the required specifications for the re-calculation of the structure.

Table 1.3 synthesizes three concepts corresponding to different scales within the investigation domain and to different types of provided data, from that of the single test reading to the whole dataset of test results. The definition of these concepts is then presented in the following sections.

Test Reading

> A test reading[22] is a value of one single measure (one value of RH, UPV, pull-out force or any other NDT method).

Test Location

> A test location (TL) is a limited area selected for measures used to obtain one test result that is used in the estimation of in situ compressive strength. Several test readings can be carried out in the same test location.

---

[22]Or "observed value" according to ISO 5725.

There are some specifications regarding the size of a test location. Inside the test location, successive measurements can be taken within a close neighbourhood. RH impact may slightly damage the concrete surface, so two successive RH tests must not be closer than 25 mm. The same minimal distance of 25 mm must exist between the measurement point and any edge of the specimen, as well as from any rebar. The dimensions of the test location must be chosen to satisfy the minimal distance between RH measures and from the rebar. EN 12,504–2 suggests that the size of TLs cannot exceed 300 mm × 300 mm. For pull-out tests, a minimum distance of 200 mm between the centres of test positions is recommended.[23] For UPV tests, there is no constraint (see also Sect. 10.3 for defining the minimal distance between TLs).

Test Result

A test result (Tr) is a value representative of a test location. It can be a single test reading, or a mean or a median of N values of test readings. N_{Read} is the minimum number of repetitions to be sure that the test result is inside a predefined confidence interval defined from the measure uncertainty (see Chap. 10).

Test results will be used for the determination of compressive strength alone or combined with other test results obtained from different techniques (NDT or not).

Test Region

The test region (TR) is a given volume of concrete, which is homogeneous. It may be a part of a component, a whole component, a set of components or a larger zone, like a whole floor in a building or a whole structure. A test region can also denote one or several structural elements (or precast concrete components) assumed or known to be from the same population or equivalent to the defined volume associated with identity testing for compressive strength.

The investigation domain (ID) may contain one or several test regions (TR). A test region must contain several test locations.

---

[23] As in EN-12504-3.

All test results obtained in a test region are assumed to belong to the same population. Thus, the properties of each test region can be described through a unique pair of compressive strength values (mean and standard deviation between test results obtained at different test locations). This standard deviation (also called measure uncertainty) includes the precision of the measure and the variability of the material.

The identification of test regions can be helpful to properly identify the conversion models (see Sect. 1.4). It is also relevant in case of structural assessment for the determination of zones with different concrete strengths, which are not representative of the whole structure.

Precision, Accuracy and Variability

Precision corresponds to the closeness of agreement between measures obtained by replicate measurements. It has to be put in relation with repeatability and reproducibility at the same test location TL (including a displacement in the immediate vicinity, for instance, inside a rebar mesh). This issue will be addressed in Sect. 1.5.2.

Accuracy corresponds to the closeness of agreement between one measure and the reference value.

Variability is an intrinsic consequence of both material heterogeneity and evolution of environmental conditions. It has to be put in relation with measures at different TLs in a TR or in the investigation domain.

Uncertainty is the combination of precision (repeatability and reproducibility) and variability.

These concepts are presented and illustrated in Chap. 10.

1.3.3.3 Discrimination of Test Regions in Case of a Large Investigation Domain

The delimitation of the investigation domain (ID) can be done by a visual inspection of the structure and/or by a priori considerations provided by former investigations or a good knowledge of the structure.

Structural considerations cannot be neglected during the definition of the ID.

If the investigation domain covers a whole structure or a large part of a whole structure, a specific procedure can be applied in order to discriminate it into test regions (TR). This procedure is aimed at reducing the variability inside each TR (see RILEM TC 249-ISC Recommendations Sect. 6.6 and Task 4), test regions being identified such that the concrete is assumed to be homogeneous in each of them. Therefore, one pair of values (mean and standard deviation) is considered as representative of the compressive strength population in a single test region.

While the underlying logic is not-case-specific, it is not possible to follow exactly the same process or algorithm to identify relevant test regions. However, the main key-stages are:

(a) assuming a tentative determination of test regions (TR) based on prior knowledge of the structure and on a first analysis of the set of test results at all TLs. This analysis can be based on a statistical analysis and it can be facilitated by plotting a map of the test results if the ID is a large domain. If the map reveals significant contrasts in the distribution of the test results inside the ID, test regions (TR) can be tentatively defined according to these contrasts.

Grouping data on the basis of statistics only is irrelevant if it does not correspond to some logic regarding the history of the structure (casting process...) or the type of components (beams, columns, slabs...).

A group of similar components should be considered as a test region.

(b) checking the validity of the tentative determination of test regions. This comes to check that each test region has their own concrete strength mean and standard deviation that are statistically different than those of other test regions.

This process must be iterative, since the checking can lead either to group two tentative test regions that have the same properties, or to subdivide a test region whose variability remains too large.[24] Chapter 4 of this book details several examples of how test regions can be identified.

---

[24] A good example of what can be done in practice is given by A. Masi, L. Chiauzzi, V. Manfredi, Criteria for identifying concrete homogeneous areas for the estimation of in-situ strength in RC buildings. Constr Build Mater 121, 576–587 (2016).

1.4 Cores

1.4.1 Location of Cores

1.4.1.1 Recommendations from Standards

Core locations have to be in zones representative of the average conditions of concrete, taking into account casting and ageing effects. They must also consider the following:

- the diversity of structural members and stress conditions.
- coring must not endanger structural integrity and so must be performed at a sufficient distance from joints and corners.
- coring must avoid reinforcement. Before coring it is important to detect rebar positions using a covermeter and/or a ground penetrating radar (chipping the concrete cover remains a possibility).
- the coring location must also avoid cold, construction joints or macro-cracked areas.

It can be specified[25] that if the core contains a reinforcement bar along its transversal direction, the value of compressive strength can be lower than the true one. There are equations for correcting the effect of the rebar in the core.[26] Cores containing a bar in the longitudinal direction have to be discarded.

1.4.1.2 Predefined or Conditional Selection

Cores must be chosen in order to provide values which make it possible to build a representative picture of the structure without compromising the structural safety. Cores are considered as giving reference strength values (see RILEM TC 249-ISC Recommendations Sect. 6.7 and Task 6).

Once the number of cores is decided, the core location must be defined either independently of NDT measurements (predefined coring) or based on prior information provided by NDT measurements (conditional coring).

Figure 1.2 illustrates these two options. If the predefined coring is chosen (left

---

[25]EN 13791.
[26]BS EN 12504–1.

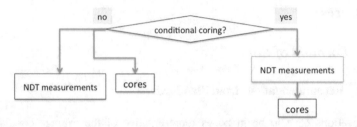

Fig. 1.2 Predefined ("no" option) or conditional ("yes" option") coring

option), the location of cores and that of NDT measurements are chosen independently. Attention must be paid to correct coverage of the structure for both. If some contrast is visible within the structure (like damage, cracking or moisture) or can be suspected for any reason (different batches, different exposure conditions, loading history...) the core selection must be such as to cover this range of variation, by taking cores in various parts of the structure.

Conditional coring (right option on Fig. 1.2) involves defining the location of cores after an analysis of NDT test results. For instance, one can rank all NDT test results and choose core locations in order to correctly cover the whole range of values given by the statistical distribution of the NDT test results.

Conditional coring is recommended, since it is expected to lead to a more reliable conversion model. Conditional coring does not induce any additional cost, and only requires that NDT test results have been made available before the time of taking cores (it may however be only a part of the full NDT program).

Conditional coring is even more preferable when the number of cores is low (3 to 8 cores), since it ensures a better coverage of the concrete condition / strength range. With such a small number of cores, the risk with predefined coring is that most cores (or even all) are selected at locations where the strength is higher (or lower) than the average strength. This will remain unknown and cannot be corrected. This risk decreases, for simple statistical reasons, when the number of cores increases.

However, the correct definition of core locations based on NDT test results requires that these data are themselves reliable. A consequence is that conditional coring cannot be based on poor test result precision (TRP).

1.4.2 Number of Cores

Defining the adequate number of cores is Task 5 in the RILEM TC 249-ISC Recommendations. A minimum number of cores is required in order to get a representative picture of the concrete in the structure and the components under investigation.

This number may be defined either: (a) as a minimum density, in terms of cores per m^2 for slabs and walls, (b) in terms of percentage of similar components to be investigated for columns or beams at the same floor level, (c) or in terms of number of cores for a given volume of concrete.

The choice of the « best» number of cores is a complex issue which does not have a simple answer. To optimize the balance between the number of NDT test locations (N_{TL}) and the number of cores (N_C), one must understand that:

(a)   the cost of cores is much higher than that of nondestructive tests,
(b)   because nondestructive tests are less expensive, they are usually the only way in practice to investigate properly the spatial variability,
(c)   NDTs always require calibration, thus core data is mandatory, for identifying the conversion model (see Sect. 1.4) and for getting reliable strength estimates,
(d)   the conversion model error (see Sect. 1.5.6) is highly dependent on the number of cores.

Thus, finding the best compromise is the answer to these opposing objectives, since:

– increasing N_{TL} is the only way to cover properly the material strength variability,
– increasing N_c ensures a better conversion model and a better precision in strength estimates,
– the total cost or the investigation time is limited.

If the objective of the assessment is to identify variability or low values, NDT brings a notable added-value.

The main criterion for defining the minimum number of cores is therefore its influence on the conversion model error (see RILEM TC 249-ISC Recommendations, Sect. 4.1 and Fig. 1.2). Identifying the conversion model means estimating the values of its N_{par} parameters (or degrees-of-freedom, dof), as will be explained at Sect. 5.5.

This identification is possible if:

$$N_c \geq N_{par}$$

From a strict mathematical point of view, the critical number of cores equals the number of degrees of freedom of the conversion model. This strict minimum number of cores may however induce large uncertainties on the assessed strength, which is not compatible with uncertainty expectations and the target EQL of Table 1.1. The way the precision of strength estimates increases with N_c is described below. More information about the practical minimum number of cores is provided in Sect. 1.6.2 and Chap. 2 of the book is devoted to this specific issue.

1.4.3 Dimensions

The dimensions of cores have an influence on compressive strength. There are different practices all over the world. However, the minimum diameter is also defined as a function of the size of the coarse aggregate. An essential point is about the ratio of the size of the coarse aggregate to the core diameter, which must not be higher than 1:3.[27]

The preferred diameter of cores in standards is about 100 mm. This diameter guarantees a good precision of strength measurement, with a low coefficient of variation in standard tests (about 2%). Using a smaller diameter may seem interesting as more cores can be taken for the same budget, making it possible to have a better spatial coverage of the structure. However, the measured strength is more scattered (at least two times larger for a diameter equal to 50 mm) and the precision of the conversion model will be affected. In any case, the core diameter must not be less than 50 mm.[28]

The height-to-diameter ratio can vary from 1 to 2. The reference is a ratio of 2 but correction coefficients are available in various standards if the height-to-diameter ratio is different from 2.[29]

Moreover, testing a core with a nominal diameter of about 100 mm and equal length (L/D = 1) gives a strength value equivalent to the strength value of a 150 mm cube.

According to European standards, testing a core with a nominal diameter at least 75 mm and not larger than 150 mm and with a length to diameter ratio equal to 2.0 gives a strength value of a 150 mm by 300 mm cylinder manufactured and cured under the same conditions (cylinder strength).

1.4.4 Extraction, Conservation and Preparation of Cores

The machine used for extracting cores must be rigid enough and must be securely fixed to the structure during the extraction.

In order to avoid damage to the core when extracting an incomplete core, it would be preferable to take a core that crosses the full thickness of the element to be investigated, but in some cases partial coring should be considered.[30]

At the end of coring, the core must be directly washed off and dried with cloth. The core must be referenced. The orientation of the core (vertical or horizontal) must be also specified. If the visual inspection reveals defects (like cracks, significant

---

[27] EN 12504–1.

[28] EN 13791.

[29] BS and ASTM C42 propose an equation for correcting the variation of height to diameter ratio. EN13791 uses a correction factor of 0.80 to convert a 1:1 core to a 2:1 core.

[30] In the case of partial coring, the core can be removed by a shear effort (inserting a screwdriver or small chisel and tapping smartly with a hammer will usually be sufficient to snap the core at its base).

voids, etc.), they must also be documented. The cores must be stored in a rigid box to protect them during the transport.

Moisture content is an important factor. The strength measured on cores depends on the moisture content at the time of the test. Any change in moisture content, including the preparation process, will modify the value of the compression test result.

Once cores have been taken, it is recommended to keep them in a condition which is as close as possible to what it was in the structure, specifically regarding the moisture content. The preferred option is to keep them in a sealed plastic bag, in order to avoid any variation of internal moisture until the compression test.

Two other options are possible in different countries, each option having its own peculiarities and leading to different results:

- *trying to keep the cores at the same level of humidity as in the structure, but this has probably been modified during drilling and cutting,*
- *wetting the cores by keeping them under water, but the measured strength will be different from that in the structure,*

If this is not possible for technical reasons or if another storage option is chosen (for instance cores kept at normal temperature and air drying in the laboratory, or cores kept under water until the compression test), correction factors can be applied in order to transform the measured strength into "on-site strength".

If this is not possible for technical reasons or if another storage option is chosen (for instance cores kept at normal temperature and air drying in the laboratory, or cores kept under water until the compression test), correction factors can be applied in order to transform the measured strength into "on-site strength".

As an example, a study performed by Bartlett[31] proposed the following relationships for correcting the effect of core moisture on compressive strength (range of compressive strength: 20 – 80 MPa):

$$f_{c,dry} = 1.114\, f_{c,wet}$$

$$f_{c,drilled} = 1.09\, f_{c,wet}$$

where f_c,dry is the compressive strength measured on air-dried cores, f_c,drilled the compressive strength measured on cores as they were drilled, f_c,wet the compressive strength measured on saturated cores.

While describing a significant effect, these relations must however be considered cautiously, since the correcting coefficients are probably imprecise and depend upon a variety of factors.

---

[31] F.M. Bartlett, J.G. MacGregor, Effect of moisture condition on concrete core strengths. ACI Mater J 91(3), 227–236 (1994).

However, according to the European Standard EN 13,791, cores are recommended to be exposed to a laboratory atmosphere for some time prior to testing.

Before testing the core, the length-to-diameter ratio must be controlled, assuming that the carbonated layer of core is cut off. Once the core has the proper length, the end surfaces must be properly prepared with a suitable machine in order to obtain parallel and flat surfaces. In the end, the dimensions of cores (length, diameter)[32] must be measured and they must be weighed for calculating their density. The density comparison can give helpful information on the differences of concrete quality.

1.4.5 Core Testing (Mechanical Testing, NDT)

To control the damage that could result from the coring process, NDTs (especially direct UPV measurement) can be performed on cores before compressive testing and compared with measures obtained on site prior to coring. A significant change could be linked to a core defect: damaged cores would depart from the general regression plot of test results on cores against test results on the structure before coring.

Before placing the cores on the loading plates of the compressive machine, it must be checked that the surfaces of the cores are clean and dry. The speed of loading must be equal to 0.6 MPa/s $\pm$ 0.2 MPa/s.

Results must be reported in a table with all the information for each core.[33] As an example, the following items could be proposed for the report.

(a) description and identification of the test specimen;
(b) estimated maximum size of aggregate;
(c) date of coring;
(d) visual inspection, noting any abnormalities identified;
(e) reinforcement (when appropriate): diameter, in mm, position(s) in mm;
(f) method used for the preparation of specimen (cutting, removal of carbonated layer, grinding, or capping);
(g) length and diameter of the core as received;
(h) length/diameter ratio of prepared specimen;
(i) surface moisture condition at time of the test and moisture condition of the specimen;
(j) date of the test;

---

[32] EN 12504–1 recommends to measure core diameter within 1%, from pairs of measurements taken at right angles, at the half and quarter points of the length of the core. It is also underlined that the length must also be assessed within 1%.

[33] The extensive procedure is described in European standards (EN 12504–1:2012, Testing concrete in structures—Part 1: Cored specimens—Taking, examining and testing in compression, European Committee for Standardization (CEN), Brussels, Belgium, 2012—EN 12390–3:2019, Testing hardened concrete - Part 3: Compressive strength of test specimens, European Committee for Standardization (CEN), Brussels, Belgium, 2019).

(k) core compressive strength, to nearest 0.1 MPa (N/mm²);

(l) any deviations from the standard method of examination or compression testing;

(m) a declaration by the person responsible for the examination and testing that these were carried out according to the standard.

The report may include:

(n) mass of the specimen, in kg;

(o) apparent density of specimen, to the nearest 10 kg/m³;

(p) condition of the specimen on receipt;

(q) curing conditions since the core was received;

(r) date of test (if appropriate);

(s) age of specimen at the time of test;

(t) other relevant information e.g. void volume.

1.5 Identification of the Conversion Model

1.5.1 Main Steps and Principles

Available information at the beginning of the assessment process is, as shown in Fig. 1.3, provided by:

- a large number of NDT test results obtained at N_{TL} test locations,
- a limited number N_c of compression strength test results measured on cores which have been drilled at some test locations, following either a predefined pattern or according to NDT test results (i.e. conditional coring).

> The model identification stage consists in processing the information at locations where both core strength $f_{c\,i}$ and NDT test results Tr_i are available, which can be written as a series of pairs (Tr_i, f_{ci}), with $i = 1, N_c$. The $f_{c\,i}$ strengths are « **reference strength**» values.

Fig. 1.3 Illustration of test locations on a test region (black "x" = NDT locations, blue crosses = core locations, here $N_c = 5$)

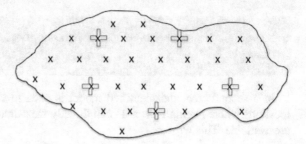

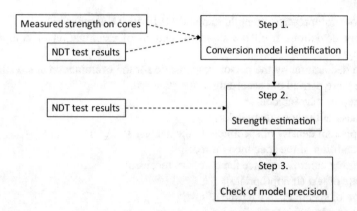

Fig. 1.4 Steps to be followed in the conversion model identification process

In some cases, several cores can be taken at the same test location or several specimens can be taken from a single core, and thus several strength values can be available at the same test location. In such a case, the mean value of all measured strengths is considered as the reference strength.

When several NDT methods are used, each (Tr_i, f_{ci}) pair is replaced with a dataset containing one strength value and all NDT test results at the same test location.

Figure 1.4 describes the conversion model identification process, which is identical whatever the type of semi-destructive or nondestructive method, among those described at Table 1.2.

In a first step, a number N_c of (Tr_i, f_{ci}) pairs is available corresponding to test locations where both measured core strength and NDT test results have been obtained. This dataset, named "identification set" is used for identifying the values of all parameters of the conversion model M. This is commonly ensured by finding the values of model parameters that minimize the discrepancies between estimated strength $f_{c,est\,i}$ and measured strength $f_{c\,i}$ at the same locations. To do that, the user must:

– choose a model type. The variety of possible conversion models is very large. Models are defined by their type (explicit *vs* implicit), their mathematical expression and number of parameters N_{par} (cf Sect. 1.5.4),
– choose a suitable model approach (cf Sect. 1.5.5).

Identifying the best model comes down to identify the values of the N_{par} model parameters.

In a second step, the identified model is used to estimate strength at all test locations where strength has not been directly measured and only NDT test results are available. This writes:

$$f_{c,est\,i} = M\left(par_j, Tr_i\right) \quad j = 1, N_{par}$$

where par_j, are the values of the model parameters identified during the first step and Tr_i are additional NDT test results, which have not been used during the identification step.

A conversion model can be identified and applied:

(a)   with a specific model for each test region
(b)   or with teh same model covering several or all test regions.

The choice between these options is based on:

- *the consideration of prior knowledge about the structure and the concrete,*
- *information provided by NDT test results and their statistical distribution,*
- *the number of cores that will be used for identifying the conversion model(s), since the uncertainty of strength estimates decreases when the number of cores increases while each model requires a minimum number of cores.*

Chapter 4explains how this choice can be done in relation with the definition of test regions.

The last step is to evaluate the precision of the estimation with the identified model (or models if different models are used for different test regions) over the whole investigation domain (cf Sect. 5.6).

The following sections (Sects. 1.5.2–1.5.6) detail the process of the conversion model identification.

1.5.2   NDT Test Result Precision

Test results are the only data that are used for identifying the conversion model and estimating the quantities corresponding to each EQL. The precision of these input data has a major effect on the uncertainty of the strength estimates. Thus, controlling the test result precision is mandatory (see RILEM TC 249-ISC Recommendations Task 3 and Sect. 6.4).

1.5.2.1   Definition of NDT Test Result Precision (TRP)

Precision includes the repeatability and the reproducibility. The repeatability is obtained by repeating the measurement at a TL with the same operator and the same device. The reproducibility corresponds to the case where the operator or/and the device is changed.

If we consider the repeatability, as explained at Sect. 1.3.3, one must differentiate between the NDT test reading (result of one single measurement) and the NDT test result (representative of the property at a given test location), which is obtained from a given number of repetitions N_{read} of the measurement at a given test location or at a close vicinity. Deriving the NDT test result from the series of individual test readings is usually done through averaging, after having eliminated local extreme values. This process is described in detail in some technical standards.[34]

If several operators make the measurements with different devices, reproducibility must be considered.

The input data for the analysis is the NDT test result.

Whenever test result precision is considered in the following, it is related with the NDT results (Tr) and not with the individual readings. The test result precision at a given test location can be easily quantified by repeating N_{rep} times the process which provides a NDT result (from N_{read} readings) and estimating the standard deviation and/or coefficient of variation between these N_{rep} test results. A lower value iselect a few (2 to 5) test locationsndicates a better repeatability.

It is recommended, during the investigation stage, to devote a specific part of the program to assess the magnitude of the test result precision (TRP), which is a key factor in the next stages of compressive strength assessment.

The value of TRP has a major impact on the recommended number of cores at each EQL (see Chap. 2).

---

[34]For rebound hammer for instance, it is the case in European standards (EN 12504–2:2012, Testing concrete in structures—Part 2: Non-destructive testing—Determination of rebound number, European Committee for Standardization (CEN), Brussels, Belgium, 2012) and in North-American standards (ASTM C805:2013, Standard test method for rebound number of hardened concrete, 2013where it is said to discard readings differing from the average of ten readings by more than six units and determine the average of the remaining readings. If more than two readings differ from the average by six units, the entire set of readings must be discarded.

1.5.2.2   Assessment of NDT Test Result Precision

TRP is assessed through several repetitions of the measurement process at some test locations over the investigation domain in similar conditions.

One simple way to assess test result precision is:

- to select a few (2 to 5) test locations,
- at each test location, to repeat N_{rep} times the measurement process,[35]
- to quantify the standard deviation of test results sd_{rep} or the coefficient of variation COV_{rep} between the N_{rep} test results at this test location,
- to process these few values of sd_{rep} or COV_{rep}, one value being derived from each test location, in order to assess the repeatability of the test over the investigation domain.

A TRP value is derived which is representative of the repeatability of the test result over the investigation domain. This repeatability can be the same over the ID or it can vary and be larger in some regions. A cautious estimate of the TRP must be finally derived.

If reproducibility has to be evaluated, the same set of measurements must be performed at the same TL by other operators and/or with other devices. Standard deviation of the test results obtained by all operators on the same test location must be computed and used to evaluate the reproducibility.

The comparison between sd_{rep} or COV_{rep} obtained during the tests with values of Table 1.4 leads to qualifying the NDT precision as being high (TRP1), medium (TRP2) or poor (TRP3).

Some precision levels are difficult to obtain in practice due to the intrinsic nature of the test and material response (for instance, high precision level is difficult to obtain for RH test results or micro-cores). Poor test result precision may prevent the identification of any accurate conversion model and, as a consequence, can drastically limit the precision of strength estimate.

---

[35]One must point out that the whole measurement process is repeated, which requires a total number of test readings equal to N_{read} x N_{rep} where N_{read} is the number of individual readings that are repeated for obtaining a test result.

Table 1.4 Indicative order of magnitude of test result precision (TRP)

		High precision TRP1	Medium precision TRP2	Poor precision TRP3
Usual NDT				
RH	Coefficient of variation COV_{rep}	$COV_{rep} \leq 3\%$	$3\% < COV_{rep} \leq 7\%$	$COV_{rep} > 7\%$
	Standard deviation sd_{rep}	$sd_{rep} \leq 1$	$1 < sd_{rep} \leq 3$	$sd_{rep} > 3$
UPV	Coefficient of variation COV_{rep}	$COV_{rep} \leq 1\%$	$1\% < COV_{rep} \leq 3\%$	$COV_{rep} > 3\%$
	Standard deviation sd_{rep}	$sd_{rep} \leq 50$ m/s	50 m/s $< sd_{rep} \leq 125$ m/s	$sd_{rep} > 125$ m/s
Other non-destructive methods				
Coefficient of variation of Pull Out[a]		$COV_{rep} \leq 5\%$	$5\% < COV_{rep} \leq 8\%$	$COV_{rep} > 8\%$
Coefficient of variation of Windsor probe penetration test[b]		$COV_{rep} \leq 2\%$	$2\% < COV_{rep} \leq 5\%$	$COV_{rep} > 5\%$
Coefficient of variation of strength on micro-cores[c]		$COV_{rep} \leq 15\%$	$15\% < COV_{rep} \leq 25\%$	$COV_{rep} > 25\%$

[a]For individual tests, from comparison of results obtained by various researchers, in ACI 228.1R-95, In-place methods for determination of strength of concrete, chair N.J. Carin, 1995
[b]For penetration depth measured with $N_{read} = 6$ (ACI 228.1R-95, In-place methods for determination of strength of concrete, chair N.J. Carino, 1995)
Variability on micro-cores is much larger than that on cores. It is also highly influenced by the size of the largest aggregates. See A.O. Celik, K. Kilinc, M. Tuncan, A. Tuncan, Distributions of compressive strength obtained from various diameter cores. ACI Mater J 109(6), 597–606 (2012)

Table 1.5 Task definition specific for each EQL

Task	EQL1	EQL2	EQL3
Check test result precision (TRP)—*T3*	Recommended	Mandatory (TRP value must be given)	Mandatory (TRP value must be given, and must correspond to medium or high precision)
Conditional coring—*before T6*	Recommended	Mandatory	Mandatory
Check model prediction error—*T9*	Recommended	Recommended	Mandatory (RMSE on strength must be reported)

Increasing the number of test results could appear to be a partial solution but it is often preferable, both for cost and statistical reasons, to try first to improve the test result precision. A drastic control of all uncontrolled influential factors (irregular surfaces…) is the best way to improve the test result repeatability.

If it is planned to combine several NDT methods, the precision must be quantified for each technique, since any method with a poorer precision will constitute a handicap for an efficient combination.

1.5.3   Data Processing: The Case of Outliers

1.5.3.1   What Are Outliers?

When analyzing real data sets, it is often found that some observations are different from the majority of the data. Such observations are usually called outliers and can be defined as individual data values that are numerically distant from the rest of the sample, thus masking its probability distribution.

Outliers always deserve a careful attention, either because they may have a significant impact in concrete strength estimation or, in some cases, they may in fact represent a different concrete population that deserves a separate assessment.

An outlier analysis can usually be seen as a two-step process. In the first step, several methods can be used to identify potential candidate outliers of the data. In the second step, the candidate outliers need to be handled to account for their effect in subsequent statistical analyses of the data. The following subsections summarize these two steps. Chapter 5 is dedicated to the analysis of outliers, with its theoretical bases and examples.

1.5.3.2   Possible Identification of Outliers

Outliers can occur in both univariate and multivariate data, i.e. in a series of NDT test results and when these test results are considered with the corresponding core strength at the same test location.

For univariate data, the identification of outliers is usually based on the analysis of the statistical distribution of the data and on how the individual values depart from the main body of the data.

For multivariate data, outlier identification is often carried out by analyzing the correlation between the variables and how it departs from the expected characteristics or by analyzing the distance between data points and groups of data points. To illustrate possible multivariate situations that can occur, Fig. 1.5a presents a scenario where two NDTs are carried out at the same TL of a given TR and for which the results indicate the existence of a single data point departing from the correlation trend observed for the remaining test results. In another example, Fig. 1.5b presents also a scenario where two NDTs are carried out at the same TL of a given TR but, in this case, the results indicate the existence of a smaller group of results departing from the bulk of the remaining test results.

As far as possible, it is recommended to analyze the NDT measurement results while on the field, in order to be able to perform additional measurements if suspicious data are detected.

1.5.3.3 Possible Alternatives for Outliers

After having identified candidate outliers, the subsequent step is to deal with their presence and/or handle their effects on the data set. Most available options to deal with outliers revolve around three main possibilities:

- correcting an outlier value (if an outlier has been identified as an error and if it is possible to replace that observation by a correct value),
- removing an outlier value from the data set,
- accommodating an outlier value by reducing its effect on the subsequent statistical analysis of the data (e.g. using outlier-resistant techniques such as robust statistics to work with the data set).

The ideal way is to proceed to additional measurements that will confirm if its presence is due to an error that can be eliminated. If this is the case, correcting its value can be carried out by using data from the additional measurements. However, errors are not often identified and one must check if any physical or technical reason could explain the outlier origin (by checking the casting history, the reinforcement location, local damage or evolution of properties such as those resulting from carbonation…). In any case, outlier values be either removed from the data set or accommodated using appropriate statistical methods. The first option is not recommended herein as long as keeping outlying data corresponds to a conservative option regarding the strength assessment. It must be remembered, however, that determining whether or not an observation is an outlier is ultimately a subjective exercise. With respect to outlier accommodation, procedures such as those based on robust statistics[36] should be followed to handle outliers and reduce their effects (see Chap. 5).

> It is mandatory to record all information about measurements that have been considered as outliers and removed from the original data set. Such data (number of outliers, location, values, reasons for excluding the outliers) must be mentioned in the investigation report.

---

[36]Robust statistics corresponds to a set of statistical methods whose efficiency is not strongly affected by outliers or any deviation from common assumptions, like those regarding the normal distribution of the variables.

1.5.4 Types of Conversion Models

The choice and identification of a conversion model correspond to Tasks 7 and 8 in the concrete strength assessment process, as defined in the RILEM TC 249-ISC Recommendations. The general expression of a mathematical model used as a conversion model can be written as:

$$f_{c,est, I} = M(par_j, Tr_{k, i})$$

where $Tr_{k, i}$ (with $i = 1, N_{TL}$) is the i-th test result of the k-th NDT method and par_j (with $j = 1, N_{par}$) is the value of the j-th model parameter. This expression works both for univariate models (single NDT, $k = 1$) and bivariate models (combined NDT, $k > 1$).

Once the model identification parameter stage is completed, the value of the N_{par} model parameters par_j is known, and the model can be written:

$$f_{c,est} = M (Tr_k).$$

The input data of a conversion model are exclusively non-destructive test results Tr.

This excludes any empirical model that would use additional material data (like for instance mix properties: aggregate size, water to cement ratio), that cannot be assessed on site by nondestructive techniques.

Former RILEM Recommendations[37] were based on the identification of iso-strength curves. Because of many influencing factors on these curves, a series of six coefficients had to be identified to cover the effects of cement type, cement content, aggregate type, proportion of fines, maximum aggregate size and admixtures. A specific (and heavy) experimental program was required for each coefficient. The result was a case-specific conversion model. However, the problem remains open when on-site strength is addressed, even once these curves are available, since knowing the value of each of these influencing factors may be difficult and the correction of the estimates is therefore quite difficult.

Various types of conversion models can be considered and are briefly discussed in the following.

---

[37]RILEM TC 43-CND, I. Facaoaru (chair), Draft recommendation for in situ concrete strength determination by combined non-destructive methods. Mater Struct 26, 43–49 (1993).

1.5.4.1 Explicit Versus Implicit Models

Explicit models are those defined by an explicit equation M (par_j, $Tr_{k,i}$), whatever its level of complexity. Identifying the model consists of identifying the N_{par} values of the model parameters par_j by finding the best agreement between measured strengths and estimated strengths on the identification set of pairs (Tr_i, f_{ci}), with $i = 1$, N_c.

Implicit models cover all kinds of models for which this best agreement is found without providing an explicit expression.

An example is that of neural networks that have been used by several authors for this purpose. Other examples are those provided by metaheuristic algorithms like particle-swarm-optimization (PSO), Bayesian networks (BN), genetic algorithms (GA) or similar methods. Such models are cited in Chap. 3 devoted to the state-of-the-art about combination.

The identification of the N_{par} model parameters of implicit models follows the same general rules (minimizing an error), but the values of the parameters are usually not explicit and the calibrated model is used as a "black box" to estimate strength from nondestructive test results. Such models may have in fact a very large number of internal parameters which is a source of risk. The model may deliver estimates which are very close to reference values on the identification set, while having poor performances when applied for estimating strength for new NDT test results. If implicit models are used, it is mandatory to evaluate their prediction error.

> We recommend the use of explicit models whose main interest is that peer-review and control is easy.

"Black-box" models are used at the client's own risk. If implicit models are used, it is strongly recommended that the analysis report provides all information (including the value of all internal model parameters), thus making possible a peer-review and external validation.

1.5.4.2 Single Versus Combined NDT Models

The general expression of a conversion model makes possible the use of several NDT methods and thus the estimation of concrete strength by combining the values of several NDT test results (from different techniques) in a single equation.

The interest in combining several methods has been controversial for many years. The potential interest of such a combination lies in the fact that when some uncontrolled influential factors are suspected, several NDT methods can be beneficial because they have a different sensitivity to this factor. A well-known example is the combination of RH and UPV for assessing concrete strength (both rebound and UPV increase with strength) and eliminating the effect of changes in moisture concrete condition (an increase in moisture content increases the UPV and decreases the rebound number). Such a combination has been promoted in former RILEM Recommendations[32] and is commonly used in practice. Despite this theoretical interest, the effective benefit of combination remains a matter of discussion.

The use of combined methods requires that:

- these methods are affected in different ways by influential factors, as it is the case regarding the influence of concrete moisture on RH and UPV test results,
- the precision of each of the methods to be combined is similar: combining a high precision method with a poor precision method is useless,
- since the number of parameters of the conversion model is larger for combined methods, the minimum number of cores must increase accordingly.

The basic principles of combining several NDT methods are discussed in Chap. 3. How it can be done in practice is explained in Chap. 9, where the added-value of combination is discussed.

1.5.4.3   Mathematical Shape of the Model

Explicit models are defined by a mathematical expression. While on one hand any mathematical expression would be theoretically convenient, on the other hand, the number of cores must increase with the number of model parameters in order to keep the same precision of the strength estimate. Thus, it is recommended to follow a parsimony rule, the number of model parameters being restricted to the strict minimum.

> **It is recommended to limit the number of model parameters to 2 for univariate models (single method) and to 3 parameters when two methods are combined.**

Regarding univariate models, three (empirical) mathematical expressions are commonly used to describe the dependence of the ND test result to concrete strength, which enable a description of the different sensitivity of strength to Tr variations:

- linear conversion model M_L:

$$f_c = a + b \, Tr$$

where Tr is the test result provided by one NDT method.
- power law conversion model M_P:

$$f_c = a \, Tr^b$$

- exponential conversion model M_E:

$$f_c = a \, e^{b \, Tr}$$

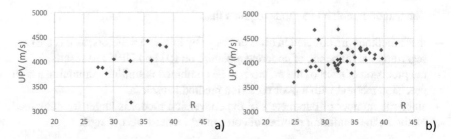

Fig. 1.6 Bivariate outlier scenarios: **a** existence of a single data point departing from the correlation trend observed for the remaining test results; **b** existence of a smaller group of results departing from the bulk of the remaining test results

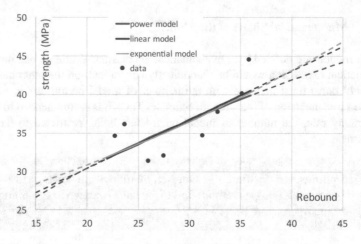

Fig. 1.6 Illustration of model identification from the identification set (here $N_c = 8$)

Figure 1.6 illustrates how three different model shapes can be used for the same data set.

Identification result for the three model shapes defined at Sect. 1.5.4.3 (linear, power and exponential).[38]

The shape of the mathematical model is not the most influencing factor regarding the accuracy of the concrete strength assessment. When the fitting quality of various mathematical models is compared on a given dataset, it may slightly change between models (Fig. 1.6), but this is not decisive.

---

[38]The identification process with the three model shapes respectively leads to: (a) $f_c = 0.632 \, R + 17.78$ ($r^2 = 0.543$) for the linear model, (b) $f_c = 7.672 \, R0.460$ ($r^2 = 0.480$) for the power model, (c) $f_c = 22.09 \exp (0.0167 \, R)$ ($r^2 = 0.532$) for the exponential model. While the three models are very close in the interval corresponding to the identification set, they disagree when they are used for extrapolation.

- Linear models are preferable as long as the dataset does not exhibit a clear nonlinear tendency.
- Nonlinear models may be better when the strength range covered by the model is very large.

Whatever the model, extrapolation, i.e. using the model outside the range of values used for fitting, must be avoided.

The same principles apply to bivariate models. When combined methods are considered, the number of possible expressions widely increases, but it seems preferable to keep some homogeneity, with similar functions for all test results. One must then consider:

- linear conversion model M_L:

$$f_c = a + b\ Tr_1 + c\ Tr_2$$

where Tr_1 and Tr_2 are the test results provided by two NDT methods.
- power law conversion model M_P:

$$f_c = a\ Tr_1^b T\ r_2^c$$

- exponential conversion model M_E:

$$f_c = a\ e^{b\ Tr1}\ e^{c\ Tr2}$$

All these expressions require the identification of three parameters at the model identification stage.

When the goodness of fit of several univariate or bivariate mathematical models is compared for a given data set, the fitting error may slightly change between the models, but this is not decisive, since the prediction error (Sect. 5.6) must be the only relevant measure of the model error.

Particular attention must be paid to the prediction error of nonlinear and bivariate models which are usually more sensitive to random uncertainties in the data set. This issue is illustrated in Chap. 9.

1.5.5 Model Identification Approach

An existing conversion model must not be used for directly deriving strength estimates for NDT test results. The identification of a specific model is mandatory in all cases.

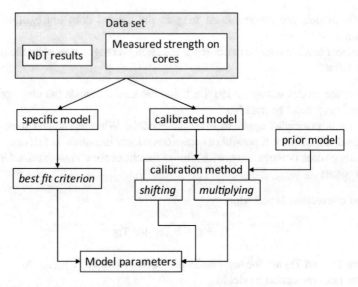

Fig. 1.7 The two main options for identifying model parameters

1.5.5.1 Specific Identification or Calibration Options

The flowchart of Fig. 1.7 details the first step of Fig. 1.4 and illustrates what are the main options regarding the model identification approach, on the basis of available data, i.e. from the N_c pairs $(f_{c, i}, NDT_i)$.

Identifying a specific model or calibrating a prior model are the two main options for identifying model parameters.

1.5.5.2 Specific Model Identification, Bi-objective Approach

The most common way of identifying the conversion model is to directly identify the N_{par} model parameters par_j by using regression analysis. This only requires choosing a specific mathematical shape for the model expression (see Sect. 1.5.4.3) and identifying its parameters.

Since $N_C \geq N_{par}$, two situations exist:

- if N_c is exactly identical to N_{par}, the identification problem has a unique and exact solution, with a perfect fit between measured strength and estimated strength (fitting error is zero). However, the prediction error in this case can be very large. This situation exists when one has only two cores for a univariate model, or three cores for a bivariate model.

– If N_c is strictly larger than N_{par}, solving the identification problem through a minimization algorithm leads to a best-compromise solution, which minimizes the gap between estimated and measured strengths. A least-square criterion is the most commonly used. The prediction error decreases as N_c increases.

It can be shown that, in practice, each new identification set would lead to a specific set of model parameters which is not the exact solution that would be obtained without measurement uncertainties (all values of the identification set contain some uncertainties, coming from instance from test result precision). However, there is some consistency between each of these sets, corresponding to a certain trade-off between the model parameters. This point is addressed at Sect. 12.4.

> However, the common specific regression approach is not adapted to capture the variability of concrete strength, for which a specific bi-objective approach must be used.

The bi-objective approach addresses two targets simultaneously by fitting both the experimental mean strength $f_{c, mean}$ and the strength standard deviation $sd(f_c)$. For univariate conversion models, this leads to solve a problem with two equations and two unknowns, whose solution is straightforward.[39] The detailed process for identifying the model parameters depends on the mathematical shape of the model.

For instance, in the case of a linear model ($f_{c\,est} = a + b\,Tr$), the two conditions can be respectively written as:

$$f_{c\,estmean} = a + b\,Tr_{mean} = f_{c\,mean}$$

and

$$sd(f_{c\,est}) = b\,sd(Tr) = sd(f_c)$$

where $f_{c\,mean}$ and Tr_{mean} are, respectively, the mean values of measured core strength and NDT test results, and where $sd(f_c)$ and $sd(Tr)$ are, respectively, the standard deviations of measured core strength and NDT test results.

Consequently, the values of the two model parameters become:

$$b = sd(f_c)/sd(Tr)$$

and

$$a = f_{c\,mean} - b\,Tr_{mean}$$

---

[39]M. Alwash, Z.M. Sbartaï, D. Breysse, Non-destructive assessment of both mean strength and variability of concrete: a new bi-objective approach. Constr Build Mater 113, 880–889 (2016).

The same principles apply to bi-objective power law model and bi-objective exponential model (detailed information is available in Chap. 12).

The bi-objective approach does not imply any additional experimental cost as it uses the same data (test results) of other identification approaches.

1.5.5.3 Model Identification Through Calibration

An alternative approach is the identification of the conversion model using a prior model which is just calibrated to fit test results. The prior model can be written as:

$$f_{c, est, o} = M(par_j, Tr)$$

where the value of all model parameters (par_j) is fixed.

The prior model must be documented and must have been established in a similar context, including the same range of strengths and the same type of concrete (e.g. fiber reinforced concrete, precast concrete, etc.).

The prior estimated strength is calculated on all points of the identification set from the Tr_i test results:

$$f_{c, est, o, i} = M(par_j, Tr_i)$$

The calibrated model can be written:

$$f_{c, est, cal} = M(par_j, Tr, C)$$

where C is a calibration factor, e.g. an additional parameter that is identified during the calibration stage. The calibration stage consists of identifying the C value (C is thus the unique degree-of-freedom of the model) in order to get the best fit between estimated calibrated strength and measured strengths.

In practice, there are two ways for performing this calibration, both comparing the mean values of prior strength $f_{c, est, o, mean}$ with the mean value of measured strengths on cores $f_{c, mean}$.

- shift calibration, where the difference between the two means is quantified:

$$C_s = f_{c, mean} - f_{c, est, o, mean}$$

leading after calibration to:

$$f_{c, est, cal} = f_{c, est, o, mean} + C_s$$

- scaling calibration, where the ratio between the two means is quantified:

$$C_M = f_{c, mean} / f_{c, est, o, mean}$$

leading after calibration to:

$$f_{c, est, cal} = C_M x f_{c, est, o, mean}$$

1.5.5.4 Possible Problems with Model Identification

In some situations, the model identification approach may result in a conversion model which is not physically sound. Such situations are most common when the number of cores is very low and/or when the NDT test result precision is poor. As a consequence, it can happen that the strength to NDT test result sensitivity is inconsistent (for instance negative slope in a linear model), or that the estimated strength is negative for some NDT test results.

In such cases, the conversion model must be rejected and the available data must be carefully analyzed. This can lead to reject some data (considered as outliers, Sect. 5.3), or to proceed to additional measurements (NDT and/or cores), or to rely only on core strength measurements (thus limiting the assessment to mean strength).

1.5.5.5 Recommendations Regarding the Model Identification Approach

Chaps. 8 and 9 will illustrate how conversion models can be identified in practice from data sets and will compare the efficiency of several types of models and of several calibration approaches. There is no general guidance for choosing the best model identification approach, which may depend on the situation.

An important criterion is the number of cores N_c, since the lower N_c is, the lower the number of free model parameters must be. This can lead to choosing a calibration approach (only one free parameter, C) when N_c is very low (namely equal to 3 or 4), while a specific model is preferable when N_c is larger.

When concrete variability has to be assessed (i.e. for EQL 2 and EQL 3, see Table 1.1), **the bi-objective approach is recommended, since it is the only one to correctly address it.**

1.5.6   Using the Conversion Model and Quantifying Its Error

1.5.6.1   Assessing Local Strength, Mean Strength and Strength Standard Deviation with the Conversion Model

Local strength is the first parameter that needs to be estimated, with a targeted accuracy which depends on the estimation quality level EQL, as defined in Table 1.1. Once the conversion model is identified, this assessment is possible. At each test location where only a NDT test result is available (see Fig. 1.3) the model can be applied for deriving the estimation of local strength:

$$f_{c\ est\ I} = M(Tr_i)$$

This operation can be repeated for all test locations (i.e. at $N_{TL} - N_c$ points), thus leading to a series of ($N_{TL} - N_c$) local strength estimates. These estimates can be combined with strength values directly measured on the N_c cores.

A possibility is to combine N_{TL} estimated strength values with N_c measured strength values in order to obtain the strength distribution. If so, it is also possible to consider a different weighting coefficient for directly measured strengths (supposed to be more reliable) and for estimated strength (supposed to be less reliable because of the use of a correlation).

1.5.6.2   Fitting Error and Prediction Error

Whatever the option used for identifying the model parameters (specific or calibrated), it is recommended to check the accuracy of strength estimation (RILEM TC 249-ISC Recommendations, Task 9). This is mandatory for EQL3.

The most common measurement of the fitting error is related to the value of the determination coefficient R^2, which measures the goodness of fit of the theoretical model to the data set. This coefficient does not provide information about the effective capacity of the model to estimate the strength for a new dataset, which has not been used at the model identification stage. This is issue discussed in Chaps. 8 and 9.

The recommended measure of model error is the root mean square error (RMSE) which provides the statistical lack of fit, expressed in MPa, i.e. the "mean distance" between the reference (measured) value and the value which is estimated by using the model. The RMSE values directly quantify (in MPa) the mean statistical error in any estimation of the local strength value.

The RMSE value can be obtained by:

$$RMSE = \sqrt{\frac{\sum_{i=1}^{n}(f_{c,est\,i} - f_{ci})^2}{n}}$$

where n is the number of points where $f_{c,est\,i}$ is calculated.

One must understand the difference between the fitting error (RMSE$_{fit}$) and the prediction error (RMSE$_{pred}$). The former relates to the dataset used for model identification (n = N$_c$), while the latter relates to a new data set (n = N$_{TL}$ − N$_c$). The fitting error is calculated on the identification data set, by comparing $f_{c\,i}$ with $f_{c,est\,i}$ for all N$_c$ points at which test results Tr$_i$ and core strengths $f_{c\,i}$ are used in order to calibrate the model. The prediction error is calculated by comparing measured values and strength estimates at points which have not been used to calibrate the model. Prediction error is always larger than fitting error, which is due to the random nature of the estimated parameters of the conversion model. Any estimation of the model error based only on the fitting error, as it is common in practice, is meaningless.

Chapter 11 details the definitions of RMSE and Chaps. 8 and 9 illustrate how RMSE can be calculated and analyzed in practice.

1.5.6.3 Influence of the Number of Cores on the Fitting and Prediction Errors

Figure 1.8 illustrates the general trend describing how both RMSE$_{fit}$ and RMSE$_{pred}$ vary with the number of cores (i.e. the size of the calibration set). When the number

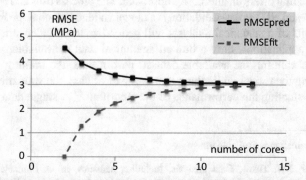

Fig. 1.8 RMSE trend with the number of cores, for a 2-parameter model (values are illustrative) (The difference between the fitting and prediction errors has been analyzed and explained as well with synthetic data (Non destructive strength evaluation of concrete: analysis of some key factors using synthetic simulations, M. Alwash, D. Breysse, Z.M. Sbartaï, Constr Build Mater 99, 235–245, 2015) as with real data (Analysis of the single and combined non-destructive test approaches for on-site concrete strength assessment: general statements based on a real case-study, K. Ali-Benyahia, Z.M. Sbartaï, D. Breysse, S. Kenai, M. Ghrici, Case Stud Constr Mater 6, 109–119, 2017))

of cores is very large, both curves converge towards the true value of the error. However, $RMSE_{fit}$ and $RMSE_{pred}$ curves show opposite trends: $RMSE_{fit}$ increases with the number of cores while $RMSE_{pred}$ decreases. If the analysis is limited to the first one, the (wrong) idea is that the best method is to reduce the number of cores (in fact, there is a perfect fit and $RMSE_{fit} = 0$ when the number of cores is equal to the number of model parameters, and the model cannot capture all random variations when the number of cores increases). In fact, the real ability of the model increases with the number of cores, as illustrated by the $RMSE_{pred}$ curve.

The magnitude of the prediction error depends on a variety of factors, which include the test result precision, and the range of variation of concrete strength over the investigation domain. The number of cores necessary to get the target precision for strength estimation is thus controlled by these same factors. In common practice, the number of cores is often small, inducing very large differences between both RMSE values.

1.5.6.4 Quantifying the Prediction Error

There are two main possibilities for quantifying $RMSE_{pred}$: (a) direct assessment on a control dataset, (b) cross-validation analysis.

Direct checking consists of keeping a number of cores for validation, i.e. not using the corresponding strength data to fit the conversion model and, once the model has been identified, to estimate the concrete strength from NDT values of this new dataset, and to compare the estimated strength to the measured strength on this control set. The main drawback is that cores being expensive, their number is limited, and it is usually more useful to use all core data to improve the model identification.

Cross-checking offers an alternative approach. The process uses all N_C data pairs. Both strength and NDT cross-validation is a model validation technique for assessing how the results of a statistical analysis will generalize to an independent data set. In most real applications, only a limited amount of data is available, which leads to the idea of splitting the available dataset: part of data, the training set (TS), is used for fitting the model while the remaining part of data, the validation set (VS), is used for evaluating the performance of the algorithm.[40] A single data split yields

---

[40]Relevant information is available in:

- F. Mosteller, J.W. Tukey, Data analysis, including statistics, in G. Lindzey, E. Aronson (eds.), *Handbook of Social Psychology*, Vol. 2. Res. Methods (Addison-Wesley, Reading, Massachusetts, 1968), pp. 80–203;
- M. Stone, Asymptotics for and against cross-validation. Biometrika 64, 29–35 (1977). doi:10.1093/biomet/64.1.29;
- S. Geisser, A predictive approach to the random effect model. Biometrika 61, 101–107 (1974). doi:10.1093/biomet/61.1.101;
- S. Arlot, A. Celisse, A survey of cross-validation procedures for model selection. Stat Surv 4, 40–79 (2010). doi:10.1214/09-SS054.

a validation estimate of the risk, and averaging over several splits yields a cross-validation estimate. There are several ways of dividing the dataset into TS and VS by varying the size of the two sets. The most used ways are:

- leave one out cross validation (in which one observation is left out at each step),
- leave k-out cross validation (in which k observations are left out at each step),
- K-fold cross validation (where the original sample is randomly partitioned into k subsamples and one is left out in each iteration).

1.6 Overall Assessment Methodology and Recommended Number of Cores

The accuracy of the estimation of concrete strength (mean value, standard deviation, local values) depends on two types of contributing factors:

- those corresponding to the intrinsic characteristics of concrete, namely mean strength and standard deviation, and to the technical aspects of the NDT method,
- those which can be modified according to choices made during the investigation, namely, the number and location of cores, the precision of test results, and additional variables such as the choice between predetermined or conditional coring, the type of conversion model and the choice of the methodology for identifying the conversion model parameters.

These factors have an influence on each of the targets of estimation: mean strength, standard deviation and error on local strength (RMSE). However, these influences may be complex and they are difficult to quantify.

1.6.1 Organization of the Evaluation Approach

1.6.1.1 Recommended Sequence of Tasks

The flowchart in Fig. 1.1 summarized the main steps of the evaluation process. A series of important improvements have been introduced in the RILEM TC 249-ISC Recommendations, and the corresponding assessment process is detailed in the flowchart of Fig. 1.9. These improvements are:

- the definition of the estimation quality level (EQL) prior to any investigation,
- the fact that NDT measurements are carried out at a preliminary stage, as a way to perform a screening of the whole investigation domain,
- the quantification of NDT test result precision (TRP),
- conditional coring, i.e. the definition of core location based on NDT test result distribution,

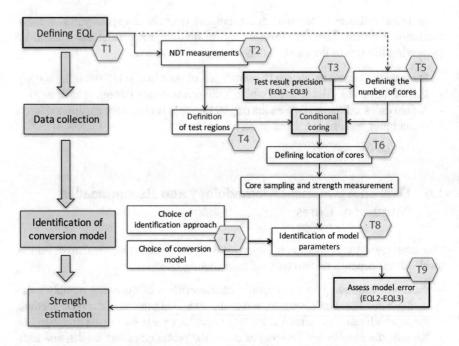

Fig. 1.9 Detailed steps of the recommended concrete assessment process

– the quantification of the model error.

All steps of the flowchart have been described in the previous sections. The dashed line in the upper part of the flowchart corresponds to the case where the number of cores is predefined by external constraints. Table 1.5 specifies additional tasks that are recommended or prescribed according to the target EQL.

1.6.1.2   Target Tolerance Intervals, Number of Cores and Risk of Wrong Assessment

The last issue is the proper justification of a relevant number of cores. For each of the three prescribed targets, each EQL corresponds to a target tolerance interval on the parameter to be assessed, as indicated in Table 1.1.

The tolerance interval can be expressed either in relative terms or in absolute terms:

– with a relative uncertainty $\pm x\%$, this means that the assessed value X_{est} satisfies:

$$X_{est}/(1 + x) \leq X_{true} \leq X_{est}/(1-x)$$

– with an absolute uncertainty $\pm$ X MPa, this means that the assessed value X_{est}
satisfies:

$$X_{est} - X \leq X_{true} \leq X_{est} + X$$

In both cases, X_{true} and X_{est} are, respectively, the true value and the estimated value
of the parameter of interest (mean strength, standard deviation).

However, due to many random influences, whatever the magnitude of the accepted
tolerance interval, which depends on the EQL, there is always some probability that
the assessed value falls outside the accepted interval.

An adequate definition of the number of cores must correspond, for each of these
targets, to an accepted risk (or maximum probability) that the assessed value will
miss the prescribed tolerance interval. It is on this basis that a recommended number
of cores can be provided.

Regarding the error on local strength (RMSE), which is the third criteria defining
EQL in Table 1.1, the target value defines an upper bound, which can be also expressed
either in relative terms or in absolute terms. Therefore, it can be written either as a
percentage x of the true mean strength or as an absolute quantity X in MPa:

$$RMSE_{true} \leq x.f_{c\,true} \quad \text{or} \quad RMSE_{true} \leq X$$

However, due to many random influences, whatever the magnitude of this accepted
tolerance interval, which depends on the EQL, there is always a risk for $RMSE_{true}$ to
be larger than this accepted upper bound. These random influences are analysed in
the RILEM TC 249-ISC Recommendations (Sect. 4.1) and in Chap. 2 of this book.

1.6.2 Recommended Number of Cores

The number of cores is among the main influencing factors regarding the strength
estimation quality. Thus, the recommended number of cores must be in agreement
with the magnitude of the target tolerance interval, for a given accepted risk.

The number of cores necessary to obtain test results in conformity with a
prescribed EQL has to be in agreement with three criteria:

(a) to have enough cores to achieve an acceptable assessment of the mean strength,
(b) to have enough cores to achieve an acceptable assessment of the standard
deviation of strength (at EQL2 and EQL3 only),
(c) to have enough cores to achieve an acceptable error on local strength (i.e. RMSE
below a given threshold, at EQL2 and EQL3 only).

For each criterion, the strength estimation quality is expressed in Table 1.1 in terms
of tolerance intervals.

The accepted risk for the first two targets is taken at 10%, which comes to a
10% chance for the assessed value to be outside the target tolerance interval. This

corresponds, for a symmetrical risk, to have a 5% chance of under/overestimating the mean strength and the concrete variability. For the third target (local strength estimation), the risk is taken at 5%, which comes to a 5% probability that the threshold RMSE is exceeded.

There is no way to formally derive a recommended number of cores covering all possible situations from a theoretical model or from real engineering practice. The numbers provided in this section are derived from calculations carried out on synthetic data. Such data allow for a much more extensive analysis of possible scenarios. They are also the only way to address the assessment issue in terms of risk. Details regarding these calculations are given in Chap. 2.

The recommended number of cores are supposed to apply in most real situations for on-site investigation of concrete. They must be taken as a general guidance, which corresponds to the three EQL defined in Table 1.1.

The recommended number of cores N_c is given by:

$$N_c = max(N_{c1}, N_{c2}, N_{c3})$$

where N_{c1}, N_{c2} and N_{c3} denominate the minimum number of cores required for each criterion.

For each criterion and each EQL, these numbers depend on three inputs:

- the non-destructive Test Result Precision (TRP) as defined in Sect. 1.5.2.1 and Table 1.5,
- the concrete mean strength,
- the coefficient of variation (COV) of concrete strength.

The TRP is denominated "high precision", "medium precision" and "poor precision", in agreement with reference values given in Table 1.5. **If TRP has not been assessed, the numbers given for "poor precision" must be taken.**

Since concrete mean strength and COV are usually unknown, the most unfavorable combination (regarding possible mean strength and COV of concrete strength) should be considered, for the selected TRP.

If the investigator has some reliable information, including that provided by NDT screening or the documentation on the structure, which allows him to reduce the possible range of variation of any of these two parameters, he can use this information to define the recommended number of cores. This information must be documented.

Chapter 2 provides full details on the recommended number of cores. It explains first on what bases the RILEM TC 249-ISC identified these numbers. Thus, it provides synthetic information (Chap. 2, Sects. 2.4 and 2.5) about how to define this number in practice.

Acknowledgements The 16th author would like to acknowledge the financial support by Base Funding—UIDB/04708/2020 of CONSTRUCT—Instituto de I&D em Estruturas e Construções, funded by national funds through FCT/MCTES (PIDDAC).

References

Standards

1. ACI228.1R-19, Report on methods for estimating in-place concrete strength, Reported by ACI Committee 228, A.J. Boyd (chair) (2019)
2. ASTM C42:2016, Standard test method for obtaining and testing drilled cores and sawed beams of concrete (2016)
3. ASTM C803:2010, Standard test method for penetration resistance of hardened concrete (2010)
4. ASTM C805:2013, Standard test method for rebound number of hardened concrete (2013)
5. ASTM C823:2017, Standard practice for examination and sampling of hardened concrete in constructions (2017)
6. BS 6089:2010, Assessment of in-situ compressive strength in structures and precast concrete components—Complementary guidance to that given in BS EN 13791, British Standard (2010). ISBN: 978 0 580 67274 3
7. EN 12504–1:2012, Testing concrete in structures - Part 1: Cored specimens—Taking, examining and testing in compression, European Committee for Standardization (CEN), Brussels, Belgium (2012)
8. EN 12504–2:2012, Testing concrete in structures—Part 2: Non-destructive testing - Determination of rebound number, European Committee for Standardization (CEN), Brussels, Belgium (2012)
9. EN 12504–3:2005, Testing concrete in structures—Part 3: Determination of pull-out force, European Committee for Standardization (CEN), Brussels, Belgium (2005)
10. EN 12504–4:2004, Testing concrete—Part 4: Determination of ultrasonic pulse velocity, European Committee for Standardization (CEN), Brussels, Belgium (2004)
11. EN 13791:2007, Assessment of in-situ compressive strength in structures or in precast concrete products, European Committee for Standardization (CEN), Brussels, Belgium (2007)
12. EN 1998–3:2005, Eurocode 8: Design of structures for earthquake resistance—Part 3: Assessment and retrofitting of buildings, European Committee for Standardization (CEN), Brussels, Belgium, (2005)

Scientific Publications

14. Ali-Benyahia, K., Sbartaï, Z.M., Breysse, D., Kenai, S., Ghrici, M.: Analysis of the single and combined non-destructive test approaches for on-site concrete strength assessment: general statements based on a real case-study. Case Study Constr. Mater. **6**, 109–119 (2017)

15. Alwash, M., Breysse, D., Sbartaï, Z.M.: Non destructive strength evaluation of concrete: analysis of some key factors using synthetic simulations. Constr. Build. Mater. **99**, 235–245 (2015)
16. Alwash, M., Sbartaï, Z.M., Breysse, D.: Non-destructive assessment of both mean strength and variability of concrete: a new bi-objective approach. Constr. Build. Mater. **113**, 880–889 (2016)
17. Arlot, S., Celisse, A.: A survey of cross-validation procedures for model selection. Stat. Surv. **4**, 40–79 (2010) https://doi.org/10.1214/09-SS054
18. Bartlett, F.M., MacGregor, J.G.: Effect of moisture condition on concrete core strengths. ACI Mater. J. **91**(3), 227–236 (1994)
19. Biondi, S.: The knowledge level in existing building assessment. In: 14th World Conference on Earthquake Engineering, Beijing, China, 12–17 October 2008
20. Breysse, D.: Nondestructive evaluation of concrete strength: an historical review and a new perspective by combining NDT methods. Constr. Build. Mater. **33**, 139–163 (2012)
21. Brozovsky, J., Zach, J.: Influence of surface preparation method on the concrete rebound number obtained from impact hammer. In: 5th Pan American Conference for NDT, Cancun, Mexico, 2–6 Oct. 2011
22. Bungey, J.H., Millard, S.G., Grantham, M.G.: Testing of Concrete in Structures. CRC Press, Boca Raton (2018)
23. Celik, A.O., Kilinc, K., Tuncan, M., Tuncan, A.: Distributions of compressive strength obtained from various diameter cores. ACI Mater. J. **109**(6), 597–606 (2012)
24. Felicetti, R.: The drilling resistance test for the assessment of fire damaged concrete. Cem. Concr. Compos. **28**(4), 321–329 (2006)
25. Geisser, S.: A predictive approach to the random effect model. Biometrika **61**, 101–107 (1974). https://doi.org/10.1093/biomet/61.1.101
26. Machado, M.D., Shehata, L.C.D., Shehata, I.A.E.M.: Correlation curves to characterize concretes used in Rio de Janeiro by means of non-destructive tests. Rev IBRACON Estrut. Mater. **2**(2), 100–123 (2009)
27. Maierhofer, C., Reinhardt, H.-W., Dobmann, G.: Non-Destructive Evaluation of Reinforced Concrete Structures. Woodhead Publishing (2010)
28. Malhotra, V.M., Carino, N.J. (eds.): Handbook of Nondestructive Testing of Concrete. CRC Press, Boca Raton, Florida (2003)
29. Masi, A., Chiauzzi, L.: An experimental study on the within-member variability of in situ concrete strength in RC building structures. Constr. Build. Mater. **47**, 951–961 (2013)
30. Masi, A., Chiauzzi, L., Manfredi, V.: Criteria for identifying concrete homogeneous areas for the estimation of in-situ strength in RC buildings. Constr. Build. Mater. **121**, 576–587 (2016)
31. Moczko, A.T., Carino, N.J., Petersen, C.G.: CAPO-TEST to estimate concrete strength in bridges. ACI Mater. J. **113**(6), 827–836 (2016)
32. Mosteller, F., Tukey, J.W.: Data analysis, including statistics. In: Lindzey, G., Aronson E. (eds.), Handbook of Social Psychology, Vol. 2. Res. Methods, pp. 80–203. Addison-Wesley, Reading, Massachusetts (1968)
33. Nguyen, N.T., Sbartaï, Z.M., Lataste, J.F., Breysse, D., Bos, F.: Assessing the spatial variability of concrete structures using NDT techniques—laboratory tests and case study. Constr. Build. Mater.s **49**, 240–250 (2013)
34. Pascale, G., Di Leo, A., Carli, R., Bonora, V.: Evaluation of actual compressive strength of high strength concrete by NDT. In: 15th World Conference on NDT, Rome, Italy, 15–21 Oct. 2000, idn527
35. RILEMTC 207-INR, Non-destructive assessment of concrete structures: reliability and limits of single and combined techniques. In: Breysse, D. (ed.), State-of-the-Art Report of the RILEM Technical Committee 207-INR (2012). https://doi.org/10.1007/978-94-007-2736-6
36. RILEMTC 43-CND, I. Facaoaru (chair), Draft recommendation for in situ concrete strength determination by combined non-destructive methods. Mater. Struct. **26**, 43–49 (1993)
37. Stone, M.: Asymptotics for and against cross-validation. Biometrika **64**, 29–35 (1977). https://doi.org/10.1093/biomet/61.1.101

Chapter 2
How to Identify the Recommended Number of Cores?

**Jean-Paul Balayssac, Denys Breysse, Maitham Alwash,
Vincenza Anna Maria Luprano, Xavier Romão, and Zoubir Mehdi Sbartaï**

Abstract The concrete strength assessment process is influenced by uncertainties at many levels, including random measurement errors, sampling uncertainty and identification of the conversion model parameters. Therefore, instead of estimating the true value of the concrete strength, it is preferable to say that the objective of the assessment process is to predict a strength value ranging at a tolerable distance from the true strength value. This implies a deep revision of the assessment paradigm, in which both the acceptable tolerance interval and the risk of a wrong assessment must be given at the very beginning of the investigation. A large series of simulations has been carried out in order to understand and quantify how, for a given tolerance on the strength estimation, the risk value varies as a function of the precision of measurements, the number of cores and the strength distribution. Empirical models have been identified from the simulation results. These models have been finally used to calculate how many cores are required in various situations, to achieve the accuracy corresponding to three different estimation quality levels. This chapter describes the principles of the simulation, and how their results were used in order to build a series of tables where the recommended number of cores is made available in a variety of situations.

J.-P. Balayssac (✉)
LMDC, Université de Toulouse, INSA/UPS Génie Civil, Toulouse, France
e-mail: jean-paul.balayssac@insa-toulouse.fr

D. Breysse · Z. M. Sbartaï
University Bordeaux, I2M-UMR CNRS 5295, Talence, France

M. Alwash
Department of Civil Engineering, University of Babylon, Babylon, Iraq

V. A. M. Luprano
ENEA - Italian National Agency for New Technologies, Energy and Sustainable Economic Development, Department for Sustainability - Non Destructive Evaluation Laboratory, Brindisi, Italy

X. Romão
CONSTRUCT-LESE, Faculty of Engineering, University of Porto, Porto, Portugal

© RILEM 2021   57
D. Breysse and J.-P. Balayssac (eds.), *Non-Destructive In Situ Strength Assessment of Concrete*, RILEM State-of-the-Art Reports 32,
https://doi.org/10.1007/978-3-030-64900-5_2

2.1 Introduction

The information provided in the first chapter (Guidelines) has more details that what is written in the RILEM TC 249-ISC Recommendations [1]. The Guidelines in the first chapter of this book provide a more extensive description of the context and the background, and they also provide additional information that is useful for engineers who want to apply the Recommendations in practice. However, one very practical issue was not developed in the Guidelines: the recommended minimum number of cores that is needed to identify the conversion model. The number of cores is a governing factor regarding the final precision of the strength assessment. Increasing this number has a beneficial influence on the precision of the strength estimation, but it also has a direct impact on the investigation cost. The problem is complex because the precision of the strength estimation can be influenced by many factors, such as the test result precision (TRP), the concrete properties (mean strength and strength variability) and choices made at the various steps of the concrete assessment process (see Chap. 1, Fig. 1.9).

Faced with this complexity, the RILEM TC 249-ISC has chosen to explain into detail how the whole assessment process is organized. This chapter is specifically dedicated to the issue of the number of cores needed for the strength assessment. It is divided into five main sections. The reader who only wants to identify the recommended number of cores could skip Sects. 2.2–2.3 and go directly to Sects. 2.5 and 2.6 where practical information is made available in a series of tables corresponding to specific cases that may be found in real strength assessment situations. On the other hand, the first three sections explain in more detail the procedures that led to these tables. Section 2.2 describes the theoretical bases that justify how relationships can be identified between the number of cores, the target precision on strength assessment and a variety of influencing factors. It also introduces the concept of risk curves for each parameter that is estimated (mean strength, strength variability, local strength) which are a very practical mean to make these relationships more explicit. Sections 2.3 and 2.4 describe the work developed by the RILEM TC 249-ISC to identify these relationships and to synthesize the results in a series of tables. A large set of synthetic simulations were carried out and their outputs were processed in order to quantify how the various influencing factors impact the final precision of strength assessment (Sect. 2.3). A further data processing step leads to empirical models (Sect. 2.4) which enable calculating the recommended number of cores as a function of all the influencing parameters. In a last stage, the robustness of the recommended number of cores is discussed.

The reader can now directly go to Sect. 2.5 for a practical application of the RILEM TC 249-ISC Recommendations or follow the logical development of this chapter, in order to fully understand the consequences of his different choices in the assessment process.

2.2 Theoretical Principles—Strength Assessment Precision and Risk Curves

2.2.1 Considering Risk in Non-destructive Concrete Strength Assessment

Analyzing the non-destructive strength assessment issue requires a paradigm change [2] that can be described as follows. Figure 2.1 reproduces the strength assessment process that is commonly carried out by practitioners. In this process, a given conversion model (which can be either a pre-existing model or a specific one adapted to a certain context) is used to transform the non-destructive test (NDT) results into an estimated strength, which is the final result. However, when the process is more closely analyzed, it is seen to be oversimplified, thus requiring different types of uncertainties to be considered, as described in Fig. 2.2. This new flowchart divides the process into two stages: model identification and model use.

This flowchart is divided into two stages, corresponding respectively to model identification and to model use. When some attention is paid to what happens during the full process of strength estimation, it is easy to understand that uncertainty (or

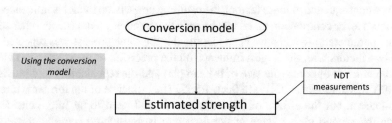

Fig. 2.1 Usual strength estimation process

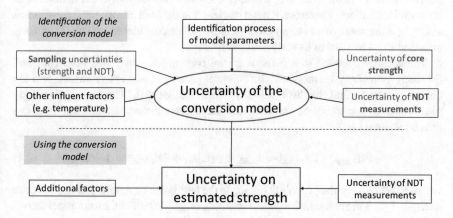

Fig. 2.2 Uncertainties arising in the different stages of the strength estimation process

errors) can impact the process at many stages. Significant research efforts have been recently devoted to a more systematic analysis of all the degrees of freedom of the non-destructive investigation and assessment process [3–5]. In the first main stage (i.e. the conversion model identification), the influence of the following effects has been highlighted:

- statistical (sampling) uncertainty, due to the limited size of the dataset upon which the model is calibrated, i.e. typically the number of cores,
- measurement uncertainty, on core strength measurements and on NDT measurements, which depends mostly on the measurement technique itself, but also on the device, on the expertise of who takes the measurement and on the environmental context. This also includes all biases that can appear during the drilling and the core preparation,
- factors related to the model identification process itself (from data to model parameters), such as the choice of the mathematical shape of the model, or the method that is used to select the location of cores. These factors have a large number of degrees of freedom and a large potential for improvement,
- additional uncontrolled factors, not considered in the analysis and that can have some influence on the core strength values, on the NDT results, or on their relationship (e.g. temperature or carbonation).

As a consequence of these factors, the resulting conversion model is influenced by a random error component, and it can be seen that another model (i.e. another set of model parameters) would have been identified if the same process was repeated [6]. The two factors with the largest influence on the precision of the conversion model are the number of cores (the size of the sample) and the repeatability and precision of NDT measurements (TRP, see Sect. 1.5.2). The influence of the former is already well-known, but the effects of the latter are not and need to be fully considered. When the second global stage of the flowchart is considered (i.e. the conversion model use), one has now an "uncertain" conversion model that will be used with new data (new NDT results) that have possibly been influenced by different or additional uncontrolled factors. Therefore, it must be clear that the final output, i.e. the estimated strength, is the result of a random process, and must be considered as such. Therefore, attention must be paid to its statistical distribution.

These statements led to a revision of the framework which has been followed for many years by practitioners. The classical paradigm was deterministic, and the challenge was to find the "true value" of concrete strength. The revised paradigm must consider uncertainties and risk, and the corresponding strength estimate can thus be written as.

$$P(f_{c,true} - Df_{c,true}) < f_{c,est} < P(f_{c,true} + Df_{c,true}) = 1 - \alpha \qquad (2.1)$$

where $P()$ stands for the probability, $f_{c,true}$ is the true value (unknown) of the concrete strength, $f_{c,est}$ is the estimated concrete strength, $\Delta f_{c,true}$ is half of the tolerance interval defined for the true strength and $(1 - \alpha)$ is the confidence level of the estimation (where α is the risk of a wrong assessment, i.e. the risk of obtaining an estimate outside the prescribed tolerance interval). In Eq. (2.1), the property to be estimated can be the mean strength, the local strength, but the same expression is also applicable for the standard deviation of strength. According to the revised paradigm, the challenge is no longer to identify the true value of the concrete property but, instead, to estimate it within some tolerance interval, and at a given (accepted) risk. Furthermore, it can be seen that Eq. (2.1) can be adapted if the tolerance interval is defined in relative terms (i.e. in percentage) instead of absolute ones. In this case, Eq. (2.1) can be rewritten as:

$$P\big(f_{c,true} \times (1 - U)\big) < f_{c,est} < P\big(f_{c,true} \times (1 + U)\big) = 1 - \alpha \qquad (2.2)$$

where $U = \Delta f_{c,true} / f_{c,true}$.

2.2.2   Risk of a Wrong Estimation

As explained in the introduction, a common way to express the precision of the assessment using NDTs is to provide the estimated value with a $\pm U\%$ error (tolerance interval). In fact, due to many sources of uncertainty, if the whole estimation process is repeated many times, different estimated values are obtained and their cumulative distribution function (CDF) can be plotted. The reliability of the estimation improves as the scatter of the data reduces. Therefore, factors affecting the scatter also control the reliability. For instance, increasing the number of cores or reducing the within test repeatability (TRP) of NDT results reduces the scatter and improves the precision. To correlate these factors with the precision of the assessment, it is preferable to proceed in two steps:

(a)   Using the CDF curve of multiple estimates, by fixing a tolerance interval $\pm U\%$ around the true reference/target value,
(b)   Quantify the probability (or risk) that the estimated value lies outside the prescribed tolerance interval, i.e. the risk that the estimated quantity is either more than $T \times (1 + U)$ or lower than $T \times (1 - U)$, T represents the true value of the quantity.

The way of deriving the risk values (i.e. the probability of a wrong estimate that lies outside the prescribed tolerance interval) from the CDF curve is illustrated in Fig. 2.3.

This concept of risk is slightly adapted for the third possible objective of the estimation process, i.e. the local strength assessment, since the prediction error for local strength $RMSE_{pred}$ is always positive. Thus, Eqs. (2.1) and (2.2) are transformed into Eqs. (2.3) and (2.4) that refer to absolute and relative one-sided tolerance intervals, respectively:

$$P\big(RMSE_{pred} > X\big) = 1 - \alpha \qquad (2.3)$$

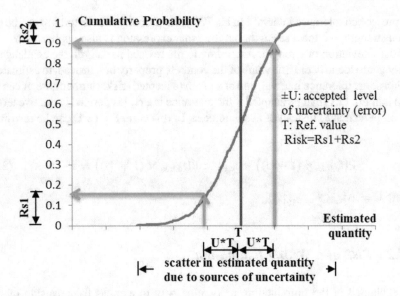

Fig. 2.3 Proposed concept to derive a risk value that corresponds to a given tolerance interval using the CDF curve of multiple estimation processes

$$P(RMSE_{pred} > a.f_{c,mean}) = 1 - \alpha \tag{2.4}$$

If the risk level α is given, X in Eq. (2.3) corresponds to the $1 - \alpha$ percentile of the distribution of local errors $RMSE_{pred\ 1-\alpha}$, and a in Eq. (2.4) expresses this percentage as a fraction of the concrete mean strength.

2.2.3 Most Influencing Factors

In light of the previous arguments, it is therefore mandatory to better understand and quantify the respective role of all uncertainties described in Fig. 2.2. The number of cores is the first (and best known) factor, and its role has been identified by all standards, which usually prescribe a "minimum number of cores" [7, 8]. Within the new paradigm, which is risk-based, one can assume that the larger the number of cores, the more accurate the estimation (i.e. the smaller the tolerance interval for a given risk level). However, because of the combined effects of the several uncertainty factors on the final estimation, there is no simple mathematical expression that defines an explicit relation between the number of cores and the final accuracy.

This problem has recently been addressed with more detail in studies that considered either real data or synthetic data [3, 9]. For instance, Table 2.1 was synthesized in [3] after a meta-study that analyzed more than 2500 test result pairs (rebound number R, cube strength f_c) from 17 previously published datasets of various origins [10].

Table 2.1 Effects of influencing factors on the reliability of the assessment, using the bi-objective approach for calibrating the conversion model

	Effect of the factors on the reliability of the	
Influencing factors (when they decrease)	Mean strength assessment	Strength variability assessment
Number of cores used to identify the conversion model (N_C)	Reduces	Reduce
Within-test variability of NDT results (TRP)	Improves	Improves
Concrete strength variability, $sd(f_c)$	Improves	Reduce

This meta-study addressed the ability to assess both mean strength and strength standard deviation.

The study confirmed that the two most influencing factors are the number of cores and the TRP of NDT results. However, the mean strength assessment is more accurate the smaller the concrete variability is, whereas the variability assessment is more accurate when this variability is large.

The major role played by the number of cores justifies the concept of risk curves.

2.2.4 Risk Curves

Risk values can be plotted as a function of the number of cores (N_C) to establish curves called "Risk curves". These curves usually show how the risk value decreases when the number of cores is increased, which reduces the consequences of uncertainties. This way of analyzing the efficiency of the strength assessment process was first promoted in [11] where risk curves for the assessment of the mean strength and of the strength standard deviation are developed to analyze the role of several influencing factors (number of cores, choice of the conversion model and its calibration method, use of single NDT technique or of a combination of NDTs). Recent studies [12, 13] have used risk curves for local errors with real datasets obtained from existing buildings to identify how these errors depend on the number of cores and on how the conversion model should be identified and calibrated.

Risk curves, as illustrated on Fig. 2.4 (see Sect. 2.3.3.1), quantify the relationship between the number of cores and the risk level on a given quantity (i.e. local strength, mean strength or strength standard deviation) for a given concrete, defined through its mean strength and strength variability, and a given within-test variability of NDT results.

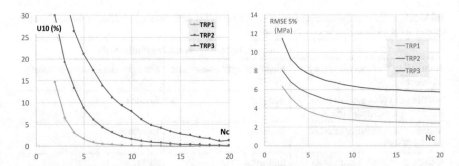

Fig. 2.4 Risk curves for U_{10} (left) and $RMSE_5$ (right) for the three TRPs, considering rebound test results and a specific linear conversion model (C 30–20 concrete)

2.2.5 Multi-objective Risk Curves and Recommended Number of Cores

In the RILEM Recommendations, three quantities are addressed at EQL2 and EQL3 (Chap. 1, Table 1.1), which represent the level of accuracy of the concrete strength estimation. A risk curve can be built for each of these quantities, once the tolerance interval and the risk level are defined. In Sect. 2.3.4, formulas and tables are provided for deriving the recommended number of cores. The values were determined considering the following assumptions:

(a) The tolerance intervals were fixed (at an indicative level) by RILEM TC 249-ISC in Table 1.1. For instance, regarding the mean strength assessment, they correspond to ±15% at EQL1 and EQL2 and to ±10% at EQL3. It can be noted that, if different tolerance intervals were considered, there would be no obstacle to derive different numbers of cores corresponding to these new precision requirements.

(b) The admissible risk level was taken as 10% for the first two quantities and as 5% for the third one (i.e. that of the $RMSE_{pred\ 95}$ value). This difference comes from the fact that, with a symmetrical distribution of the error, the 10% two-sided risk for the first two quantities is equivalent to the 5% one-side risk for the third quantity.[1] The 5% risk thus corresponds:

- to a 5% risk of overestimating the concrete mean strength,
- to a 5% risk of underestimating the concrete strength variability,
- to a 5% risk of underestimating the error on the local strength.

---

[1] From a mathematical point of view, the 10% two-sided risk is not equivalent to the 5% one-sided risk. However, regarding the structural safety purpose, the real risk is that of underestimating the concrete strength (the risk of overestimating strength only addresses financial issues, and no safety issues). Regarding variability, there is some risk when it is assumed to be lower than its real value, as the resulting characteristic strength (defined as a lower percentile) may be overestimated.

(c) At EQL2 and EQL3, each of the three quantities leads to a different number of
     cores. Therefore, the largest value among these three numbers is considered the
     reference minimum number of cores since it is the smallest value that satisfies
     the three objectives.

2.3   How Risk Curves Are Built

2.3.1   The Principles of Synthetic Simulations

The RILEM TC 249-ISC has chosen to carry out a series of synthetic simulations
in order to obtain the necessary data to build the risk curves. The approach based on
synthetic simulations had been used in recent research and had shown a good level
of efficiency for the strength assessment of concrete [4, 14]. Since it is impossible
to cover a large variety of assessment situations with real data only (i.e. with data
coming from the field), the following advantages of synthetic simulations (SS) were
put forward:

- SS are a quick and low-cost process,
- SS enable the analysis of a large variety of contexts regarding the concrete
   properties, the precision of test results or the investigations strategy,
- SS provide the possibility of deriving a large quantity of results, as it is very easy
   to simulate different conditions for the assessment, e.g. vary the number of cores
   or the conversion model identification method; the efficiency of different methods
   can be tested for each dataset that is made of simulated core strength values and
   NDT results.

However, the major drawback of SS is that results will obviously depend on
the quality of the assumptions made for the simulations. Assumptions are required
regarding the concrete strength distribution, the effect of some hidden influencing
factors (like the level of carbonation or concrete moisture), and the magnitude of all
input parameters (concrete strengths, within-test variability of test results). Before,
performing these analyses, it was first checked that SS can lead to results that reproduce all the main features known by experts (such as the effect of influencing factors,
as described in Table 2.1) and that they provide self-consistent outputs. Therefore,
the range of variation of each input parameter was carefully chosen, in order to cover
a large domain of interest, for which the final results can be validated.

This chapter will not detail how SS were performed in order to quantify the efficiency of the assessment strategy (covering the data collection stage, i.e. investigation
stage, as well as the post-processing of data, i.e. the conversion model identification process), since this has been described in detail in previous publications [15].
Whereas the simulation methodology principles remained unchanged for the analyses that were performed, a MATLAB code was nonetheless developed to increase
its computational efficiency.

2.3.2 Application Domain and Assumptions

2.3.2.1 Input Parameters

The parameters that are supposed to have the largest influence on the precision of the strength assessment belong to three groups:

(a) Concrete properties that are **independent of the investigation**: the concrete mean strength $f_{c\,mean}$, the concrete strength coefficient of variation $cv(f_c)$. These parameters correspond to the target of the concrete strength estimation strategy.

(b) Parameters that define the **strategy of data collection**: the number of cores N_c, the selected NDT technique, the choice between having only one technique or a combination of techniques, the choice regarding the selection of the location of cores (predefined, conditional), the repeatability of the NDT results ε_{REL}. The influence of N_c and ε_{REL} will be privileged in this first draft.

(c) Parameters that correspond to choices during the **post-processing of the test results**, and different model identification approaches, typically the mathematical shape of the conversion model and the method for identifying the model parameters. Fifteen identification approaches were compared, which are summarized in Table 2.2. In terms of the NDT, they correspond to the use of the rebound hammer (R), of the ultrasonic pulse velocity method (UPV), or to a combination of both (R + UPV). Table 2.2 also shows that three mathematical shapes (linear, power and exponential) were tested. Lastly, several model identification approaches are compared, as S stands for SPECIFIC (i.e. identification of a specific conversion model), C stands for CALIBRATED (with two options: drift Δ or multiplying factor k), S-bi and C-bi stand for the bi-objective approach. The specific model that was used for the approach combining NDTs is the common SonReb model, with a double power law.

The simulation strategy was designed such as to enable multiple comparisons, namely:

Table 2.2 Definition of the 15 model identification approaches (IA)

Model	R			UPV		R + UPV
Linear	S	C	Δ k	S		
	S-bi			S-bi		
Power	S	C	Δ k	S		S
	S-bi			S-bi		
Exponential				C	Δ k	

- the use of rebound test results and that of UPV test results, with everything else being kept identical (with S or S-bi),
- SPECIFIC and SPECIFIC BI-OBJECTIVE models (for R and UPV, for linear and power models),
- SPECIFIC and CALIBRATED (two options) models, for R using linear and power models,
- single NDT technique and combined NDT techniques (using a SPECIFIC power model).

2.3.2.2 Targets to Be Assessed

Three targets are defined as a function of the EQL, which correspond to the possible different targets of the assessment process. During the SS process, each configuration (i.e. concrete type, number of cores, within-test-variability of test results...) is simulated 10 000 times and generates the corresponding assessment results. The RISK value is calculated for each target and corresponds to the frequency of having a wrong estimation for the assessed quantity, i.e. an assessed value which is outside the prescribed tolerance interval. As such, the following are calculated:

- For mean strength, RISK is determined for tolerance intervals of $\pm10\%$, $\pm15\%$, $\pm20\%$, $\pm30\%$.[2] The results corresponding to the first two targets are termed U_{m10} and U_{m15} hereon.
- For standard deviations, RISK is determined for tolerance intervals of $\pm20\%$, $\pm30\%$ and $\pm50\%$.[3, 4] The results corresponding to the last two targets are termed U_{sd30} and U_{sd50} hereon.
- For RMSE, the outputs are directly the RMSE values whose exceedance risk is 5 and 10%. These results are termed $RMSE_5$, $RMSE_{10}$ hereon.

2.3.2.3 Simulation Sets

In the SS, the following are considered:

- concrete strength is considered to follow a Gaussian distribution with a mean strength $f_{c\,mean}$ and a coefficient of variation $cv(f_c)$.

---

[2]The first two values are those kept in the TC 249-ISC Recommendations (Table 1). The last two values correspond to less ambitious tolerance intervals. Simulation results have shown that the first two objectives are easy to satisfy in most cases with a small number of cores. Therefore, these two values ($\pm10\%$ and $\pm15\%$) were kept in the RILEM TC 249-ISC Recommendations.

[3]The simulation results have shown that the most ambitious objective ($\pm20\%$) is likely to be very difficult to satisfy in most cases with a reasonable number of cores. This is the reason why only the two other objectives were kept in the RILEM TC 249-ISC Recommendations.

[4]The tolerance intervals can be expressed either in relative terms (e.g. $\pm50\%$) or in absolute terms (e.g. ±4 MPa). Both were considered since the way the tolerance interval is expressed has an impact on the minimum number of cores (see Sect. 2.4).

Twenty-five combinations of concrete properties are simulated involving $f_{c\ mean}$ values ranging from 10 to 50 MPa and $cv(f_c)$ values ranging from 10 to 30%.[5] **This is the range of concrete properties based on which all our conclusions will apply**.

- an extensive NDT screening before the selection of cores, with a number of test locations $N_{TL} = 100$.
- three TRPs corresponding to high precision (TRP1), medium precision (TRP2) and low precision (TRP3) conditions and that are defined by ε_{REL} values of 1, 2 and 4 units, respectively, for R test results, and of 50 m/s, 100 m/s and 200 m/s, respectively, for UPV test results. These values are consistent with those given in Table 4 of the RILEM TC 249-ISC Recommendations.
- a number of cores N_c varying between 2 and 20.
- the definition of the core locations can be either random/predefined (PC) or conditional (CC),
- a single NDT or a combination of NDT techniques can be used, as defined in Table 2.2, with a total of 15 options.

2.3.3 Illustration of Risk Curves in a Specific Case and Illustration of the Influencing Factors

This first stage of the analysis is only qualitative. As seen in the following, the level of dependency between the results and the input parameters is comparable for all concretes, whatever their mean strength and variability. For this qualitative analysis, a reference concrete (C 30–20 with $f_{c\ mean} = 30$ MPa and $cv(f_c) = 20\%$) located at the center of the investigation range is used.

The simulation outputs are also analysed in the following regarding the influence of:

- the NDT within-test variability,
- the choice of the mathematical shape of the conversion model,
- using conditional coring to select the location of cores (if it is not mentioned, the simulation considers predefined coring),
- using the bi-objective approach,
- using a combination of NDT methods.

---

[5]Each concrete is denominated "C N1-N2", where N1 indicates the mean strength in MPa and N2 refers to the coefficient of variation of strength, in %.

2.3.3.1   Effect of the NDT Within-Test Variability (TRP)

Figure 2.4 presents typical risk-curves and illustrates how risk varies (i.e. decreases) as the number of cores increases, from 2 to 20 cores. Figure 2.4 (left) corresponds to the first quantity (mean strength), and the risk is given in percent while Fig. 2.4 (right) corresponds to the third quantity, i.e. the local error in MPa, and corresponds to a 5% exceedance probability. This means there is a 5% probability that the local error (absolute difference between the reference local strength at a certain test location and the strength estimated from the NDT result at the same location) is larger than the value on the y-axis. Three curves are drawn for the different TRPs where green, blue and red correspond to TRP1 (high precision), TRP2 (medium precision) and TRP3 (low precision), respectively. The regular decrease of risk from TRP3 to TRP1 is visible, given the large influence of this factor.

The three curves for the mean strength assessment converge towards zero-risk but, even with a very large number of cores, the $RMSE_5$ cannot decrease below a certain threshold value which is correlated to the TRP (for 20 cores, this threshold is about 2.5 MPa, 4 MPa and 5.5 MPa for the three TRPs, respectively).

2.3.3.2   Effect of the Conversion Model

Figure 2.5 shows risk curves for the same concrete, for $RMSE_5$ and for TRP2 rebound test results, and compares the RMSE values for three options of the conversion model. These options consider conversion models obtained by identifying a specific model through a regression (involving cases where the model might be linear or defined by a power law) or by calibrating a predefined linear conversion model with a simple shift k.

The differences between these curves are much smaller than those seen in Fig. 2.4 and all converge to the 4 MPa threshold. The differences are only in the rate of

Fig. 2.5 Risk curves for $RMSE_5$ for three options regarding the conversion model identification, TRP2 rebound test results and specific linear conversion model

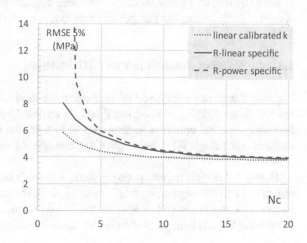

Fig. 2.6 Risk curves for
$RMSE_5$, considering the
three TRPs, rebound test
results and a specific linear
conversion model:
predefined coring (solid)
versus conditional coring
(dotted)

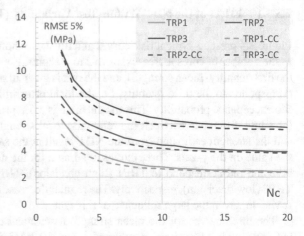

convergence, with the calibrated model and the specific power model being the best
and the worst option, respectively. This is due to the effect of the uncertainty about
the model parameters discussed in Sect. 2.3.3.1. The calibration option has only one
free parameter and is more robust when a low number of cores is considered, whereas
the power law is less robust since the uncertainty on the exponent parameter can have
a very negative influence when only 2 or 3 cores are involved.

2.3.3.3 Effect of Conditional Coring

Figure 2.6 shows how the use of conditional coring can change the $RMSE_5$ risk
values for different TRPs. The difference between the results obtained for the 3 TRP
levels is kept but conditional coring slightly reduces the RMSE values, as shown by
the dotted line curves which lay below the solid line curves. Therefore, if one wants
to limit the risk to a given value, for instance 5%, the use of conditional coring is
able to slightly reduce the number of cores that is necessary.

2.3.3.4 Effect of Combining Two NDT Methods

Figure 2.7 represents the effect of combining rebound and ultrasonic pulse velocity
test results on $RMSE_5$, considering TRP2. To ensure the consistency of this comparison and since the common SonReb model is a double power law, the risk curves
obtained with single NDT techniques and considering specific power laws are also
drawn.

Figure 2.7 reveals an interesting pattern. Although the RMSE reduces when the
number of cores is large enough (10 or more), there is a controversial effect when the
number of cores is low (NC < 6) since the RMSE value is much larger for the model
combining the NDT methods than for the models using a single NDT method. This

Fig. 2.7 Risk curves for RMSE$_5$ and TRP2 considering single techniques (rebound and UPV) and a specific power model, and for a combination of techniques and the SonReb model

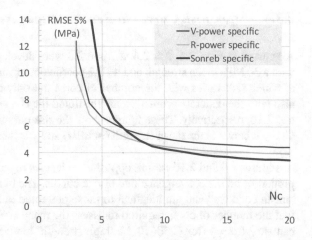

is a direct consequence of the uncertainty of the conversion model parameters since the SonReb model has 3 parameters and the others conversion models only have 2.

2.3.3.5   Effect of Using a Bi-Objective Approach

The bi-objective approach was proposed to better capture the concrete variability [14]. To analyse the effect of using this approach, it is therefore logical to look at the risk curves regarding the variability assessment.

Figure 2.8 (left) shows that the risk regarding the concrete variability assessment is significantly reduced thanks to the use of the bi-objective approach. A 5% risk level can be reached with 20 cores, which is not possible when considering the usual model identification approach. However, the performance of the bi-objective approach for estimating local strength values is slightly worse than that of the conventional model identification approach (Fig. 2.8 right).

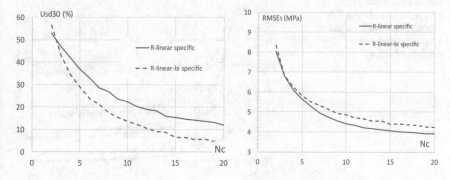

Fig. 2.8 Risk curves for U$_{sd30}$ (left) and RMSE$_5$ (right), with TRP2 (rebound and UPV) and a linear specific model

2.3.4 From Risk Curves to Recommended Number of Cores

All curves in Figs. 2.4, 2.5, 2.6, 2.7 and 2.8 were developed for a specific concrete with a 30 MPa mean strength and a 20% coefficient of variation (C 30–20). The rate at which risk varies with the number of cores not only changes with the TRP, but also with the concrete properties, as illustrated for U_{10} and for $RMSE_5$ in Figs. 2.9 and 2.10, respectively. These figures show the risk curves for five concretes termed C-x–y where x is the mean strength (in MPa) and y is the coefficient of variation (in %).

Figures 2.9 and 2.10 exhibit opposite tendencies regarding the effect of concrete properties on the convergence rate of risk curves. The blue curves correspond to the "mean concrete" and are identical to those presented in Fig. 2.8. Figure 2.9 shows that the number of cores required to assess the mean strength and ensure a 5% risk can vary from 4 (for C 50–10, i.e. high strength, low variability) to more than 20

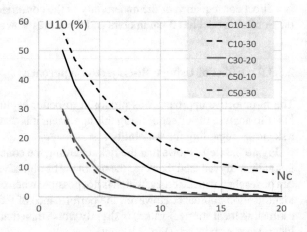

Fig. 2.9 Risk curves for U_{10} for 5 different concretes considering TRP2, rebound test results, a linear specific model and predefined coring

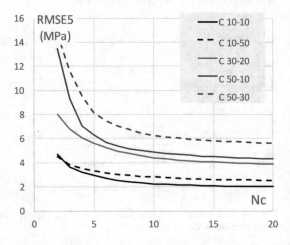

Fig. 2.10 Risk curves for $RMSE_5$ for 5 different concretes considering TRP2, rebound test results, a linear specific model and predefined coring

(for C 10–30, i.e. low strength, large variability). The same type of dependency is visible for the local error $RMSE_5$, but with a reversed sensitivity, i.e. smaller errors are obtained for low strength concretes.

All these results confirm the ability of synthetic simulations to capture the main features governing the precision of the strength assessment strategy, but they also confirm that the problem is a complex one, which does not follow simple rules. The major effects of the number of cores N_c and of the precision of NDT measurements (TRP) are confirmed. They are systematically the most influencing factors, whatever the assessed target (mean strength, variability, RMSE of local values). It has also been checked (but not shown here) that two NDT techniques, namely rebound and ultrasonic pulse velocity, have a very close efficiency when considered with the same TRP level. This is explained by the values defining the TRP categories in the RILEM TC-ISC 249 Recommendations (Table 1.4), as these values were chosen in order to obtain this result.[6]

2.4 Modeling Risk Curves and Deriving the Required Number of Cores

2.4.1 Modelling Risk Curves

As all risk curves (Figs. 2.4, 2.5, 2.6, 2.7, 2.8, 2.9 and 2.10) show how risk decreases monotonically with the increase of the number of cores, i.e. Risk $= f(N_c)$, they can also be considered as a way to identify the minimum number of cores that is required in order to have a risk value below a given acceptable threshold. However, it was shown that the risk value depended not only on the number of cores, but also on a series of other input variables, among which the most influent are seen to be the within-test repeatability and the concrete strength characteristics. The main objective is, therefore, to identify an empirical model that is able to describe these relationships. Given the simulation-based approach that was followed, this model was identified using an experimental dataset containing the results of all synthetic simulations, in order to cover the application domain defined in Sect. 2.2. Once established, the accuracy of the model was checked by comparing real values of outputs (i.e. risk values) with those predicted by the model.

For each quantity of interest that has to be estimated (mean, standard deviation, RMSE of local value) and each tolerance interval, the idea is to establish an empirical model that would be valid in most cases (i.e. for a large variety of concretes) and that quantifies the output (risk-value for the mean and standard deviation, RMSE of

---

[6]This issue will be analyzed further in this document (Sect. 12.3). Still, referring that a "medium precision" R measurement has the same efficiency of a "medium precision" UPV measurement does not mean that, in practice, R measurements have the same efficiency of UPV measurements. The threshold values defining the TRP categories are such that real UPV measurements will most often be TRP1 or TRP2 while R measurements will more probably be TRP2 or TRP3.

local value) as a multi-variable function such as:

$$Y = c_0'(N_c)^{C1}(\varepsilon_{NDT})^{C2}(f_{c\,mean})^{C3}(cv(f_c))^{C4} \qquad (2.5)$$

where N_c is the number of cores, ε_{NDT} is the within-test repeatability, $f_{c\,mean}$ is the mean concrete strength, $cv(f_c)$ is the coefficient of variation of the strength and where c_0 to c_4 are empirical model coefficients. These coefficients are calculated using a least-square minimization process on $\ln(Y)$, which leads to:

$$\ln(Y) = c_0 + c_1\ln(N_c) + c_2\ln(\varepsilon_{NDT}) + c_3\ln(f_{c\,mean}) + c_4\ln(cv(f_c)) \qquad (2.6)$$

In order to make the model analysis clearer, it is preferable to work with standardized variables, i.e. variables which vary in the $[-1, +1]$ interval, whatever their original unit. The standardizing process is the following:

(a) X being the original variable, $\ln(X)$ is calculated
(b) define m and M as the minimum and maximum values taken by $\ln(X)$, respectively,
(c) [X] is the standardized variable, with:

$$[X] = a.\ln(X) + b \qquad (2.7)$$

with $a = 2 / (M - m)$ and $b = (m + M) / (m - M)$.

In practice, this corresponds to values given in Table 2.3.

By introducing the standardized logarithms, the model of Eq. (2.6) becomes:

$$\ln(Y) = a_0 + a_1[\ln(N_c)] + a_2[\ln(\varepsilon_{NDT})] + a_3[\ln(f_{c\,mean})] + a_4[\ln(cv(f_c))] \qquad (2.8)$$

Once the model coefficients are identified, it is possible to come back to the original value of Y from the logarithm $\ln(Y)$.

Table 2.3 Standardization of the model variables

	Original variable	Standardized values
N_c	2, 3, 4, … 20	$-1, -0.648, -0.398, …, +1$
ε_{NDT} for rebound	1/40, 2/40, 4/40	$-0.333, +0.333, +1$
ε_{NDT} for UPV	50/4000, 100/4000, 200/4000	$-1, -0.333, +0.333$
$f_{c\,mean}$	10, 20, 30, 40, 50	$-1, -0.139, +0.365, +0.723, +1$
$cv(f_c)$	10, 15, 20, 25, 30	$-1, -0.262, +0.262, +0.668, +1$

2.4.2   Identification and Validation of Risk Models

2.4.2.1   Risk Model Identification Process and Results

The models are identified in order to fit the simulation results for the whole application domain, and were developed to focus on risk values corresponding to the practical range of interest. The full set of simulation results (i.e. all concretes, all N_c values, all TRPs, two NDT methods) corresponds to $25 \times 19 \times 3 \times 2 = 2\,850$ risk (or RMSE) values for each assessed quantity and each choice regarding the model identifications strategy (i.e. choice of mathematical function for the conversion model and choice of calibration method). Among these 2 850 values, only those corresponding to values close to the range of interest were kept for the risk model calibration, which corresponds to:

- the risk values belonging to the [2.5%, 15%] interval, which contains the 5% risk for the assessment of the mean strength and of the standard deviation of the strength,
- the RMSE values that are lower than 80% of the concrete strength standard deviation, as larger RMSE values would be useless for a reliable assessment.

Tables 2.4 and 2.5 provide the results (Table 2.4 for rebound and Table 2.5 for UPV) obtained from the identification of all the risk model parameters, where N refers to the number of results kept for the model calibration. These correspond to the identification approach with a *univariate linear specific model*. Since the bi-objective approach is expected to be the only efficient approach for the estimation of concrete variability, models for the *single linear bi-objective specific* identification approach were also identified for the concrete variability target, both for an assessment objective expressed in relative values (i.e. 30 and 50%) and absolute values (i.e. 2 and 4 MPa).

Results in Tables 2.4 and 2.5 show that all empirical risk models are highly relevant, as the relative uncertainty on all model parameters is very small. The sign of the model coefficients confirms that the risk decreases for higher N_c values ($a_1 < 0$) and increases for higher TRP values ($a_2 > 0$). The sensitivity to concrete properties (mean strength and strength standard deviation) is more complex and requires further analysis (Sect. 2.3.2.3).

2.4.2.2   Risk Model Validation

Once the risk models are identified, they can be checked by comparing the predicted risk value (from the model) with the risk value resulting from the synthetic simulations. Figures 2.11 and 2.12 represent the measured values against the predicted value for the risk for the assessment of the standard deviation U_{sd30} (i.e. with a tolerance interval of $\pm30\%$) and for the local error $RMSE_5$, respectively. As explained previously, the models were identified from the simulation outputs after removing

Table 2.4 Synthesis of the identified empirical models considering rebound measurements (standard deviation of the coefficients between parentheses, *for the bi-objective approach)

Target	Tolerance interval	N	a_0	a_1	a_2	a_3	a_4	r^2
U_m	±10%	421	1.492 (0.017)	−1.963 (0.045)	1.543 (0.039)	−1.088 (0.039)	0.634 (0.021)	0.826
	±15%	241	0.642 (0.052)	−1.770 (0.075)	1.121 (0.055)	−0.899 (0.043)	0.366 (0.030)	0.710
$U_{sd\,rel}$	±30%	264	4.272 (0.057)	−1.999 (0.053)	2.523 (0.066)	−1.302 (0.035)	−1.596 (0.075)	0.869
	±50%	339	2.567 (0.028)	−1.793 (0.064)	2.231 (0.079)	−1.121 (0.043)	−1.355 (0.051)	0.719
$U^*_{sd\,rel}$	±30%	418	3.363 (0.035)	−1.853 (0.046)	1.454 (0.035)	−0.898 (0.025)	−1.357 (0.033)	0.828
	±50%	390	2.010 (0.023)	−1.475 (0.064)	0.987 (0.047)	−0.775 (0.038)	−1.150 (0.050)	0.616
$U^*_{sd\,abs}$	±2 MPa	539	1.922 (0.024)	−0.790 (0.052)	0.722 (0.044)	0.677 (0.038)	0.0311 (0.023)	0.407
	±4 MPa	265	0.184 (0.061)	−1.654 (0.056)	0.958 (0.044)	1.410 (0.060)	0.101 (0.022)	0.774
$RMSE_5$ (MPa)		596	1.2593 (0.0055)	−0.3479 (0.0058)	0.6664 (0.0057)	0.4239 (0.0045)	0.1173 (0.0044)	0.986
$RMSE_{10}$ (MPa)		669	1.1859 (0.0052)	−0.3097 (0.0054)	0.6683 (0.0059)	0.4203 (0.0048)	0.1154 (0.0046)	0.983

(a) the risk values ranging outside the [2.5%, 15%] interval for the strength standard deviation and (b) the RMSE values exceeding 80% of the concrete strength standard deviation, which corresponds to an upper limit of 12 MPa in the most critical case ($f_{c\,mean}$ = 50 MPa, cv = 30%).

These two figures show that points corresponding to the two NDT techniques (R and UPV) have the same trend. Furthermore, the results also show that the main feature is the effect of TRP, which clearly appears in Fig. 2.12, where TRP1 leads to lower $RMSE_5$ values whereas TRP3 has the poorest performance. This effect is not directly visible in Fig. 2.11, where it is combined with the effect of the number of cores, as an identical risk level can be reached for two different N_c values according to TRP, for instance with 8 cores with for TRP1 but also with 13 cores with TRP2.

The scatter of the clouds of points of the two figures is small. This is particularly true for $RMSE_5$, as the models (Tables 2.4 and 2.5) are very good predictors of the effective performance of the investigation strategy. Regarding the estimation of the concrete standard deviation, Fig. 2.11 shows that a predicted risk of 5% (which will be the objective for which the minimum number of cores will be defined in Sect. 2.3.4) corresponds to similar effective results which belong to the [3%, 7%] interval, whatever the combination of number of cores, type and precision of NDT results, and whatever the concrete properties (mean strength and variability). This

Table 2.5 Synthesis of the identified empirical models, considering UPV measurements (standard deviation of the coefficients between parentheses, *for the bi-objective approach)

Target	Tolerance interval	N	a_0	a_1	a_2	a_3	a_4	r^2
U_m	±10%	463	2.618 (0.023)	−1.851 (0.049)	1.406 (0.040)	−0.795 (0.024)	0.652 (0.024)	0.767
	±15%	264	1.426 (0.021)	−2.013 (0.067)	1.369 (0.053)	−0.764 (0.032)	0.536 (0.029)	0.782
$U_{sd\,rel}$	±30%	241	5.822 (0.107)	−1.719 (0.057)	2.377 (0.069)	−0.757 (0.027)	−1.142 (0.040)	0.847
	±50%	331	4.378 (0.080)	−1.722 (0.061)	2.399 (0.079)	−0.711 (0.031)	−1.358 (0.049)	0.744
$U^*_{sd\,rel}$	±30%	441	4.298 (0.055)	−1.795 (0.047)	1.253 (0.032)	−0.556 (0.021)	−1.249 (0.031)	0.816
	±50%	414	2.665 (0.038)	−1.317 (0.062)	0.871 (0.046)	−0.608 (0.034)	−1.033 (0.048)	0.563
$U^*_{sd\,abs}$	±2 MPa	485	2.471 (0.034)	−0.898 (0.052)	0.696 (0.041)	0.883 (0.042)	0.092 (0.024)	0.493
	±4 MPa	310	1.436 (0.041)	−1.040 (0.065)	0.627 (0.055)	0.739 (0.063)	0.102 (0.031)	0.467
$RMSE_5$ (MPa)		499	1.259 (0.0055)	−0.3479 (0.0058)	0.666 (0.0057)	0.4239 (0.0045)	0.1173 (0.0044)	0.986
$RMSE_{10}$ (MPa)		570	1.6281 (0.0097)	−0.2763 (0.0062)	0.6057 (0.0077)	0.5627 (0.0048)	0.1666 (0.0053)	0.983

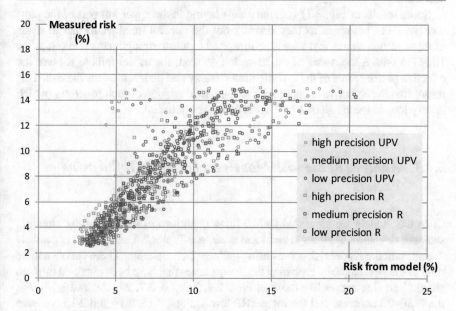

Fig. 2.11 Simulation results and risk model for U^*_{sd30} (all concretes, bi-objective approach, different colour markers correspond to different TRPs)

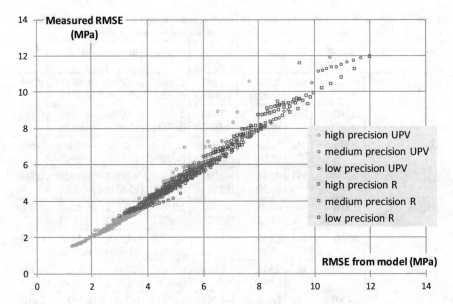

Fig. 2.12 Simulation results and model for RMSE$_5$ (all concretes, different colour markers correspond to different TRPs)

confirms that using the identified risk models for deriving the number of cores is adequate.

Some results of Fig. 2.11 (i.e. the points laying in the upper left part of the plot) deserve a last comment as they depart from the general trend. A further analysis of these results shows that they all correspond to a low strength concrete ($f_{c\,mean}$ = 10 MPa) with a low value of cv. In such a context, the model fails to capture the effective performance of the investigation strategy and underestimates the risk. This means that the objective of capturing the standard deviation of such concretes (which is very low, in the range of 1 or 2 MPa) cannot be satisfied.

2.4.3 Analysis of the Risk Models and of Their Influencing Factors

Once the models are identified for the three quantities of interest (mean strength, standard deviation of strength and local error, see Tables 2.4 and 2.5), it is possible to represent the model risk curves using their explicit equation. These curves will be used in the next steps, by replacing the many curves that have been obtained through simulations (i.e. curves like those of Figs. 2.4, 2.5, 2.6, 2.7, 2.8, 2.9 and 2.10). For the C 30–20 concrete and the three TRP levels, Figs. 2.13, 2.14 and 2.15 compare the model curves with the simulation results in terms of mean strength assessment (Fig. 2.13), standard deviation assessment (Fig. 2.14) and local error (Fig. 2.15).

Fig. 2.13 Simulation results and model estimates for mean strength U_{10} with UPV test results (C 30–20, 3 TRPs)

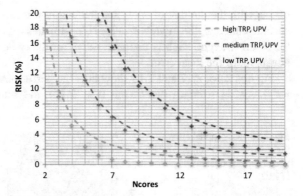

Fig. 2.14 Simulation results and model estimates for U^*_{sd30} with UPV test results (C 30–20, 3 TRPs)

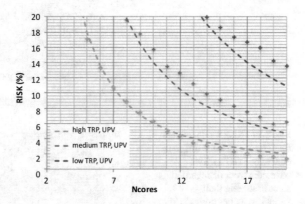

Fig. 2.15 Simulation results and model estimates for $RMSE_5$ with UPV test results (C 30–20, 3 TRPs)

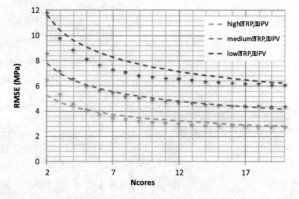

The same plotting conventions are used for the three figures, where dashed lines are used for the model predictions and markers are used for the simulation results, and each of the colours green, blue and red is assigned to one of the three TRPs. These figures confirm the capacity of the risk models to fully capture the variation of risk with the number of cores and the within-test repeatability of NDTs. For this specific concrete, the models provide risk estimates that are slightly conservative for the mean strength, slightly unconservative for the standard deviation, and perfectly adequate for the local error.

Figures 2.16, 2.17 and 2.18 show how the performance of the assessment varies for the mean strength assessment (Fig. 2.16), the standard deviation assessment

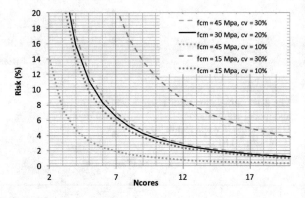

Fig. 2.16 Risk model curves for mean strength U_{10} and five different concretes (TRP2)

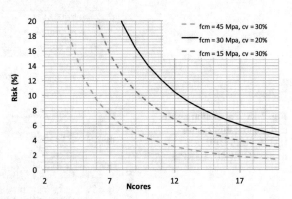

Fig. 2.17 Risk model curves for U^*_{sd30} and three different concretes (TRP2). Two concretes were disregarded since assessing their variability is not possible with an acceptable risk

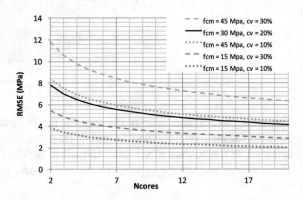

Fig. 2.18 Model for RMSE$_5$ and five different concretes (TRP2)

(Fig. 2.17) and the local error (Fig. 2.18), for five specific concretes that cover a large part of the domain of the simulations and considering medium test result precision (TRP2) using UPV measurements. The black curves in these figures correspond to the blue curves in Figs. 2.13, 2.14 and 2.15, i.e. to the C 30–20 concrete. Figure 2.17 is somehow specific as the two concretes with a cv of 10% lead to high values of risk and were disregarded: assessing their variability with an acceptable risk is not possible.

Figure 2.16 shows that decreasing the mean strength and increasing the concrete strength variability has a negative impact on the assessment strategy, which is related to the negative values of parameter a_3 and the positive values of parameter a_4 in Tables 2.4 and 2.5. Reducing the concrete strength variability has also a negative impact on the assessment of the variability itself, as can be seen in Fig. 2.17, where the two curves for the low variability concretes (C 15–10 and C 45–10) are not represented since their risk values are out of the limits of the plot.[7] Figure 2.18 confirms that the $RMSE_5$ value (local error) increases both with the increase of the mean strength and of the variability (this is also due to the positive values of parameters a_3 and a_4 in Tables 2.4 and 2.5). From the results of the three figures, it appears that the required number of cores for reaching a given assessment performance is significantly affected by the concrete properties, which are of course unknown during a real investigation process.

2.4.4 Identification of the Required Number of Cores

2.4.4.1 Identification Process for Each Assessment Target

Equation (2.8) describes the model for all possible targets of the assessment (i.e. mean strength, strength variability and local error). It is therefore straightforward to derive the minimum number of cores that is required in order to reach the same target with the prescribed accuracy (tolerance interval for mean strength or standard deviation, or 5%-risk RMSE):

$$a_1[\ln(N_c)] = \ln(Y) - (a_0 + a_2[\ln(\varepsilon_{NDT})] + a_3[\ln(f_{c\,mean})] + a_4[\ln(cv(f_c))]) = A \tag{2.9}$$

which leads to:

$$N_{c\,req} = e(A/a_1) \tag{2.10}$$

---

[7]This comment is however valid when the tolerance interval for the standard deviation is expressed in relative terms (i.e. represented by the cv). The conclusion would be different for a tolerance interval expressed in absolute terms, as shown in the next section.

Fig. 2.19 Required number of cores for reaching a ±10% tolerance interval on the mean strength with a 5% risk, for the three TRPs, with UPV test results and a cv of 20%

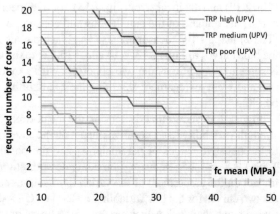

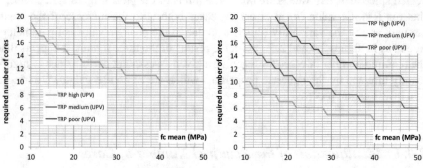

Fig. 2.20 Required number of cores for estimating the concrete strength standard deviation with a 5% risk, for the three TRPs, with UPV test results and a cv of 20%: left—with a ±30% tolerance interval; right—with a ±50% tolerance interval

where A stands for the expression defined in Eq. (2.9).[8]

Figures 2.19, 2.20 and 2.21 illustrate how this required number of cores varies according to the mean strength when assessing the mean strength (Fig. 2.19), the standard deviation (Fig. 2.20) and the local error (Fig. 2.21) and for the 3 TRP levels (here with cv defined as 20%). Figure 2.19 illustrates that this number is smaller for concretes with a larger strength and that it is highly dependent of TRP. It can also be seen that, for $f_{c\,mean} = 30$ MPa (i.e. C 30–20), the results obtained are those of Fig. 2.13 (i.e. a minimum number of cores of 5, 9, 15 for the three TRPs). Figure 2.20 (left and right) address the concrete variability assessment, which is more difficult than that of the mean strength, thus requiring a larger number of cores. For a target tolerance interval of ±30%, the curves on the left provide the same number of cores

---

[8]For RMSE, the method is slightly different, since the model (Eq. (2.5), examples of curves in Figs. 2.15 and 2.18) directly provides the magnitude of RMSE for a given accepted risk (typically = 5%). Thus, to derive $N_{c\,req}$, only the maximum accepted value of this 5%-risk RMSE needs to be specified, either in absolute terms (in MPa) or in relative terms (as a percentage of the mean strength).

Fig. 2.21 Minimum number of cores for reaching a maximum (relative) value of RMSE$_5$ (defined as 15% of f$_{c\,mean}$), with UPV test results and a cv of 20%

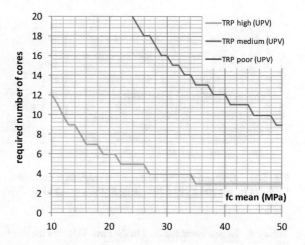

for C 30–20 as those of Fig. 2.14, i.e. 12, 20, >20 for the three TRPs. These numbers change to 5, 9 and 14 if the tolerance interval is increased to ±50%. Figure 2.21 shows the results for RMSE$_5$ than can also be compared to those of Fig. 2.15 for C 30–20. Furthermore, Fig. 2.15 also shows that the target (15% f$_{cmean}$ = 4.5 MPa) can be reached with TRP1 and TRP2 (considering 4 and 16 cores, respectively), but that it cannot be reached with TRP3.

An important issue must be pointed out, which concerns the fact that the tolerance interval on two of the quantities (standard deviation and local error RMSE) can be prescribed either in terms of relative precision or in terms of absolute precision. For instance, the two ways of prescribing the target RMSE$_5$ are ±0.15 $f_{c\,mean}$ or, alternatively, ±4.5 MPa. These two ways are equivalent for the C 30–20 concrete, as 0.15 $f_{c\,mean}$ = 4.5 MPa, but this is not true in other situations. Therefore, the way of expressing these two targets (standard deviation and local error RMSE) may be more or less constraining. The absolute way is less severe for concretes with a lower strength, whereas the relative way is less severe for concretes with a larger strength.[9] Figure 2.22 (left and right) illustrate the minimum numbers of cores for the local error when it is expressed in absolute terms. The effect of how the target is expressed is clearly visible since all curves increase monotonically in this case, while they were decreasing in Fig. 2.21 (and provide the same results for the C 30–20 concrete). Figure 2.22-right corresponds to a less severe target (6 MPa) which requires a smaller number of cores.

This statement has important consequences, since the prescribed number of cores required to reach a given 5%-risk RMSE is highly dependent on how the target is expressed, either in terms of absolute precision or in terms of relative precision. The same statement also applies to the assessment of the standard deviation.

---

[9]The same reasoning also applies to the concrete variability assessment but it is not developed here.

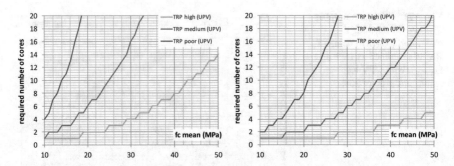

Fig. 2.22 Minimum number of cores for reaching a maximum (absolute) value of RMSE$_5$, with UPV test results and a cv of 20% (left: RMSE$_5$ = 4.5 MPa, right: RMSE$_5$ = 6 MPa)

Table 2.6 Targets corresponding to the three EQL, for the three objectives

	f$_{c\,mean}$ (%)	sd (% or MPa)		RMSE (% or MPa)	
		Realtive	Absolute	Relative	Absolute
EQL1	15				
EQL2	15	50	4	20	6
EQL3	10	30	2	15	4, 5

2.4.4.2   Application to a Multi-objective Assessment

Table 2.6 reproduces how the targets corresponding to the three Estimation Quality levels (EQL) were expressed in the RILEM TC 249-ISC Recommendations, with two options for sd and RMSE, i.e. relative or absolute.

The previous sections have shown how the models defined by Eqs. (2.9) to (2.10) and the parameters identified in Tables 2.4 and 2.5 are able to lead to the number of cores that is required for reaching each of the three targets independently. Based on these results, it is straightforward to determine the minimum number of cores that simultaneously satisfy all three targets defined according to Table 2.6. Figure 2.23 illustrates this for a case considering TRP1 UPV test results and a cv of 30%. It can be seen that, because of the multi-objective requirement, the variation of the minimum number of cores with the mean strength is not monotonic.

It is also possible to compare the required number of cores for the same assessment requirements. For instance, Figs. 2.16, 2.17 and 2.18 show these results for EQL3, considering TRP2 UPV test results, and for five different concretes that were studied. Table 2.7 synthesizes some of these results and, on the last line of the table, show that all three objectives can only be reached for three of the five concretes, and that the necessary number of cores varies between 16 and 20, which are very large values. This simply means that TRP2 test results are not adapted to an EQL3 requirement and that TRP1 test results need to be used instead.

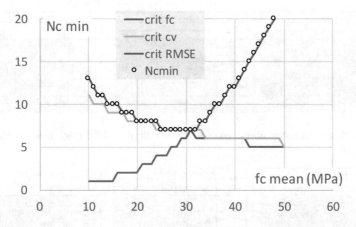

Fig. 2.23 Minimum number of cores for satisfying all three objectives (UPV test results, TRP1, cv = 30%)

Table 2.7 Minimum number of cores—multi-objective assessment, EQL3 and TRP2 UPV test results ("–" when the number exceeds 20, the less severe of the two criteria is kept for RMSE$_5$)

		C 15–10	C 15–30	C 30–20	C 45–10	C 45–30
$f_{c\,10}$		17	8	9	4	9
sd*$_{30}$	±30%	15	–	20	–	9
RMSE$_5$	15% $f_{c\,mean}$	–	15	16	5	16
	4.5 Mpa	5	2	16	–	–
Multi-objective		17	–	20	–	16

The most important conclusion at this stage is that the final "minimum number of cores" is not obtained by a simple relationship and is not so easy to determine. Therefore, the RILEM TC 249-ISC has chosen to propose an alternative and more efficient way to define the necessary minimum number of cores, adapted to the context of on-site assessment by experts and engineers that is explained is the next section.

2.5   Recommendations Regarding the Prescribed Minimum Number of Cores

2.5.1   Preliminary Statements

The objectives of the on-site strength assessment of structures was specified in the RILEM TC 249-ISC Recommendations and correspond to three main quantities that

High precision (TRP1)						Medium precision (TRP2)						Poor precision (TRP3)				
fc mean	cov (%)					fc mean	cov (%)					fc mean	cov (%)			
	10	15	20	25	30		10	15	20	25	30		10	15	20	25	30
10	3	3	3	3	3	10	4	5	6	6	7	10	7	8	9	10	11
15	3	3	3	3	3	15	4	4	4	5	5	15	6	6	7	7	8
20	3	3	3	3	3	20	4	4	4	4	4	20	6	6	6	6	7
25	3	3	3	3	3	25	4	4	4	4	4	25	6	6	6	6	6
30	3	3	3	3	3	30	4	4	4	4	4	30	6	6	6	6	6
35	3	3	3	3	3	35	4	4	4	4	4	35	6	6	6	6	6
40	3	3	3	3	3	40	4	4	4	4	4	40	6	6	6	6	6
45	3	3	3	3	3	45	4	4	4	4	4	45	6	6	6	6	6
50	3	3	3	3	3	50	4	4	4	4	4	50	6	6	6	6	6

Fig. 2.24 Recommended number of cores at EQL1 for the three TRPs (high, medium, poor)

High precision (TRP1)						Medium precision (TRP2)						Poor precision (TRP3)				
fc mean	cov (%)					fc mean	cov (%)					fc mean	cov (%)			
	10	15	20	25	30		10	15	20	25	30		10	15	20	25	30
10	4	4	4	4	4	10	5	5	6	6	7	10	8	8	9	10	11
15	4	4	4	4	4	15	5	5	5	5	5	15	8	8	8	8	8
20	4	4	4	4	4	20	5	5	5	5	5	20	8	8	9	11	12
25	4	4	4	4	4	25	5	5	5	5	5	25	9	11	14	16	18
30	4	4	4	4	4	30	5	5	5	5	6	30	12	15	19	22	25
35	4	4	4	4	4	35	5	5	6	7	8	35	15	20	24	28	32
40	4	4	4	4	4	40	5	6	7	9	10	40	19	25	31	36	41
45	4	4	4	4	4	45	6	7	9	10	12	45	23	31	37	44	50
50	4	4	4	4	4	50	7	9	11	12	14	50	28	37	45	52	60

Fig. 2.25 Recommended number of cores at EQL2 for the three TRPs (high, medium, poor)

are the concrete mean strength, the standard deviation of concrete strength and the mean error on local strength. Three levels of requirements (EQL) were defined, that correspond to different tolerance intervals for the first two objectives and to different magnitudes of the accepted uncertainty for the third objective (see Sect. 1.1).

As the number of cores needed to reach these objectives depends on many parameters and is not obtained by simple laws, the results of a large series of synthetic simulations were processed and empirical models were identified (see Sects. 2.2–2.4 of this chapter for more details). This section is based on these results, but is written in such a way that it can be used directly, without requiring additional knowledge.

Since the type of NDT method (rebound R or ultrasonic pulse velocity UPV), the within-test repeatability (TRP) and the concrete properties (mean strength and standard deviation) are the most influent parameters, the prescribed minimum number of cores can be read directly from synthetic tables corresponding to each specific case.[10] All the tables are presented in Sect. 2.6 and the following subsections illustrate the main features that govern the results given in the tables.

2.5.2 Illustrating the Effect of EQL and TRP

Figures 2.24, 2.25 and 2.26 show how the required number of cores varies with EQL and TRP. In each figure, the results for the three possible EQLs are compared, from high precision (TRP1) on the left side to poor precision (TRP3) on the right

---

[10] All numbers in the tables are derived from those provided by the models described in Sect. 2.3.

High precision (TRP1)						Medium precision (TRP2)				
fc mean	cov (%)					fc mean	cov (%)				
	10	15	20	25	30		10	15	20	25	30
10	4	5	7	8	8	10	7	10	12	13	15
15	4	4	5	6	6	15	6	7	9	10	11
20	4	4	4	5	5	20	5	6	7	8	9
25	4	4	4	4	4	25	6	7	8	10	11
30	4	4	4	4	4	30	8	9	11	13	15
35	5	5	5	5	5	35	9	12	15	17	19
40	5	6	6	6	6	40	11	15	18	21	24
45	6	6	7	7	7	45	14	18	22	26	30
50	7	7	7	8	9	50	17	22	27	31	36

Fig. 2.26 Recommended number of cores at EQL3 for the TRP1 and TRP2 (high, medium)

side. Results of Figs. 2.24, 2.25 and 2.26 are related to EQL1, EQL2 and EQL3, respectively. The poor precision (TRP3) test results are omitted for EQL3 since more repeatable test results are mandatory (see Table 1.5). All these tables are built for rebound test results and absolute targets for both the concrete standard deviation and the local error.[11] Each double entry table provides the required number of cores as a function of the concrete mean strength (line) and concrete strength coefficient of variation (column). In each table, a black bold line defines a range of concrete properties (mean and standard deviation) which corresponds to "most favourable configurations". This issue is explained and discussed further in Sect. 2.4.3.

From these figures, it is possible to confirm the combined effect of EQL and TRP. The required number of cores increases with the severity of the assessment objectives (from EQL1 to EQL3) and with the within-test repeatability of the NDT results (TRP). It is also possible to see that:

- for the EQL1 requirement (i.e. the only target is the mean strength) 3, 7 and 11 cores, for TRP1 to TRP3 respectively, are enough to address the problem irrespective of the concrete properties,
- for EQL2, a poor TRP leads to an excessive number of cores in most cases and should never be considered. For high and medium TRP, 4 and 14 cores, respectively, are enough to address the problem irrespective the concrete properties. For TRP2, this number can be reduced if one has prior information about the concrete properties, for instance 9 cores are enough if $f_{c\,mean} \leq 40$ MPa and $cov(f_c) \leq 25\%$,
- for EQL3, a medium TRP may lead to a very large number of cores in some cases. With high TRP, 9 cores are enough to address the problem irrespective the concrete properties.

---

[11] All values in the tables depend on the type of NDT, but rebound and UPV measurements lead to very close values for similar TRPs.

2.5.3 Influence of How the Targets Are Expressed

Regarding the second and third quantities to be estimated at EQL2 and EQL3, i.e. the concrete strength standard deviation and the local error, the RILEM TC 249-ISC Recommendations proposes two options (see RILEM TC 249-ISC Recommendations, Table 1), corresponding to the cases where an absolute (A) or a relative (R) accuracy are considered. This means there are four possible combinations, A–A, A–R, R–A and R–R to assess these two quantities. Since the sensitivity of the required number of cores given the different concrete properties (mean strength and variability) depends on the way the target is expressed, Table 2.8 summarizes these sensitivities as they were identified from the analysis of the results obtained from the synthetic simulations (see Sects. 2.2 and 2.3 for full details).

These sensitivities are illustrated in Fig. 2.27 where the required number of cores are compared for the four possibilities of combinations, when considering UPV measurements and the EQL2 assessment level.

These figures confirm that the way the target criterion is expressed has a large influence on the minimum number of cores that is required. This can be easily seen if one considers for each of these figures, for instance, the case where the mean strength is equal to 30 MPa and the coefficient of variation is equal to 20% (centre of the table), for which all options give 6 cores. This is normal since, for this specific concrete:

– the third criterion (on local error) is the most demanding and
– the values of absolute and relative criteria for the local uncertainty on local strengths provided in the RILEM TC 249-ISC Recommendations, Table 1, i.e. 6 MPa and 20% of the mean strength have the same value.

However, when the mean strength or the coefficient of variation change, one of the two criteria becomes more severe and leads to a larger number of cores. This is the reason why, in the following, a different number of cores must be considered for the different options.

Table 2.8 Sensitivity of the required number of cores given different concrete properties (+ means that the output increases as the input increases)

Target	Influencing factor	
	$f_{c\ mean}$	$sd(f_c)$
Mean strength U_m	−	+
Standard deviation (relative) $U_{sd\ rel}$	−	−
Standard deviation (absolute) $U_{sd\ abs}$	+	+ (weak)
RMSE$_5$ (relative)	−	+
RMSE$_5$ (absolute)	+	+

Medium precision (TRP2)				
fc mean	cov (%)				
	10	15	20	25	30
10	5	5	5	6	7
15	5	5	5	5	5
20	5	5	5	5	5
25	5	5	5	5	6
30	5	5	6	7	9
35	5	6	8	11	13
40	5	8	12	15	18
45	7	11	16	20	25
50	9	15	21	27	33

a

Medium precision (TRP2)				
fc mean	cov (%)				
	10	15	20	25	30
10	8	13	18	23	28
15	5	8	12	15	18
20	5	6	9	11	14
25	5	5	7	9	11
30	5	5	6	7	9
35	5	5	5	6	8
40	5	5	5	6	7
45	5	5	5	5	6
50	5	5	5	5	5

b

Medium precision (TRP2)				
fc mean	cov (%)				
	10	15	20	25	30
10	29	15	10	7	7
15	22	12	7	5	5
20	18	10	6	5	5
25	16	8	5	5	6
30	14	8	6	7	9
35	13	7	8	11	13
40	12	8	12	15	18
45	11	11	16	20	25
50	10	15	21	27	33

c

Medium precision (TRP2)				
fc mean	cov (%)				
	10	15	20	25	30
10	29	15	18	23	28
15	22	12	12	15	18
20	18	10	9	11	14
25	16	8	7	9	11
30	14	8	6	7	9
35	13	7	5	6	8
40	12	6	5	6	7
45	11	6	5	5	6
50	10	6	5	5	5

d

Fig. 2.27 a–d Required number of cores with four options for the criteria on sd(f_c) and RMSE$_5$, EQL2 assessment level and medium precision UPV test results (TRP2) (**a** A–A, **b** A–R, **c** R–A, **d** R–R)

2.5.4 How Are the Tables Giving the Minimum Number of Cores Organized and How Can They Be Used?

To use the tables providing the required number of cores (Sect. 2.6), an answer to a series of five questions must first be defined, as shown in the flowchart of Fig. 2.28. These questions address:

– the *choice of the assessment quality level*, EQL (RILEM TC 249-ISC Recommendations, Sect. 2.5), since the EQL defines the target uncertainty on each quantity to be estimated,
– the *choice of the NDT method*, that will deliver Rebound or UPV measurements,[12]
– the *TRP level*, that must be identified after carrying out the NDTs and by processing the test results (RILEM TC 249-ISC Recommendations, Sect. 6.4),

---

[12]If the two techniques are used, it is required to take the highest number among those required by each technique independently. If the assessor wants to use a conversion model combining the two methods, he must also consider the consequences of having a higher number of free parameters in this model (see Sect. 12.9).

Fig. 2.28 Flowchart with
the five questions to answer
before using the tables

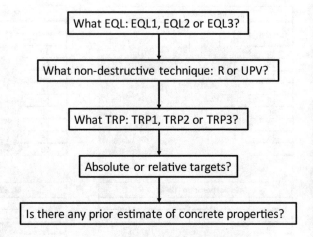

- the *choice of absolute or relative targets*, knowing that this choice may have a
  large influence on the number of cores (see Sect. 2.4). Since four options can be
  considered, given the combination of absolute or relative assessment for each of
  the two targets, they are termed A–A, A–R, R–A and R–R in the following,
- the consideration of *prior knowledge about the range of concrete properties*,
  whose origin may be preliminary considerations about the concrete or the structure
  (RILEM TC 249-ISC Recommendations, Sect. 6.1) or a first analysis of on-site
  NDT results. In any case, if such assumptions are made at this stage to reduce
  the required number of cores, their validity must be checked at the end of the
  assessment process, once the concrete properties are estimated.

2.6 Tables Providing the Recommended Minimum Number of Cores

Table 2.9 recalls, for the three Estimation Quality Levels, how targets can be defined
(with two possibilities for the standard deviation and the error on local strength), and
the target tolerance intervals (it is similar to RILEM TC 249-ISC Recommendations,

Table 2.9 Relation between the Estimation Quality Levels (EQL) and the target tolerance intervals
on strength assessment, as defined in the RILEM TC 249-ISC Recommendations, Table 1

Estimated property		EQL1	EQL2	EQL3
Mean value of local strengths		±15%	±15%	±10%
Standard deviation of local strengths	Relative (R)	Not addressed	±50%	±30%
	Absolute (A)		4 MPa	2 MPa
Error on local strength	Relative (R)		20%	15%
	Absolute (A)		6 MPa	4.5 MPa

Table 2.10 References of the relevant figure in order to identify the required number of cores

sheet	EQL	Non-destructive technique	Expression of targets on sd(f_c) and RMSE_5			
			A–A	A–R	R–A	R–R
I	EQL1	R	Fig. 2.29			
		UPV	Fig. 2.30			
IIa	EQL2	R	Fig. 2.31	Fig. 2.32	Fig. 2.33	Fig. 2.34
IIb		UPV	Fig. 2.35	Fig. 2.36	Fig. 2.37	Fig. 2.38
IIIa	EQL3	R	Fig. 2.39	Fig. 2.40	Fig. 2.41	Fig. 2.42
IIIc		UPV	Fig. 2.43	Fig. 2.44	Fig. 2.45	Fig. 2.46

Table 1). Table 2.10 synthesizes to which sheet (from Sheet I to Sheet IIIb) the reader must refer to in order to find the relevant information regarding the recommended number of cores.

Each figure contains the recommended number of cores for the different TRP levels as a function of the concrete properties. If some assumptions are made about the concrete properties (i.e. the mean concrete strength and/or the concrete variability), the number of cores is taken as the maximum of values corresponding to the hypothetic range of properties. More generally, Table 2.11 details the applicable limiting conditions regarding the concrete properties, and Table 2.12 represents the corresponding required number of cores, thus replacing the data of Sheets I–IIIb. However, *at the end of the estimation process, once the concrete properties are estimated, it is mandatory to check that the considered limiting conditions are in fact applicable.* If the answer is negative, the assessment is not valid.

Sheet I. Recommended Number of Cores for EQL I
See Figs. 2.29 and 2.30.

Sheet IIa. Recommended Number of Cores for EQL 2, With Rebound (R) Test Results
See Figs. 2.31, 2.32, 2.33 and 2.34.

Table 2.11 Limiting conditions that can apply to concrete properties, as a function of the targets on sd(f_c) and RMSE_5 (these conditions correspond to the area defined by a bold line in Figs. 2.29, 2.30, 2.31, 2.32, 2.33, 2.34, 2.35, 2.36, 2.37, 2.38, 2.39, 2.40, 2.41, 2.42, 2.43, 2.44, 2.45 and 2.46)

expression of targets on sd(f_c) and RMSE_5	Condition on $f_{c\ mean}$	Condition on $cov(f_c)$
Absolute–absolute (A–A)	$f_{c\ mean} \leq 40$ MPa	$cov(f_c) \leq 25\%$
Absolute–relative (A–R)	$f_{c\ mean} \geq 20$ MPa	$cov(f_c) \leq 25\%$
Relative–absolute (R–A)	15 MPa $\leq f_{c\ mean} \leq 40$ MPa	$cov(f_c) \geq 15\%$
Relative–relative (R–R)	$f_{c\ mean} \geq 20$ MPa	$cov(f_c) \geq 15\%$

Table 2.12 Values for the minimum number of cores that can apply if the limiting conditions of Table 2.11 are valid

Rebound (R) test results

	High precision TRP 1				Medium precision TRP 2				Poor precision TRP 3			
	A–A	A–R	R–A	R–R	A–A	A–R	R–A	R–R	A–A	A–R	R–A	R–R
EQL 1	3				6				10			
EQL 2	4	4	7	6	9	10	12	11	Excessive numbers			
EQL 3	8	8	15	13	Excessive numbers				Impossible			

Ultrasonic pulse velocity (UPV) test results

	High precision TRP 1				Medium precision TRP 2				Poor precision TRP 3			
	A–A	A–R	R–A	R–R	A–A	A–R	R–A	R–R	A–A	A–R	R–A	R–R
EQL 1	4				6				10			
EQL 2	4	4	7	6	15	11	(25)	14	Excessive numbers			
EQL 3	10	11	17	14	Excessive numbers				Impossible			

High precision (TRP1)						Medium precision (TRP2)						Poor precision (TRP3)				
fc mean	cov (%)					fc mean	cov (%)					fc mean	cov (%)			
	10	15	20	25	30		10	15	20	25	30		10	15	20	25	30
10	3	3	3	3	3	10	4	5	6	6	7	10	7	8	9	10	11
15	3	3	3	3	3	15	4	4	4	5	5	15	6	6	7	7	8
20	3	3	3	3	3	20	4	4	4	4	4	20	6	6	6	6	7
25	3	3	3	3	3	25	4	4	4	4	4	25	6	6	6	6	6
30	3	3	3	3	3	30	4	4	4	4	4	30	6	6	6	6	6
35	3	3	3	3	3	35	4	4	4	4	4	35	6	6	6	6	6
40	3	3	3	3	3	40	4	4	4	4	4	40	6	6	6	6	6
45	3	3	3	3	3	45	4	4	4	4	4	45	6	6	6	6	6
50	3	3	3	3	3	50	4	4	4	4	4	50	6	6	6	6	6

Fig. 2.29 Rebound test results recommended number of cores at EQL1 for the three TRPs (high, medium, poor)

High precision (TRP1)						Medium precision (TRP2)						Poor precision (TRP3)				
fc mean	cov (%)					fc mean	cov (%)					fc mean	cov (%)			
	10	15	20	25	30		10	15	20	25	30		10	15	20	25	30
10	3	3	3	4	4	10	4	5	5	6	7	10	6	8	9	10	11
15	3	3	3	3	3	15	4	4	4	5	5	15	6	6	7	8	9
20	3	3	3	3	3	20	4	4	4	4	5	20	6	6	6	7	8
25	3	3	3	3	3	25	4	4	4	4	4	25	6	6	6	6	7
30	3	3	3	3	3	30	4	4	4	4	4	30	6	6	6	6	6
35	3	3	3	3	3	35	4	4	4	4	4	35	6	6	6	6	6
40	3	3	3	3	3	40	4	4	4	4	4	40	6	6	6	6	6
45	3	3	3	3	3	45	4	4	4	4	4	45	6	6	6	6	6
50	3	3	3	3	3	50	4	4	4	4	4	50	6	6	6	6	6

Fig. 2.30 Ultrasonic Pulse Velocity (UPV) test results—recommended number of cores at EQL1 for the three TRPs (high, medium, poor)

High precision (TRP1)						Medium precision (TRP2)						Poor precision (TRP3)				
fc mean	cov (%)					fc mean	cov (%)					fc mean	cov (%)			
	10	15	20	25	30		10	15	20	25	30		10	15	20	25	30
10	4	4	4	4	4	10	5	5	6	6	7	10	8	8	9	10	11
15	4	4	4	4	4	15	5	5	5	5	5	15	8	8	8	8	8
20	4	4	4	4	4	20	5	5	5	5	5	20	8	8	9	11	12
25	4	4	4	4	4	25	5	5	5	5	5	25	9	11	14	16	18
30	4	4	4	4	4	30	5	5	5	5	6	30	12	15	19	22	25
35	4	4	4	4	4	35	5	5	6	7	8	35	15	20	24	28	32
40	4	4	4	4	4	40	5	6	7	9	10	40	19	25	31	36	41
45	4	4	4	4	4	45	6	7	9	10	12	45	23	31	37	44	50
50	4	4	4	4	4	50	7	9	11	12	14	50	28	37	45	52	60

Fig. 2.31 Recommended number of cores at EQL2 for the three TRPs (high, medium, poor), A–A targets

High precision (TRP1)						Medium precision (TRP2)						Poor precision (TRP3)				
fc mean	cov (%)					fc mean	cov (%)					fc mean	cov (%)			
	10	15	20	25	30		10	15	20	25	30		10	15	20	25	30
10	4	5	6	7	8	10	11	20	24	28	32	10	63	84	102	120	135
15	4	4	4	4	4	15	8	11	13	15	17	15	34	44	54	64	72
20	4	4	4	4	4	20	5	7	8	10	11	20	22	29	35	41	46
25	4	4	4	4	4	25	5	5	6	7	8	25	15	20	25	29	33
30	4	4	4	4	4	30	5	5	5	5	6	30	12	15	19	22	25
35	4	4	4	4	4	35	5	5	5	5	5	35	9	12	15	17	20
40	4	4	4	4	4	40	5	5	5	5	5	40	8	10	12	14	16
45	4	4	4	4	4	45	5	5	5	5	5	45	8	8	10	12	13
50	4	4	4	4	4	50	5	5	5	6	6	50	8	8	9	10	11

Fig. 2.32 Recommended number of cores at EQL2 for the three TRPs (high, medium, poor), A–R targets

High precision (TRP1)						Medium precision (TRP2)						Poor precision (TRP3)				
fc mean	cov (%)					fc mean	cov (%)					fc mean	cov (%)			
	10	15	20	25	30		10	15	20	25	30		10	15	20	25	30
10	18	10	6	4	4	10	30	16	10	7	7	10	49	26	16	11	11
15	13	7	5	4	4	15	22	12	7	5	5	15	37	19	12	9	8
20	11	6	4	4	4	20	18	9	6	5	5	20	30	15	10	11	12
25	9	5	4	4	4	25	15	8	5	5	5	25	25	13	14	16	18
30	8	4	4	4	4	30	13	7	5	5	6	30	22	15	19	22	25
35	7	4	4	4	4	35	12	6	6	7	8	35	20	20	24	28	32
40	7	4	4	4	4	40	11	6	7	9	10	40	19	25	31	36	41
45	6	4	4	4	4	45	10	7	9	10	12	45	23	31	37	44	50
50	6	4	4	4	4	50	9	9	11	12	14	50	28	37	45	52	60

Fig. 2.33 Recommended number of cores at EQL2 for the three TRPs (high, medium, poor, R–A targets

High precision (TRP1)						Medium precision (TRP2)						Poor precision (TRP3)				
fc mean	cov (%)					fc mean	cov (%)					fc mean	cov (%)			
	10	15	20	25	30		10	15	20	25	30		10	15	20	25	30
10	18	10	6	7	8	10	30	20	24	28	32	10	63	84	102	120	136
15	13	7	5	4	4	15	22	12	13	15	17	15	37	44	54	64	72
20	11	6	4	4	4	20	18	9	8	10	11	20	30	29	35	41	46
25	9	5	4	4	4	25	15	8	6	7	8	25	25	20	25	29	33
30	8	4	4	4	4	30	13	7	5	5	6	30	22	15	19	22	25
35	7	4	4	4	4	35	12	6	5	5	5	35	20	12	15	17	20
40	7	4	4	4	4	40	11	6	5	5	5	40	18	10	12	14	16
45	6	4	4	4	4	45	10	5	5	5	5	45	16	9	10	12	13
50	6	4	4	4	4	50	9	5	5	5	5	50	15	8	9	10	11

Fig. 2.34 Recommended number of cores at EQL2 for the three TRPs (high, medium, poor), R–R targets

Sheet IIb. Recommended Number of Cores for EQL 2, With Ultrasonic Pulse Velocity (UPV) Test Results
See Figs. 2.35, 2.36, 2.37 and 2.38.

Sheet IIIa. Recommended Number of Cores for EQL 3, With Rebound (R) Test Results
See Figs. 2.39, 2.40, 2.41 and 2.42.

Sheet IIIb. Recommended Number of Cores for EQL 3, With Ultrasonic Pulse Velocity (UPV) Test Results
See Figs. 2.43, 2.44, 2.45 and 2.46.

High precision (TRP1)						Medium precision (TRP2)						Poor precision (TRP3)				
fc mean	cov (%)					fc mean	cov (%)					fc mean	cov (%)			
	10	15	20	25	30		10	15	20	25	30		10	15	20	25	30
10	5	5	5	5	5	10	6	6	6	6	7	10	8	8	9	10	11
15	5	5	5	5	5	15	6	6	6	6	6	15	8	8	8	8	9
20	5	5	5	5	5	20	6	6	6	6	6	20	8	8	8	11	13
25	5	5	5	5	5	25	6	6	6	6	6	25	8	11	15	19	23
30	5	5	5	5	5	30	6	6	6	7	9	30	11	17	23	30	37
35	5	5	5	5	5	35	6	6	8	11	13	35	16	25	35	45	56
40	5	5	5	5	5	40	6	8	12	15	18	40	22	35	49	64	79
45	5	5	5	5	6	45	7	11	16	20	25	45	30	48	67	86	107
50	5	5	5	6	8	50	9	15	21	27	33	50	39	63	87	114	140

Fig. 2.35 Recommended number of cores at EQL2 for the three TRPs (high, medium, poor), A–A targets

High precision (TRP1)						Medium precision (TRP2)						Poor precision (TRP3)				
fc mean	cov (%)					fc mean	cov (%)					fc mean	cov (%)			
	10	15	20	25	30		10	15	20	25	30		10	15	20	25	30
10	5	5	5	6	7	10	8	13	18	23	28	10	34	54	76	98	122
15	5	5	5	5	5	15	6	8	12	15	18	15	22	35	49	63	79
20	5	5	5	5	5	20	6	6	9	11	14	20	16	26	36	47	58
25	5	5	5	5	5	25	6	6	7	9	11	25	13	20	28	37	45
30	5	5	5	5	5	30	6	6	6	7	9	30	11	17	23	30	37
35	5	5	5	5	5	35	6	6	6	6	8	35	9	14	20	26	32
40	5	5	5	5	5	40	6	6	6	6	7	40	8	12	17	22	27
45	5	5	5	5	5	45	6	6	6	6	6	45	8	11	15	20	24
50	5	5	5	5	5	50	6	6	6	6	6	50	8	10	14	18	22

Fig. 2.36 Recommended number of cores at EQL2 for the three TRPs (high, medium, poor), A–R targets

High precision (TRP1)						Medium precision (TRP2)						Poor precision (TRP3)				
fc mean	cov (%)					fc mean	cov (%)					fc mean	cov (%)			
	10	15	20	25	30		10	15	20	25	30		10	15	20	25	30
10	18	9	6	5	5	10	29	15	10	7	7	10	47	25	16	11	11
15	14	7	5	5	5	15	22	12	7	6	6	15	36	19	12	8	9
20	11	6	5	5	5	20	18	10	6	6	6	20	30	16	10	11	13
25	10	5	5	5	5	25	16	8	6	6	6	25	26	14	15	19	23
30	9	5	5	5	5	30	14	8	6	7	9	30	23	17	23	30	37
35	8	5	5	5	5	35	13	7	8	11	13	35	21	25	35	45	56
40	7	5	5	5	5	40	12	8	12	15	18	40	22	35	49	64	79
45	7	5	5	5	6	45	11	11	16	20	25	45	30	48	67	86	107
50	6	5	5	6	8	50	10	15	21	27	33	50	39	63	87	114	140

Fig. 2.37 Recommended number of cores at EQL2 for the three TRPs (high, medium, poor), R–A targets

High precision (TRP1)						Medium precision (TRP2)						Poor precision (TRP3)				
fc mean	cov (%)					fc mean	cov (%)					fc mean	cov (%)			
	10	15	20	25	30		10	15	20	25	30		10	15	20	25	30
10	18	9	6	6	7	10	29	15	18	23	28	10	47	54	76	98	122
15	14	7	5	5	5	15	22	12	12	15	18	15	36	35	49	63	79
20	11	6	5	5	5	20	18	10	9	11	14	20	30	26	36	47	58
25	10	5	5	5	5	25	16	8	7	9	11	25	26	20	28	37	45
30	9	5	5	5	5	30	14	8	6	7	9	30	23	17	23	30	37
35	8	5	5	5	5	35	13	7	6	6	8	35	21	14	20	26	32
40	7	5	5	5	5	40	12	6	6	6	7	40	19	12	17	22	27
45	7	5	5	5	5	45	11	6	6	6	6	45	18	11	15	20	24
50	6	5	5	5	5	50	10	6	6	6	6	50	17	10	14	18	22

Fig. 2.38 Recommended number of cores at EQL2 for the three TRPs (high, medium, poor), R–R targets

High precision (TRP1)						Medium precision (TRP2)				
fc mean	cov (%)					fc mean	cov (%)			
	10	15	20	25	30		10	15	20	25	30
10	4	5	7	8	8	10	7	10	12	13	15
15	4	4	5	6	6	15	6	7	9	10	11
20	4	4	4	5	5	20	5	6	7	8	9
25	4	4	4	4	4	25	6	7	8	10	11
30	4	4	4	4	4	30	8	9	11	13	15
35	5	5	5	5	5	35	9	12	15	17	19
40	5	6	6	6	6	40	11	15	18	21	24
45	6	6	7	7	7	45	14	18	22	26	30
50	7	7	7	8	9	50	17	22	27	31	36

Fig. 2.39 Recommended number of cores at EQL3 for the two TRPs (high, medium), A–A targets

High precision (TRP1)	cov (%)					Medium precision (TRP2)	cov (%)				
fc mean	10	15	20	25	30	fc mean	10	15	20	25	30
10	9	12	14	17	19	10	38	50	61	71	81
15	5	7	8	9	10	15	20	27	33	38	43
20	4	4	5	6	7	20	13	17	21	24	28
25	4	4	4	4	5	25	9	12	15	17	20
30	4	4	4	4	4	30	8	9	11	13	15
35	5	5	5	5	5	35	9	9	10	10	12
40	5	6	6	6	6	40	11	11	11	11	12
45	6	6	7	7	7	45	12	13	13	13	13
50	7	7	7	8	8	50	14	14	14	15	15

Fig. 2.40 Recommended number of cores at EQL3 for the two TRPs (high, medium), A–R targets

High precision (TRP1)	cov (%)					Medium precision (TRP2)	cov (%)				
fc mean	10	15	20	25	30	fc mean	10	15	20	25	30
10	37	20	13	9	8	10	67	36	24	17	15
15	28	15	10	7	6	15	51	28	18	13	11
20	23	13	8	6	5	20	42	23	15	11	9
25	20	11	7	5	4	25	36	20	13	10	11
30	18	10	6	5	4	30	32	17	11	13	15
35	16	9	6	4	5	35	29	16	15	17	19
40	15	8	5	5	6	40	26	15	18	21	24
45	13	7	6	6	7	45	24	18	22	26	30
50	12	7	7	8	9	50	22	22	27	31	36

Fig. 2.41 Recommended number of cores at EQL3 for the two TRPs (high, medium), R–A targets

High precision (TRP1)	cov (%)					Medium precision (TRP2)	cov (%)				
fc mean	10	15	20	25	30	fc mean	10	15	20	25	30
10	37	20	14	17	19	10	67	50	61	71	81
15	28	15	10	9	10	15	51	28	33	38	43
20	23	13	8	6	7	20	42	23	21	24	28
25	20	11	7	5	5	25	36	20	15	17	20
30	18	10	6	5	4	30	32	17	11	13	15
35	16	9	6	4	4	35	29	16	10	10	12
40	15	8	5	4	4	40	26	14	9	9	10
45	13	7	5	4	4	45	24	13	8	7	8
50	12	7	5	4	4	50	22	12	8	6	7

Fig. 2.42 Recommended number of cores at EQL3 for the two TRPs (high, medium), R–R targets

High precision (TRP1)	cov (%)					Medium precision (TRP2)	cov (%)				
fc mean	10	15	20	25	30	fc mean	10	15	20	25	30
10	4	5	6	7	8	10	7	9	11	13	15
15	4	4	5	6	7	15	5	7	9	10	12
20	4	4	4	5	6	20	5	6	7	9	10
25	4	4	4	5	5	25	7	7	10	13	16
30	5	5	6	6	6	30	8	11	16	20	25
35	6	6	7	7	9	35	11	17	23	30	37
40	7	8	8	10	12	40	15	24	33	43	53
45	8	9	11	14	17	45	20	32	45	58	71
50	9	10	14	18	22	50	26	42	59	76	94

Fig. 2.43 Recommended number of cores at EQL3 for the two TRPs (high, medium), A–A targets

High precision (TRP1)	cov (%)					Medium precision (TRP2)	cov (%)				
fc mean	10	15	20	25	30	fc mean	10	15	20	25	30
10	6	9	12	16	19	10	23	36	51	66	81
15	4	6	8	10	12	15	15	24	33	43	53
20	4	4	6	8	9	20	11	17	24	31	39
25	4	4	5	6	7	25	9	14	19	25	30
30	5	5	6	6	6	30	8	11	16	20	25
35	6	6	7	7	7	35	10	11	13	17	21
40	7	8	8	8	9	40	12	13	14	15	18
45	8	9	9	10	10	45	14	16	17	17	18
50	9	10	11	11	12	50	17	18	19	20	21

Fig. 2.44 Recommended number of cores at EQL3 for the two TRPs (high, medium), A–R targets

High precision (TRP1)	cov (%)					Medium precision (TRP2)	cov (%)				
fc mean	10	15	20	25	30	fc mean	10	15	20	25	30
10	33	18	12	9	8	10	56	31	21	15	15
15	28	15	10	8	7	15	47	26	17	13	12
20	24	14	9	7	6	20	41	23	15	11	10
25	22	12	8	6	5	25	37	21	14	13	16
30	20	12	8	6	6	30	35	19	16	20	25
35	19	11	7	7	9	35	32	18	23	30	37
40	18	10	8	10	12	40	30	24	33	43	53
45	17	10	11	14	17	45	29	32	45	58	71
50	16	10	14	18	22	50	28	42	59	76	94

Fig. 2.45 Recommended number of cores at EQL3 for the two TRPs (high, medium), R–A targets

High precision (TRP1)					
fc mean	cov (%)				
	10	15	20	25	30
10	33	18	12	16	19
15	28	15	10	10	12
20	24	14	9	8	9
25	22	12	8	6	7
30	20	12	8	6	6
35	19	11	7	5	5
40	18	10	7	5	5
45	17	10	7	5	4
50	16	9	6	5	4

Medium precision (TRP2)					
fc mean	cov (%)				
	10	15	20	25	30
10	56	36	51	66	81
15	47	26	33	43	53
20	41	23	24	31	39
25	37	21	19	25	30
30	35	19	16	20	25
35	32	18	13	17	21
40	30	17	12	15	18
45	29	16	11	13	16
50	28	16	10	12	15

Fig. 2.46 Recommended number of cores at EQL3 for the two TRPs (high, medium), R–R targets

Acknowledgements The 4th author would like to acknowledge the financial support by Base Funding—UIDB/04708/2020 of CONSTRUCT—Instituto de I&D em Estruturas e Construções, funded by national funds through FCT/MCTES (PIDDAC).

References

1. Breysse, D., Balayssac, J.P., Biondi, S., Corbett, D., Goncalves, A., Grantham, M., Luprano, V.A.M., Masi, A., Monteiro, A.V., Sbartaï, Z.M.: Recommendation of RILEM TC249-ISC on non-destructive in situ strength assessment of concrete. Mater. Struct. **52**(4), 71 (2019)
2. Breysse, D., Balayssac, J.P.: Reliable non-destructive strength assessment in existing structures: myth or reality? RILEM Tech. Lett. **3**, 129–134 (2018). https://doi.org/10.21809/rilemtechlett.2018.73https://doi.org/10.21809/rilemtechlett.2018.73
3. Breysse, D., Balayssac, J.P., Biondi, S., Borosnyói, A., Candigliota, E., Chiauzzi, L., Garnier, G., Grantham, M., Gunes, O., Luprano, V.A.M., Masi, A., Pfister, V., Sbartaï, Z.M., Szilágyi, K., Fontan, M.: Non-destructive assessment of in situ concrete strength: comparison of approaches through an international benchmark. Mater. Struct. **50**, 133 (2017). https://doi.org/10.1617/s11527-017-1009-7https://doi.org/10.1617/s11527-017-1009-7
4. Alwash, M., Breysse, D., Sbartaï, Z.M.: Using Monte-Carlo simulations to evaluate the efficiency of different strategies for nondestructive assessment of concrete strength. Mater. Struct. **50**, 90 (2017). https://doi.org/10.1617/s11527-016-0962-xhttps://doi.org/10.1617/s11527-016-0962-x
5. Breysse, D., Fernández-Martínez, J.L.: Assessing concrete strength with rebound hammer: review of key issues and ideas for more reliable conclusions. Mater. Struct. **47**, 1589–1604 (2014)
6. Breysse, D.: Nondestructive evaluation of concrete strength: an historical review and a new perspective by combining NDT methods. Constr. Build. Mater. **33**, 139–163 (2012)
7. EN 13791:2007: Assessment of In-Situ Compressive Strength in Structures or in Precast Concrete Products. European Committee for Standardization (CEN), Brussels, Belgium (2007)
8. ACI 228.1R-03: In-Place Methods to Estimate Concrete Strength, S.P. Pessiki (chair) (2003)
9. Breysse, D., Balayssac, J.P.: Strength assessment in reinforced concrete structures: from research to improved practices. Constr. Build. Mater. **182**, 1–9 (2018)
10. Szilágyi, K.: Rebound surface hardness and related properties of concrete. Ph.D. Report. Budapest University of Technology and Economics, Budapest, Hungary (2013)
11. Alwash, M., Breysse, D., Sbartaï, Z.M., Szilágyi, K., Borosnyói, A.: Factors affecting the reliability of assessing the concrete strength by rebound hammer and cores. Constr. Build. Mater. **140**, 354–363 (2017)

12. Ali-Benyahia, K., Sbartaï, Z.M., Breysse, D., Kenai, S., Ghrici, M.: Analysis of the single and combined non-destructive test approaches for on-site concrete strength assessment: general statements based on a real case-study. Case Stud. Constr. Mater. **6**, 109–119 (2017)
13. Ali-Benyahia, K., Sbartaï, Z.M., Breysse, D., Ghrici, M. Kenai, S.: Improvement of nondestructive assessment of on-site concrete strength: influence of the selection process of cores location on the assessment quality for single and combined NDT techniques. Constr. Build. Mater. **195**, 613-622 (2019)
14. Alwash, M., Sbartaï, Z.M., Breysse, D.: Non-destructive assessment of both mean strength and variability of concrete: a new bi-objective approach. Constr. Build. Mater. **113**, 880–889 (2016)
15. Alwash, M., Breysse, D., Sbartaï, Z.M.: Non destructive strength evaluation of concrete: analysis of some key factors using synthetic simulations. Constr. Build. Mater. **99**, 235–245 (2015)

Chapter 3
Evaluation of Concrete Strength by Combined NDT Techniques: Practice, Possibilities and Recommendations

Zoubir Mehdi Sbartaï, Vincenza Anna Maria Luprano, and Emilia Vasanelli

Abstract This chapter presents a summarized state of the art of the available methods (SonReb and new methodologies) for the combination of NDT measurements. This synthesis can be helpful for selecting NDT methods for the combination. The chapter also discusses in which condition it is convenient or not to apply the combination of methods. Several possible approaches are detailed in a synthetic way. SonReb method is the most popular method for combining NDT methods, likely UPV and rebound hammer. The use of more than two NDT techniques with the objective of improving the evaluation is presented. This approach of combination can be interesting for example if moisture content varies in the tested concrete or if the quality of techniques is equivalent and if they are complementary. The results of the literature have been presented and discussed for a possible use in the RILEM recommendations. As a first result and without additional computational work, the use of the different approaches is discussed regarding the test result precision of, the TRP, and the magnitude of the final conversion model error.

3.1 Introduction

The quality of concrete strength assessment by the measurement of NDT physical parameters (as velocity of ultrasonic waves or rebound index) is affected by various sources of uncertainty due to the quality of the measurements and to the variability of the concrete. The idea of combining two or several non-destructive techniques

Z. M. Sbartaï (✉)
University Bordeaux, I2M-UMR CNRS 5295, Talence, France
e-mail: zoubir-mehdi.sbartai@u-bordeaux.fr

V. A. M. Luprano
ENEA - Italian National Agency for New Technologies, Energy and Sustainable Economic Development, Department for Sustainability - Non Destructive Evaluation Laboratory, Brindisi, Italy

E. Vasanelli
Institute of Heritage Science - National Research Council (ISPC-CNR), Lecce, Italy

© RILEM 2021

D. Breysse and J.-P. Balayssac (eds.), *Non-Destructive In Situ Strength Assessment of Concrete*, RILEM State-of-the-Art Reports 32,
https://doi.org/10.1007/978-3-030-64900-5_3

to overcome the limits of a single NDT method and improve the assessment of the concrete compressive strength has been matter of research in the last decades [1]. Many case studies exist in which different techniques have been combined, but real added value can only be obtained if the issue of combination is correctly analysed [2]. This added value can be defined in terms of (a) accuracy of estimation, (b) relevance of physical explanations and diagnosis, (c) reduction in time to reach a given answer [1]. Like it is the case for single techniques, the validity interest of a combination must be weighed against the confidence level (quality evaluation), the cost increase related to the number of measurements and to "expensive" techniques, the ease of achieving measurements under practical constraints, and the complexity in the interpretation of the results [3, 4]. It is important to underline that, as in the case of single technique, are not considered the situations in which no core has been taken from the existing structure, and are limited to situations where NDT is combined with cores.

The combination of NDT physical parameters does not prevent from the technical limitations of each individual technique. The quality of an evaluation can be affected by several factors: the characteristics of the investigated concrete, the variability of the concrete properties, the form of the used regression model, and errors related to measurements if measurements conditions are not well controlled. The efficiency of combination has been widely discussed in the literature [5]. Combination may be efficient in some cases [6–11], which correspond to cases where each technique gives alone a correct but "not perfect" result. Combination may be useless in other cases, either because a single technique gives very good results [12] or, more often, because one technique has a much poorer quality than the other [13–18]. In such case, adding a very noisy measurement to the first one, which is of acceptable quality, cannot provide added value.

Considering the case in which NDT techniques under investigation are of the same level of quality, two cases can be found:

(a) The NDT parameters are highly correlated. In this case, they can be used together in order to confirm a diagnosis [11].
(b) The NDT parameters are complementary and differently correlated to the concrete compressive strength. In this case they can be used together to improve the assessment of the concrete properties by means of different combination techniques available in the literature and will be synthetized herein.

On the basis of these considerations this chapter is structured in the three following paragraphs:

– Section 3.2 presents a synthesis of how the concrete properties influence the NDT measurements. These kinds of information can be used to decide if or not to apply the combined method depending on how concrete properties varies in concrete;
– Section 3.3 presents the methods that can be used to combine the NDT measurements for strength assessment. Several possible approaches will be detailed in a synthetic way, their interest and their limits will be pointed;
– Section 3.4 focuses on some issues that must be considered in order to combine two or several non-destructive techniques for a better concrete strength assessment, while applying the RILEM TC 249-ISC Recommendations.

3.2   How Variations of the Concrete Properties Influence the NDT Measurements

Concrete properties exhibit variability not only due to the spatial variability of structure (e.g. non-homogeneity and workmanship) during the construction, but also due to the temporal variability in exposure as temperature and humidity [19].

The final objective of this chapter is to assess when it is convenient or not, in term of time and additional cost, to combine NDT techniques in the evaluation of the concrete strength. In order to do this, it is necessary to first analyse the effect of concrete variability and the noise of the NDT techniques on the results. The underlying concept of the combined methods is, in fact, that if the two methods are influenced in different ways by the same factor, their combined use could result in a cancelling effect that improves the accuracy of the estimated strength.

For this reason, we built up two tables, one for the bulk NDT techniques (Table 3.1) and one for the surface NDT techniques (Table 3.2). In the tables, for each non-destructive test, factors that may influence the measurements and their effect are reported. The difference between bulk and surface NDT techniques lies in the thickness of material investigated: the bulk methods cover the entire thickness (or indicate a range) of the structural element investigated while the surface methods examine only few centimeters of the material from the surface. These tables were built up on the basis of scientific results from the literature or from international project as SENSO French project [20].

3.3   How to Combine NDT Measurements

Many research studies have been made since the 1960s in this area, both on laboratory specimens and (a few) on actual structures, for improving the quality of the correlation between the concrete strength and the measured parameters. Historically, the first work was cited in the paper of Samarin on the combination method [50, 51]. Most of the works in the literature were mainly focused on the combination of ultrasonic and rebound measurements, for which a higher correlation is expected if both techniques are combined instead of being used separately [52–54].

With the development of NDT in civil engineering, the different combinations (e.g. radar and ultrasound, radar and impact-echo, etc.) were proposed for different purposes [11, 20, 28, 55, 56]. Independently on the NDT used and providing test results, several methods can be used for identifying the most appropriate conversion model. The RILEM TC 249-ISC Recommendations considers explicit and implicit models (see §1.5.49). The first method, corresponding to an explicit model, is the multiple regression (§3.3.1), the two others are the artificial neural networks (§3.3.2) and the data fusion (§3.3.3) respectively.[1]

---

[1]Formally, all methods could be viewed as corresponding to explicit models. For instance, data fusion uses multiple regression for identifying the relation between NDT and concrete properties.

Table 3.1 Effect of concrete properties on NDT bulk technique; (L) is low, (I) is intermediate, (H) is high

NDT 'bulk' technique	Direct ultrasonic pulse velocity	GPR	Electrical capacity
Investigation depth (cm)	60 at 50 kHz	10 using the direct wave at 1.5 GHz 40 using reflected wave at 1.5 GHz (depending on moisture condition)	8 at 33 MHz
Investigated surface (cm^2)	10	150	50–200
Effect of increasing			
Water to cement ratio (w/c)	Decreases [21–23]	Increase of velocity and amplitude [24]	Increase of permittivity
Temperature	Between 5 and 30 °C no significant change; Decreases at 40 °C and 60 °C; Increases at temperatures below freezing [25–27]	Not documented, slight effect	Decrease of permittivity
Moisture conditions	Increases [21, 22, 25, 28–30]	Decrease of velocity and amplitude [24]	Decrease of permittivity
Porosity	Decreases [28, 30]	– Slight increase velocity and amplitude – decrease of velocity and amplitude for saturated concrete	– Slight increase of permittivity – decrease of permittivity for saturated concrete
Strength	Increases [31–35][a]	Slight increase of amplitude and velocity	Slight increase of permittivity
Load	Up to 20% of the ultimate load: increases [36]; In a medium range up to 75% of the ultimate load: constant [32, 36, 37]; For greater values: decreases [32, 36, 37]	Not documented, slight effect	Not documented, slight effect
Cost measurements class	I	I – H	I

[a] Velocity increases but with strengths greater than 35 N/mm^2 it appears almost constant

Table 3.2 Effect of concrete properties on NDT surface technique; (L) is low, (I) is intermediate, (H) is high

Concrete properties NDT 'surface' technique	Indirect ultrasonic pulse velocity	Rebound hammer	Electrical resistivity	Pull-out
Penetration depth (cm)	<5 cm depending on frequency	<2 cm [38]	From 2 to 5 cm depending on distance of electrodes	2.5
Investigated surface (cm^2)	<100 cm^2 approximately	100 cm^2	25–100 cm^2	25
If increase of				
Water to cement ratio (w/c)	Decrease	Decrease due to the decrease of strength	Decrease due to the increase of porosity [39]	Decrease due to decrease of strength
Moisture condition	Increases [11, 30, 40]	Decrease [41]	Decrease [39]	Slight decrease
Porosity	Decreases [11, 30]	Decrease due to the decrease of strength	Increase of resistivity	Decrease
Temperature	Between 5 and 30 °C no significant change; Decreases at 40 and 60 °C; Increases at temperatures below freezing [25–27]	Undocumented—no significant (10–35 °C)[a]	Decrease [42, 43]	No effect
Chlorides	No effect	No effect	Decrease of resistivity if concrete is wet or saturated [44–46]	No effect
Carbonation depth	Increase	Significant increase [47, 48]	Increase	No effect [49]
Strength	Increases [33]	Increase	Slightly increase	Increase
Cost measurements class	I	L	L	I

[a]According to EN 12504-2

However, the difference is more in common engineering practice, as neural networks are often taken as "black-boxes". Whatever the method, the user, researcher or engineer, must know and control how the model works.

3.3.1 Multiple Regression Methods

3.3.1.1 SonReb

SonReb method is a combination between ultrasonic and rebound methods, which allow performing quick and simple measurements to evaluate the compressive strength of concrete. It was proposed by RILEM TC 43-NDT [54]. This method consists in taking the average value of the ultrasonic velocity (V) obtained from two to four measurements and the average rebound value (R) from at least nine measurements on each test specimen, and then the compressive strength of the concrete is estimated in a specific chart, called nomogram. Figure 3.1 shows two examples: (a) strength curves as a function of ultrasonic velocity and rebound, (b) rebound curves as a function of strength and the ultrasonic velocity. The strength of concrete is described as a two variables equation $f_c = f(V, R)$.

Many types of equations have been used (e.g. linear, power, exponential, etc.). After having been developed mainly in Eastern European countries from the 1970s, SonReb method has now become the most common combination method for estimating the strength of concrete in structures.

SonReb method has two main limitations:

Many factors influence the measurements of ultrasonic velocity and rebound. Correction coefficients were proposed in RILEM Recommendations [54] to take into account such factors, namely the type of cement, the cement content, the type of aggregate, the volume of fine aggregates, the maximum size of the aggregates, and the admixture. The final result can thus be corrected relative to the reference strength given by the nomogram.

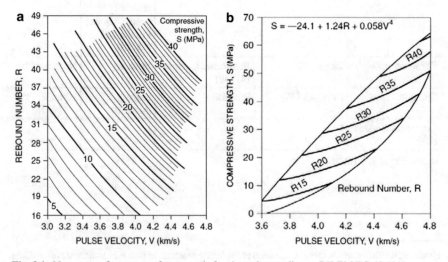

Fig. 3.1 Nomogram for curves of **a** strength, **b** rebound according to RILEM TC 43 [54]

If the proposed correction of the results remains possible in the lab, it requires a large number of information and remains very difficult on existing structures where this information is not available. However, it is possible to calibrate the SonReb model based on core tests. Several methodologies have been published based on two methods (new regression or shifting) as presented in [57–59].

3.3.1.2 Multiple Regression Analysis

Data descriptive analysis is based on statistical regression identified with different methods. The first method is to establish correlations of data pairs, and is called univariate analysis. EN 13791: 2007 [60] was devoted to the case of simple regression, ex. fc = f (V) or fc = f (R). Univariate analysis is therefore used in order to estimate concrete strength at the locations where there is only one NDT technique. The logic remains the same in the case of a multiple regression, which is a generalization of the simple regression.

The Multiple Regression Analysis (MRA) method consists in establishing correlations to estimate a variable (in our case compressive strength) with two or more independent variables (NDT techniques), and is called multivariate analysis. It is used when several techniques are used in the same areas. For example, the concrete strength can be estimated by the combination of two [10] or more observables [11, 29]. In the case of two observables given by two individual techniques (e.g. ultrasound, rebound), it is a bivariate analysis describing correlations in the form of a regression model as for SonReb method $f_c = f(V, R)$.

Many research works have been devoted to improve the quality of these correlations either by using different forms (e.g., linear, power, exponential, polynomial, etc.) or by calibrating a priori the model coefficients.

In the general case, the relationship between an estimated variable X and a series of m observables $(Y_1, Y_2 \ldots Y_m)$ can be described as a multiple linear regression model (Eq. 3.1) or multiple regression power [11, 29, 31, 56]. Equation Eq. 3.1 specifies how the dependent variable is related to explanatory variables. This equation can write in matrix shape (Eq. 3.2) or in a more compact way $X = Y + \varepsilon$.

$$X = a_0 + a_1 Y_1 + a_2 Y_2 + \ldots + a_m Y_m + \varepsilon \tag{3.1}$$

with $a_0, a_1 \cdots a_m$ as the regression coefficients; ε is the random noise representing the error term.

$$\begin{pmatrix} x_1 \\ x_i \\ x_n \end{pmatrix} = \begin{pmatrix} 1 & y_{1,1} & \cdots & y_{m,1} \\ 1 & y_{1,i} & \cdots & y_{m,i} \\ 1 & y_{1,n} & \cdots & y_{m,n} \end{pmatrix} \begin{pmatrix} a_0 \\ a_1 \\ a_i \\ a_m \end{pmatrix} + \begin{pmatrix} \varepsilon_1 \\ \varepsilon_2 \\ \varepsilon_n \end{pmatrix} \tag{3.2}$$

From Eq. 3.2, it is also possible to consider simultaneously a group of variables to estimate (several concrete properties as strength, porosity, water content, etc.), which are linked to NDT measurements. Recent studies have shown that some NDT parameters are sensitive to several properties of concrete, for example ultrasonic velocity depends on concrete strength and also the water content [30, 61, 62]. The regression models have now been developed to evaluate a pair of properties like for instance strength-saturation or porosity-saturation [11, 29].

3.3.1.3   Estimation of Model Parameters

The multiple regression models or SonReb are often estimated by the least squares method for different types of models as linear, power, exponential or polynomial. The shapes of the different models are generally compared in terms of quality. The common practice comes to keep the model that has the best fit for the set of experimental data, either because it gives the highest value of the coefficient of determination or lowest error. However, for a safer analysis, it is better to evaluate the prediction capability of the models by performing a trial and test approach.[2] This approach is based on the evaluation of compressive strength using the fitting set and the evaluation of the prediction on a new data set called testing or prediction data set [63].

The principle of the estimated regression coefficients is to minimize the sum of squared residuals between the measured value x_i and the calculated value $\hat{x}_i$ by the regression model shown in Eq. 3.3.

$$\sum_{i=1}^{n} e_i^2 = \sum_{i=1}^{n} \left(x_i - \hat{x}_i \right)^2 \tag{3.3}$$

It is important to note that a confidence interval can be calculated for each coefficient in the model. The quality of the multiple regression can be evaluated by the following main criteria (see also Chap. 11).

The Coefficient of Determination

The coefficient of determination (r^2) of the model is calculated by Eq. 3.4 and varies between 0 and 1. It provides a comprehensive view of the quality of the multiple regression. If it is close to 1, it means that the variable X is well explained by the explanatory variables Y. In general, it increases when adding a new explanatory variable.

---

[2] This is because the fitting error $RMSE_{fit}$ does not provide a relevant estimate of the model error and must be replaced by the predictive error $RMSE_{pred}$. This issue is specified in §1.5.4.3 and §1.5.6.2. It is also illustrated on a practical example in Chap. 9.

Let us consider a set of n values marked $f_{c\,i}$, $i = 1$, n, corresponding to the n measured values of strengths that have a mean value (arithmetic mean $f_{c\,mean}$). From NDT measurements and strengths measured on a calibration set, a model $f_{c,est}$ (NDT) has been identified and a set of n values $f_{c,est\,I}$ has been obtained. $f_{c\,mean}$ is the mean value (arithmetic mean) of the measured (observed) data. The most general definition of the <u>coefficient of determination r^2</u> is.

$$r^2 = 1 - SS_{res}/SS_{tot} \qquad (3.4)$$

where SS_{res} is the residual sum of squares.

$$SS_{res} = \sum_i \left(f_{c,i} - f_{c,est,i}\right)^2 \qquad (3.5)$$

and where SS_{tot} is the total sum of squares (equal to n times the variance of the data set):

$$SS_{tot} = \sum_i \left(f_{c,i} - f_{c\,mean}\right)^2 \qquad (3.6)$$

r^2 can also be seen to be the part of the total variance which is explained by the model.

The Root Mean Square Error (RMSE)

RMSE is the frequently used method for evaluating the performance of a model. It is the square root of the differences between values estimated by a model and values actually observed (Eq. 3.7). These individual differences are called residuals when the calculations are performed over the data sample that was used for the estimation. They are called prediction errors when computed out-of-sample.

$$RMSE = \sqrt{\frac{\sum_{i=1}^{n} \left(f_{c,est\,i} - f_{c\,i}\right)}{n}} \qquad (3.7)$$

3.3.2 Artificial Neural Networks

The artificial neural networks (ANN) are biologically inspired and represent a mathematical model of the functioning of the biological neuron. The ANNs offer an alternative to mathematical modelling and are part of nonparametric statistics and nonlinear models able to respond to issues of decision support, diagnosis, prediction, etc. The idea is to introduce input and output data to the ANN, and to make it learn the relationship between both through a process called training process. The objective of this process is to minimize the error between the theoretical output and that calculated by the ANN. The advantage is that the mathematical relation (based on inverse exponential form) can be nonlinear that generally increases the prediction performance.

The mathematical principle is widely described in the literature [64–66] and some software are also available as the Matlab Toolbox "Neural Network". In the field of civil engineering, the main applications of ANN concern the prediction of concrete performance, especially the respect to the compressive strength with respect to mix constituents [67, 68]. However, a limited number of works have been published regarding ANN developments in the field of the NDE of reinforced concrete structures. For instance, Hola and Schabowicz [55] have presented a neural model for the evaluation of concrete compressive strength based on the coupling of different non-destructive techniques as ultrasonic pulse velocity and impact echo. The authors found this model effective for accurate on-site prediction of concrete strength. It was applied also to a database from NDT experimental results of SENSO national French project for the evaluation of the levels of water content and the compressive strength of concrete [31].

A comparison between ANN and multiple regression analysis was discussed in [31]. The results showed that in the case of combination of NDT methods for strength evaluation using the training data set, the difference between ANN and MRA is not very significant. However, ANN is slightly better than MRA for predicting (using testing data set) concrete compressive strength and water saturation.

3.3.3   Data Fusion

Regarding data fusion, a smaller part of data combination applications concerns the concrete characterisation and more specially the strength evaluation. The different principles are already explained in [56]. The general principle is based on the identification of regression models and on the evaluation of the uncertainty of NDT measurements as presented on Fig. 3.2. This procedure is performed for all the NDT techniques used for the combination.

Data fusion consists in the combination of at least two NDT methods using a mathematical operator. The theory of beliefs proposed by Dempster Shafer can be used to segment the information and reinforce the confidence in the results, however, it is difficult to use it to evaluate a quantified value. In this respect, works has been developed with the possibilities theory [20] (Fig. 3.3) to quantify a characteristic of the concrete. Its application for on-site measurement was proposed by sampling of the inversion process with cores extracted from the evaluated structure [69]. The principle is to quantify and to decouple different characteristics of the concrete by the combination of NDT evaluation [56]. An example of compressive strength evaluation

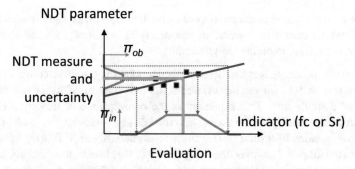

Fig. 3.2 Regression analysis and propagation of the uncertainty

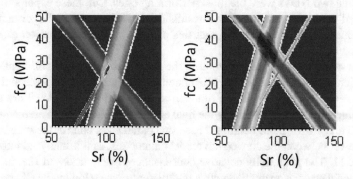

Fig. 3.3 Example of data fusion applied for the evaluation of compressive strength and saturation degree

is presented on Fig. 3.3. This figure shows two different cases, the first (left) is the case when the NDT methods converge to the same solution with high precision, and the second case (right) when the convergence of NDT is of bad quality. The method can then give information about the complementarity between NDT techniques and the possible conflict. It can be then used for evaluating if it is convenient or not to combine NDT.

3.4 Applying the RILEM TC 249-ISC Recommendations and Combining Several NDT

In scientific literature, two opposite statements can be found regarding the interest of combining several non-destructive techniques:

– some studies seem to indicate that the combination is efficient, as it leads to a better fit of experimental data (i.e. a higher coefficient of determination or a lower error),

- other studies seem to reach the opposite conclusion, with no significant improvement, which comes to a waste of resources, as combining several techniques
means more time to obtain the test results.

As a result, no consensus has been reached among the scientific community (and
the same can be told for engineers) about the real interest of using several NDT
instead of a single one. This statement is the consequence of lacking a rigorous
approach while identifying the conversion models and assessing the model error.

The combination of several NDT was also considered by experts in the benchmark
described in Chap. 6. However, the first priority of this benchmark was not to check
the validity of combining techniques. In this same chapter, Fig. 6.23 shows that, at
the most ambitious knowledge level (KL3), experts E, K, H and I who all have chosen
to combine two NDTs were the most performing ones. But, these experts have also
privileged the high precision of test results (high TRP) against a larger number of
measurements, and it was shown that this choice was the main reason explaining
their success.

When combining several techniques, the flowchart of Fig. 1.9 still applies. The
only difference lays in the type of conversion model that must be chosen and in how
its parameters are identified. The number of cores and the TRP level are the two
most influencing factors regarding the final accuracy of the strength assessment (see
Chaps. 2 and 11 for more details). These two parameters control the magnitude of
the trade-off between the conversion model parameters, as it is analysed into details
in Chap. 11. The potentially negative consequences of the trade-off (i.e. its impact
on the model error) is magnified with a multivariate conversion model Chap. 11. It is
shown in this same chapter that combining two non-destructive techniques can bring
some added-value. The conditions are that this combination must be based on high
precision test results and that the physical consistency of the conversion model must
be checked.

As a consequence, it is possible to specify the few requirements for an efficient
use of combination while applying the RILEM TC 249-ISC Recommendations:

- Task T3: use techniques that have a high TRP (if not, the poorest TRP technique
will cause the global loss of precision) and check this TRP level,
- Task T7: use a simple (explicit) conversion model, with only three model
parameters. This model can be for instance a multi-linear one, or a double
power-law,
- Task T8: calibrate the model (i.e. identify the value of the model parameters) with
a simple regression,
- Task T9: quantify the model error, using the predictive error $RMSE_{pred}$.

As Tasks T7, T8 and T9 are all related to the post-processing of test results, the
prediction error of the combined model can always be compared to that obtained if
a single technique is used. The final decision will be to keep the multi-variate model

or to come back to a univariate model, which may be more performing.[3] In any case, the final assessment error will be controlled.

3.5 Conclusion

This chapter has provided a synthesis of how the concrete properties influence the NDT measurements that can be helpful for selecting NDT methods for the combination. These kinds of information can be also used to decide if it is convenient or not to apply the combined method.

The second part summarised the methods used to combine the NDT measurements for strength assessment. Several possible approaches have been detailed in a synthetic way. SonReb method is the most popular method for combining NDT methods as UPV and rebound hammer. The results in the literature show that the combination method is not always efficient because it depends on the quality of both techniques and also on the variability of the tested material.

The use of more than two NDT techniques can improve the evaluation if for example moisture content varies in the tested concrete or if both techniques are of equivalent quality and highly complementary. For instance, the combination of rebound hammer and GPR or electrical resistivity can give more reliability in the evaluation of concrete strength independently of the way the model is identified (regression analysis or artificial neural networks). Data fusion is also an interesting tool because it takes into account the complementarity of the used NDT techniques or the possible conflict. In addition, this method takes into account the uncertainty of the NDT measurements. However, it is necessary to point out that all the presented methods of NDT combination need a calibration procedure because laboratory models do not take into account all factors that can affect NDT techniques on site.

The RILEM TC 249-ISC Recommendations can be easily adapted to the case where several NDT are used in combination, with no additional work. As it has been explained, it is thus preferable to use an explicit conversion model and the priority must be to check the precision of test results and the magnitude of the final conversion model error.

References

1. RILEM TC 207-INR: Non-destructive assessment of concrete structures: reliability and limits of single and combined techniques. In: Breysse, D. (ed.) State-of-the-Art Report of the RILEM Technical Committee 207-INR (2012). https://doi.org/10.1007/978-94-007-2736-6

---

[3]This may be the case, for instance, if one technique has a poorer TRP level than the other technique, of if the range of fluctuation of concrete properties in the investigation domain is too small.

2. Dérobert, X., Garnier, V., François, D., Lataste, J.F., Laurens, S.: Complémentarité des méthodes d'END. In: Breysse, D., Abraham, O. (eds.) Guide méthodologique de l'évaluation non destructive des ouvrages en béton armé, pp. 550. Presse ENPC, Paris, France (2005)
3. Leshchinsky, A.M.: Combined methods of determining control measures of concrete quality. Mater. Struct. **24**, 177–184 (1991)
4. Gehlen, C., Dauberschmidt, C., Nürnberger, U.: Condition control of existing structures by performance testing. Otto-Graf. J. **17**, 19–44 (2006)
5. Breysse, D.: Nondestructive evaluation of concrete strength: an historical review and a new perspective by combining NDT methods. Constr. Build. Mater. **33**, 139–163 (2012)
6. Lenzi, M., Versari, D., Zambrini, R.: Indagine sperimentale di calibrazione del metodo combinato SonReb. International Report, University, Ravenna, Italy (2010)
7. Cianfrone, F., Facaoaru, I.: Study on the introduction into Italy on the combined non-destructive method, for the determination of in situ strength. Mater. Struct. **12**(5), 413–424 (1979)
8. Oktar, O.N., Moral, H., Tasdemir, M.A.: Factors determining the correlations between concrete properties. Cem. Concr. Res. **26**(11), 1629–1637 (1996)
9. Lee, J.H.: Multivariate Analysis to Determine In-situ Concrete Strength-Beam, Bachelor Thesis., Univ. Teknologi Malaysia (2009)
10. Qasrawi, H.Y.: Concrete strength by combined nondestructive methods simply and reliably predicted. Cem. Concr. Res. **30**(5), 739–746 (2000)
11. Sbartaï, Z.M., Breysse, D., Larget, M., Balayssac, J.P.: Combining NDT techniques for improved evaluation of concrete properties. Cem. Concr. Compos. **34**(6), 725–733 (2012)
12. Domingo, R., Hirose, S.: Correlation between concrete strength and combined nondestructive tests for concrete using high-early strength cement. In: 3rd JSPS-DOST International Symposium, University Philipines Diliman, Quezon City, Philippines, 9–10 Mar 2009
13. Knaze, P., Beno, P.: The use of combined non-destructive testing methods to determine the compressive strength of concrete. Mater. Struct. **17**(3), 207–210 (1984)
14. Cumming, S.R.: Nondestructive Testing to Monitor Concrete Deterioration Caused by Sulfate Attack, Master Diss., University of Florida (2004)
15. Muniandy, H.H.H.: Multivariate Analysis to Determine In-Situ Concrete Strength—Column. Bachelor in Civil Engineering, Universiti Teknologi Malaysia (2009)
16. De Almeida, I.R.: Non-destructive testing of high strength concretes: rebound (Schmidt hammer) and ultra-sonic pulse velocity. In: Taerwe, L., Lambotte, H. (eds.) Proceedings of 2nd International RILEM/CEB Symposium on Quality control of concrete structures, pp 387–397, Ghent, Belgium, 12–14 June 1991
17. Machado, M.D., Shehata, L.C.D., Shehata, I.A.E.M.: Correlation curves to characterize concretes used in Rio de Janeiro by means of non-destructive tests. Rev. IBRACON Estrut. Mater. **2**(2), 100–123 (2009)
18. Wu, N., Han, A.: Experiment and application of combined method op pulse velocity and rebound hardness in tunnel engineering. J. Nanjing Tech. Univ. **30**(4), 93–97 (2008)
19. Nguyen, N.T., Sbartaï, Z.M., Lataste, J.F., Breysse, D., Bos, F.: Non-destructive evaluation of the spatial variability of reinforced concrete structures. Mech Ind **16**(1), 6 (2015). N°130212. https://doi.org/10.1051/meca/2014064
20. Project ANR SENSO : Stratégie d'Evaluation Non destructive pour la Surveillance des Ouvrages en béton, ANR Project Final Report, 274 pp (2009)
21. Zhu, F.Z., Zha, T.J., Guan, T.: Influence of water content on dynamic elastic modulus of concrete. Appl. Mech. Mater. **351–352**, 1605–1609 (2013)
22. Bungey, J.H., Millard, S.G.: Testing of Concrete in Structures. CRC Press, Boca Raton, Florida (2010)
23. Kaplan, M.F.: The effects of age and water to cement ratio upon the relation between ultrasonic pulse velocity and compressive strength of concrete. Mag. Concr. Res. **11**(32), 85–92 (1959)
24. Sbartaï, Z.M., Laurens, S., Balayssac, J.P., Ballivy, G., Arliguie, G.: Effect of concrete moisture on radar signal amplitude. ACI Mater. J. **103**(6), 419–426 (2006)
25. Jones, R., Facaoaru, I.: Recommendations for testing concrete by the ultrasonic pulse method. Mater. Struct. **2**, 275–284 (1969)

26. BS 1881–203:1986: Testing Concrete: Recommendations for Measurement of Velocity of Ultrasonic Pulses in Concrete. British Standards Institution (1986)
27. Naik, T.R., Malhotra, V.M., Popovics, J.S.: The ultrasonic pulse velocity method. In: Handbook of Nondestructive Testing of Concrete, Chap. 8. CRC Press, Boca Raton, Florida (2003)
28. Balayssac, J.P., Laurens, S., Arliguie, G., Breysse, D., Garnier, V., Dérobert, X., Piwakowski, B.: Description of the general outlines of the French project SENSO—quality assessment and limits of different NDT methods. Constr. Build. Mater. **35**, 131–138 (2012)
29. Kheder, G.F.: A two stage procedure for assessment of in situ concrete strength using combined non-destructive testing. Mater. Struct. **32**, 410–417 (1999)
30. Garnier, V., Piwakowski, B., Abraham, O., Villain, G., Payan, C., Chaix, J.F.: Acoustic techniques for concrete evaluation: improvements, comparisons and consistency. Constr. Build. Mater. **43**, 589–613 (2013)
31. Sbartaï, Z.M., Laurens, S., Elachachi, S.M., Payan, C.: Concrete properties evaluation by statistical fusion of NDT techniques. Constr. Build. Mater. **37**, 943–950 (2012)
32. Menditto, G., Bufarini, S., D'Aria, V., Porco, G.: Metodo combinato ultrasuoni-sclerometro (SonReb): considerazioni e riflessioni. 15o Congresso CTE Collegio dei Tecnici della, pp 305–312. Industrializzazione Edilizia, Bari, Italy, 4–6 Nov 2004
33. Shariati, M., Ramli-Sulong, N.H., Arabnejad, K.H., Shafigh, P., Sinaei, P.: Assessing the strength of reinforced concrete structures through Ultrasonic Pulse Velocity and Schmidt Rebound Hammer tests. Sci. Res. Essays **6**(1), 213–220 (2011)
34. Hannachi, S., Guetteche, M.N.: Application of the combined method for evaluating the compressive strength of concrete on site. Open J. Civ. Eng. **2**(1), 16–21 (2012)
35. Nash't, I.H., A'bour, S.H., Sadoon, A.A.: Finding a unified relationship between crushing strength of concrete and non-destructive tests. In: 3rd Middle East NDT Conference and Exhibition, Manama, Bahrain, 27–30 Nov 2005
36. Popovics, S., Rose, J.L., Popovics, J.S.: The behavior of ultrasonic pulses in concrete. Cem. Concr. Res. **20**(2), 259–270 (1990)
37. Popovics, S., Popovics, J.S.: Effect of stresses on the ultrasonic pulse velocity in concrete. Mater. Struct. **24**(1), 15–23 (1991)
38. Lam, E.S.S.: Estimating the economic life span of a reinforced concrete building. In: Grantham, M., Salomoni, V., Majorana, C. (eds.) Concrete Solutions, 2nd edn., pp. 291–298. CRC Press, Boca Raton, Florida (2009)
39. Gjørv, O.E., Vennesland, Ø.E., El-Busaidy, A.H.S.: Electrical resistivity of concrete in the oceans. In: 9th Annual Offshore Technology Conference, Houston, Texas, 2–5 May 1977, pp 581–588, paper 2803 (1977)
40. Sutan, N.M., Jaafar, M.S.: Evaluating efficiency of nondestructive detection of flaws in concrete. Russ. J. Nondestruct. Test. **39**(2), 87–93 (2003)
41. Szilágyi, K., Borosnyói, A.: 50 years of experience with the Schmidt rebound hammer. Concr. Struct., Annu. Tech. J. **10**, 46–56 (2009)
42. Bertolini, L., Polder, R.B.: Concrete resistivity and reinforcement corrosion rate as function of temperature and humidity of the environment, TNO report 97-BT-R0574. Netherlands Organisation for Applied Scientific Research, Delft, The Netherlands (1997)
43. Polder, R.B.: Test methods for on-site measurement of resistivity of concrete—a RILEM TC-154 technical recommendation. Constr. Build. Mater. **15**(2–3), 125–131 (2001)
44. Tuutti, K.: Corrosion of Steel in Concrete, CBI Research Report 4.82. Cement and Concrete Research Institute, Stockholm, Sweden (1982)
45. Andrade, C., Sanjuan, M.A., Alonso, M.V.: Measurement of chloride diffusion coefficient from migration tests. In: Corrosion 93, NACE International, Houston, Texas, paper 319
46. Sbartaï, Z.M., Laurens, S., Rhazi, J., Balayssac, J.P., Arliguie, G.: Using radar direct wave for concrete condition assessment: correlation with electrical resistivity. J. Appl. Geophy. **62**(4), 361–374 (2007)
47. Kim, J.K., Kim, C.Y., Yi, S.T., Lee, Y.: Effect of carbonation on the rebound number and compressive strength of concrete. Cem. Concr. Compos. **31**(2), 139–144 (2009)

48. Aydin, F., Saribiyik, M.: Correlation between Schmidt Hammer and destructive compressions testing for concretes in existing buildings. Sci. Res. Essays **5**(13), 1644–1648 (2010)
49. Moczko, A.T., Carino, N.J., Petersen, C.G.: CAPO-TEST to estimate concrete strength in bridges. ACI Mater. J. **113**(6), 827–836 (2016)
50. Samarin, A., Dhir, R.K.: Determination of in situ concrete strength: rapidly and confidently by nondestructive testing. In: Malhotra, V.M. (ed.) In Situ/Nondestructive Testing of Concrete, ACI SP-82, pp 77–94. American Concrete Institute (1984)
51. Malhotra, V.M., Carino, N.J. (eds.): Handbook of Nondestructive Testing of Concrete. CRC Press, Boca Raton, Florida (2003)
52. Samarin, A., Meynink, P.: Use of combined ultrasonic and rebound hammer method for determining strength of concrete structural members. Concr. Int. **3**(3), 25–29 (1981)
53. Facaoaru, I.: Non-destructive testing of concrete in Romania. In: Symposium on NDT of Concrete and Timber, pp. 39–49. Institute of Civil Engineers, London, United Kingdom, 11–12 June 1969
54. RILEM TC 43-CND, Facaoaru, I. (chair): Draft recommendation for in situ concrete strength determination by combined non-destructive methods. Mater. Struct. **26**, 43–49 (1993)
55. Hola, J., Schabowicz, K.: New technique of non-destructive assessment of concrete strength using artificial intelligence. NDT E Int. **38**(4), 251–259 (2005)
56. Ploix, M.A., Garnier, V., Breysse, D., Moysan, J.: NDE data fusion to improve the evaluation of concrete structures. NDT E Int. **44**(5), 442–448 (2011)
57. Sbartaï, Z.M., Garnier, V., Villain, G., Breysse, D.: Assessment of concrete by a combination of non-destructive techniques. In: Non-Destructive Testing and Evaluation of Civil Engineering Structures, Chap. 8, pp. 259–297. ISTE Press—Elsevier (2018)
58. Alwash, M., Breysse, D., Sbartaï, Z.M., Szilágyi, K., Borosnyói, A.: Factors affecting the reliability of assessing the concrete strength by rebound hammer and cores. Constr. Build. Mater. **140**, 354–363 (2017)
59. Breysse, D., Balayssac, J.P., Biondi, S., Borosnyói, A., Candigliota, E., Chiauzzi, L., Garnier, G., Grantham, M., Gunes, O., Luprano, V.A.M., Masi, A., Pfister, V., Sbartaï, Z.M., Szilágyi, K., Fontan, M.: Non-destructive assessment of in situ concrete strength: comparison of approaches through an international benchmark. Mater. Struct. **50**, 133 (2017). https://doi.org/10.1617/s11 527-017-1009-7https://doi.org/10.1617/s11527-017-1009-7
60. EN 13791:2007: Assessment of In-situ Compressive Strength in Structures or in Precast Concrete Products. European Committee for Standardization (CEN), Brussels, Belgium (2007)
61. Popovics, S.: Effects of uneven moisture distribution on the strength of and wave velocity in concrete. Ultrasonics **43**(6), 429–434 (2005)
62. Ohdaira, E., Masuzawa, N.: Water content and its effect on ultrasound propagation in concrete— the possibility of NDE. Ultrasonics **38**, 546–552 (2000)
63. Arlot, S., Celisse, A.: A survey of cross-validation procedures for model selection. Stat. Surv. **4**, 40–79 (2010). https://doi.org/10.1214/09-SS054https://doi.org/10.1214/09-SS054
64. Rumelhart, D.E., Hinton, G.E., Williams, R.J.: Learning internal representations by error prop-agation. In: Rumelhart, D.E., McClelland, J.L. (eds.) Parallel Distributed Processing: Explo-ration in the Microstructure of Cognition: Foundations, pp. 318–362. MIT Press, Cambridge, Massachusetts (1987)
65. Bishop, C.M.: Neural Networks for Pattern Recognition. Oxford University Press (1995)
66. Rafiq, M.Y., Bugmann, G., Esterbrook, D.J.: Neural network design for engineering applica-tions. Comput. Struct. **79**(17), 1541–1552 (2001)
67. Yeh, I.C.: Modelling of strength of high-performance concrete using artificial neural networks. Cem. Concr. Res. **28**(12), 1797–1808 (1998)
68. Ni, H.G., Wang, J.Z.: Prediction of compressive strength of concrete by neural networks. Cem. Concr. Res. **30**(8), 1245–1250 (2000)
69. Villain, G., Sbartaï, Z.M., Dérobert, X., Garnier, V., Balayssac, J.P.: Durability diagnosis of a concrete structure in a tidal zone by combining NDT methods: laboratory tests and case study. Constr. Build. Mater. **37**, 893–903 (2012)

Chapter 4
Identification of Test Regions and Choice of Conversion Models

Jean-Paul Balayssac, Emilia Vasanelli, Vincenza Anna Maria Luprano, Said Kenai, Xavier Romão, Leonardo Chiauzzi, Angelo Masi, and Zoubir Mehdi Sbartaï

Abstract The main objective of test region (TR) identification is to define an efficient conversion model. The first part of the chapter aims a difficult question, the identification of test regions (TR) because each structure is specific and so it is impossible to give a unique methodology. Here, three different possibilities are proposed. The first one is based on synthetic data obtained on a continuous structure for which TR are identified by means of k-means clustering method. The second approach concerns a real building for which TR are determined by means of two different statistical methods based on the analysis of confidence interval and ANOVA. On three real case studies, the second part of the chapter compares the performances of two scenarios, either the consideration of several TRs and so a conversion model on each one, or the consideration of a unique TR with only one model. The efficiency

---

[1]EN 13791:2007 [1].

J.-P. Balayssac (✉)
LMDC, Université de Toulouse, INSA/UPS Génie Civil, Toulouse, France
e-mail: jean-paul.balayssac@insa-toulouse.fr

E. Vasanelli
Institute of Heritage Science - National Research Council (ISPC-CNR), Lecce, Italy

V. A. M. Luprano
ENEA - Italian National Agency for New Technologies, Energy and Sustainable Economic Development, Department for Sustainability - Non Destructive Evaluation Laboratory, Brindisi, Italy

S. Kenai
Civil Engineering Department, University of Blida, Blida, Algeria

X. Romão
CONSTRUCT-LESE, Faculty of Engineering, University of Porto, Porto, Portugal

L. Chiauzzi · A. Masi
School of Engineering, University of Basilicata, Potenza, Italy

Z. M. Sbartaï
University Bordeaux, I2M-UMR CNRS 5295, Talence, France

© RILEM 2021 117
D. Breysse and J.-P. Balayssac (eds.), *Non-Destructive In Situ Strength Assessment of Concrete*, RILEM State-of-the-Art Reports 32,
https://doi.org/10.1007/978-3-030-64900-5_4

of each scenario is quantified by the error on the estimation of both mean strength and local strength.

4.1 Introduction

Test Regions (TR), are parts of the building whose data belong to the same population as it is specified in EN 13791:2007.[1] Nevertheless, the identification of TR is difficult to address since any methodology cannot be generalized to all the case studies because each structure is specific. Instead of proposing a common methodology, this appendix gives examples of different processes to identify test regions (TR). In paragraph 2.1 is described the case of a continuous structure (cylindrical tank) for which TR are identified by means of k-means clustering method. In paragraph 2.2 a real case study building is considered for which TR are determined by means of two different statistical methods. The first method is based on a statistical analysis of the data and a determination of confidence intervals, which allows the discrimination of test regions. Then a combination of T-student test and ANOVA is used for validation. The second method with two different approaches is based on the use of confidence interval.

The identification of TR has the objective of defining an efficient conversion model. Two choices can be done: one is to consider a single conversion model for each test region and the second one is to consider a unique conversion model for all the test regions. For both the case, the efficacy of the conversion models is assessed by calculating the predictive model error, in order to choose the better solution. It is important to note that revising the choice (from several TR to one TR) does not imply any additional data, thus the comparison between model errors with the two options can lead to revise easily this last choice.

In a second part of the chapter (paragraph 3), the influence of the consideration of the existence of TR on the identification of the conversion models will be analysed on three real case studies. In particular if two TR can be identified two scenarios will be studied, either the identification of a conversion for each TR or the determination of a unique model. The performances of the two scenarios will be compared.

4.2 Identification of TR

4.2.1 TR Identification in a Continuous Structure

4.2.1.1 Methodology

In this section, a methodology for the identification of TR is proposed on the basis of synthetic data. It consists in performing a pre-auscultation of the structure with NDT methods in a limited number of TL, in order to reduce the cost. Once done this pre-auscultation, test results are analysed, either separately for each technique or after a combination. For instance, rebound and UPV can be merged to assess compressive strength by using Sonreb method (see Chap. 3 for a complete description of this method).

In the example presented in the following sections, synthetic data are generated on three cylindrical tanks. In this case, the pre-auscultation results are plotted on maps, which show the variation on the tank surface of the different test results or that of the estimated strength by merging test results. The identification of the test regions is achieved by a statistical processing with k-means clustering method.

4.2.1.2   Generation of the Synthetic Data

The basic source data is density d. Once d is known, all other properties (strength, rebound, velocity …) are generated by using simple relationships, and measurements have been generated from true properties by simply adding a random measurement error.

Since the focus is on the delimitation of "homogeneous areas", several tanks have been generated which all show specific patterns. In each tank, three scales are considered regarding the concrete properties:

- at wall scale, the tanks are divided between a LEFT AREA and a RIGHT AREA which can have different properties. The physical limit between LEFT and RIGHT does not follow a vertical line: the RIGHT area contains 6 batches at ground level, and only two batches at top level (the total number of batches is $20 = 6 + 5 + 4 + 3 + 2$).
- at batch scale, for each of the two areas, some batch-to-batch variability is considered, but the volume of concrete of on given batch is supposed to be homogeneous,
- each surface corresponding to a given batch contains several possible test locations where the local concrete properties can be tested.

Thus variability exists at several scales:

- between LEFT and RIGHT AREA, (a) the density in each area is supposed to follow a Gaussian distribution (mean and SD are given), (b) some accidents are superimposed to the regular variability. An "accident" corresponds to a specific batch whose density, therefore strength, is significantly lower than expected due to fabrication or casting problems. Accidents were generated randomly, with a probability p and a magnitude d, corresponding to a strength reduction factor,
- in each batch, the density is fixed as the concrete is assumed to be statistically homogeneous, but the true strength has some low scale variability (which may be due for instance in small variations in moisture). This variability is modelled with an additional random noise (gaussian, S.D. = 2 MPa).

Table 4.1 Values used for the generation of synthetic data

Tank	Left area				Right area			
	m	COV	p	d	m	COV	p	d
1	2.35	0.04	0	–	2.35	0.04	0	–
2	2.37	0.06	0	–	2.28	0.06	0	–
3	2.34	0.04	0	–	2.31	0.04	3%	0.08

Table 4.2 Range of variation of density and of the other properties

Density	2.25	2.30	2.35	2.40
f_c (MPa)	28	31	34	37
Rebound R	34.4	36.1	37.7	39.2
UPV (m/s)	4150	4236	4316	4392

Table 4.1 provides all values used for the generation of the three tanks. The different columns indicate, for the two areas, the mean (m) and coefficient of variation (COV) of the Gaussian density distribution, the probability (p) of an accidental batch and the magnitude (d) of such an accident, as a relative decrease of density.

The values of Table 4.1 have been chosen in order to obtain density distributions, and therefore strength distributions that are in agreement with what is observed in real structures. The values of "reference" UPV and "reference" rebound are calculated from the "reference" strength with a conversion model which is a power law. Table 4.2 provides the values that are simulated when the concrete density varies in the [2.25–2.40] interval. Some measurement uncertainty is finally added to the reference values at each test location.

The data maps are presented on Fig. 4.1 for the three tanks. For each tank the measured compressive strength and the value of NDT properties for, respectively, rebound, UPV and pull-out, are plotted. For a better clarity, scales on X and Y axis are different. The different concrete batches are clearly visible, corresponding to a rectangular pattern which corresponds to the casting sequence. The small scale fluctuations correspond to the within batch variability. As expected, Tank 1 presents a rather homogeneous distribution of strength on the entire surface and only random variations. On the other hand, one can remark that for Tank 2 the two distinct areas are distinguished (with stronger values on the left and weaker ones on the right). For Tank 3 a weak zone is identified on the right part of the map. Moreover, for each tank, one can remark that the maps of NDT measurements are well correlated with the map of compressive strength. Particularly both weakest and strongest zones are localized at the same places on all the maps.

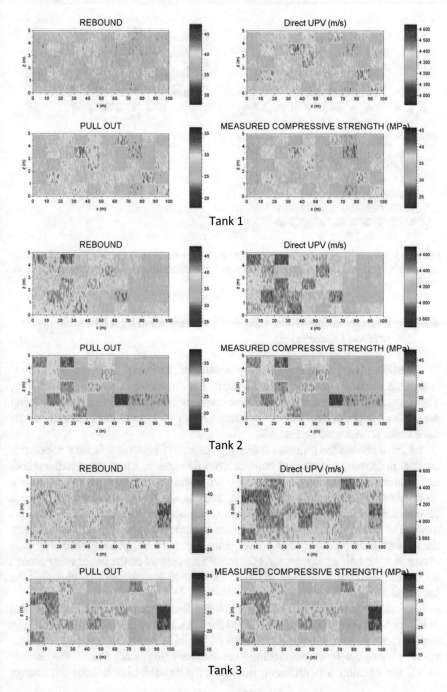

Fig. 4.1 Original maps (rebound, direct UPV, pull out and measured compressive strength)

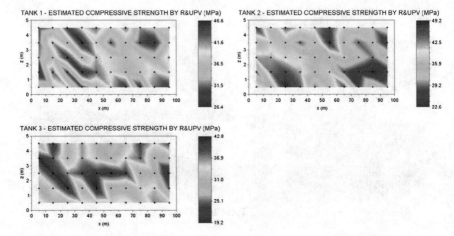

Fig. 4.2 Screening maps of estimated strength on the three tanks by combining Rebound Number (R) and Ultrasonic Pulse Velocity (UPV) test results (measurements performed at the centre of the batches)

4.2.1.3 Determination of TR After NDT Investigation by a Clustering Method (K-mean)

We considered a pre-auscultation by means of Rebound Hammer and Ultrasonic Pulse Velocity performed in TL located at the centre of the batches (50 TL). The test results are processed to estimate compressive strength in each TL by using the relation proposed by Breysse.[2] A screening map of the estimated strength for the three tanks is presented on Fig. 4.2.

We have chosen the k-means clustering method[.34] because it is very popular. It consists in defining a number of clusters k (for this reason, it is a supervised method, as the number of clusters is defined by the operator) and to randomly choose k observations from the data set and use them as the initial mean values. Then k clusters are created by progressively associating every observation with the nearest mean. The centroid of each of the k clusters becomes the new initial mean and then the procedure is repeated until convergence, i.e. until the population inside each cluster is not modified anymore. As the result may depend on the initial value chosen for the clustering, it is relevant to repeat the method several times by changing the initial value.

The clustering is applied with two, three and four clusters for each Tank 1, 2 and 3. Figure 4.3 presents the results obtained for Tank 1 in the case of two clusters. The first remark concerns the weak influence of the initialization in this case, even for the location of the TL corresponding to each cluster (Fig. 4.3c). Even if the number of TL for iteration 6 is different, the mean inside each cluster does not change

---

[2]Breysse [2].

[3]Wu [3].

[4]Javadi et al. [4].

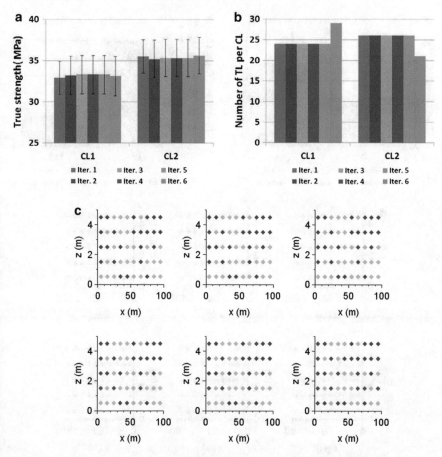

Fig. 4.3 Clustering by k-means with 2 clusters—Tank 1. **a** Mean inside each cluster for the 6 iterations. **b** Number of TL inside each cluster for the 6 iterations. **c** Location of the TL for each cluster (blue: CL1-green: CL2)

significantly (the coefficient of variation of the mean calculated in each cluster from the six iterations is less than 1%). However, considering the scatter of the results, one can conclude that the two populations are not significantly different. If three or four clusters are defined, at the end of the process, they still cannot be considered as statistically different.

So, the conclusion is that a unique TR can be considered for Tank 1 which is coherent regarding the generation of the synthetic data of Tank 1 (§4.2.1.2).

Figure 4.4 shows the results for Tank 2 when two clusters are considered. The six iterations reveal no initialization effect, neither on the mean nor on the position of the TL of the two clusters. Moreover, the weaker zone is clearly located on the right side of the map. The means obtained for the two clusters are significantly different (31 and 35 MPa) and the number of TL inside each cluster is not significantly different.

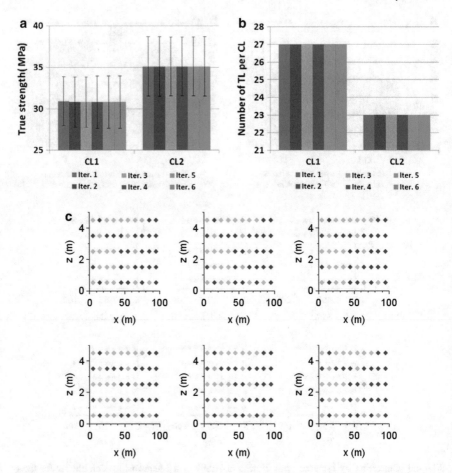

Fig. 4.4 Clustering by k-means with 2 clusters—Tank 2. **a** Mean inside each cluster for the 6 iterations. **b** Number of TL inside each cluster for the 6 iterations. **c** Location of the TL for each cluster (blue: CL1-green: CL2)

If three clusters are considered, the effect of the initialization is significant and the mean inside each cluster is respectively 28.7, 32.8 and 35.8 MPa. The number of TL inside each cluster is respectively 13, 23 and 14. Considering four clusters is not relevant because the difference of mean strength between cluster 2 and 3 is not significant (respectively 31.3 and 33.1 MPa).

So for Tank 2, two significantly different TR can be considered.

For Tank 3, the identification of the weaker zone in the right part is not easy with the screening grid at the center of the batches, whatever the number of clusters. By using a larger quantity of TL (260 regularly distributed on the surface of the tank), this identification is possible when four clusters are considered, with a compressive strength mean inside each of them of 27.9, 31.0, 32.4 and 34.8 MPa respectively. It can be noticed that the mean of the clusters 2 and 3 is not significantly different

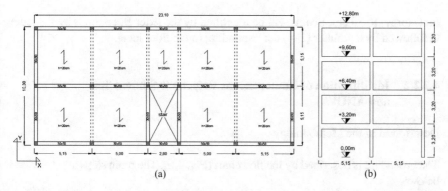

Fig. 4.5 In plan (**a**) and elevation (**b**) layout of the building. 18 columns and 21 beams per floor

(31 and 32.4 MPa respectively). Moreover, the coefficient of variation of strength inside the weaker cluster is quite large (about 21%), which is induced by a significant overlap between the strength distributions of the different clusters. Therefore, in this particular case this clustering method is not efficient to identify test regions.

4.2.2 TR Identification in a Real Case Study Building

4.2.2.1 Presentation of the Case Study

The case study is that of a building presented in a paper published by Masi et al. [5](Fig. 4.5). It concerns an existing RC framed structure designed to vertical loads only. It has a rectangular plan shape with a surface of 23.10 × 10.30 m (X and Y direction, respectively) and four storeys in elevation, with constant inter-storey height equal to 3.20 m. The original design specifications show that the structure has lateral load resisting frames only along the longitudinal dirction X. Specifically, along the X direction, three resisting frames with bay length varying in the range 2.85 m (the shorter one corresponding to the staircase width) are present. Beam section dimensions are constant throughout the structure and equal to 30 × 50 cm. Along the transversal direction Y, the structure has two bays with only the exterior frames having a rigid beam (30 × 50 cm) while, in the interior frames, columns are connected through the one-way RC slab with thickness equal to 22 cm. The staircase structure is placed in a central position and is made up of two inclined cranked beams at each storey, arranged in two adjacent frames along the Y direction. Cross-section dimensions of staircase beams are equal to 30 × 50 cm. The smallest dimension of columns cross-section is 30 cm while the other dimension depends on the storey number and the in-plane position, varying from for 45 cm (internal columns at the

---

[5]Breysse [2].

first storey) to 30 cm. Beams will be identified by the letter B, columns by the letter C followed by a number (1–4) corresponding to the floor level.

4.2.2.2   Identification of Homogeneous Regions (t-Student Test and ANOVA)

Description of the Methodology

The procedure is described by the flowchart (Fig. 4.6). The main steps are described below:

– identification of samples regarding structural aspects (beams and columns for instance),
– computation of coefficient of variation (COV) of both rebound number (R) and Ultrasonic Pulse Velocity (UPV). If the COV of both NDT results within each sample is more than 15% the sample cannot be considered as homogeneous and additional tests are needed to split it,
– computation of the confidence interval (95%) for each sample for both R and UPV. If the range of confidence interval of two samples has no intersection, it is considered that they do not belong to the same population,
– if there is an intersection between the range of confidence interval of two samples a t-Student test and an analysis of variance (ANOVA) are performed for each non-destructive test result.

Results

The process is applied to the dataset of the case study, where 54 TL have been used. The results are briefly presented here (for more details it can be referred to the paper of Masi et al.[6]). Figure 4.7 illustrates how the analysis of the range of confidence interval allows the discrimination of the samples. Here the prior assumption is that a sample is made of the beams or the columns of a single storey, based on the fact that each storey (floor) in the construction of reinforced concrete buildings is usually related to a different concrete casting phase. Moreover, this refers also to provisions given in seismic codes (e.g. EC8 for existing buildings[7]).

The comparisons show that C1 is the only sample that can be considered as a different one from all the other samples. Figure 4.8 presents the results of the confidence interval analysis and the samples which are comparable or not.

Figure 4.9 gives the results obtained after having applied t-student test.[8] The Student's t-test is a method of testing hypotheses about the mean of small samples drawn from a normally distributed population when the variance of the population

---

[6]Masi et al. [5].
[7]EN 1998–3:2005 [6].
[8]Breysse [2].

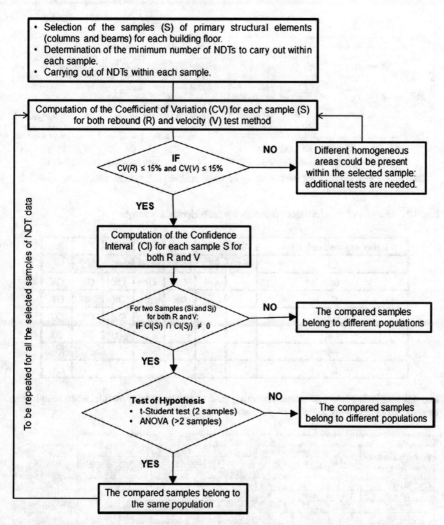

Fig. 4.6 Description of the procedure for the identification of concrete areas having homogeneous properties

is unknown and the samples are independent. Specifically, considering two samples A and B and after having verified the applicability of the Student's t-test, if X_A and X_B denote the samples' mean values of the two investigated areas, two alternative hypotheses should be verified:

- H_0: $X_A = X_B$; there is essentially no difference between the two groups;
- H_1: $X_A \neq X_B$; there is a significant difference between the two groups.

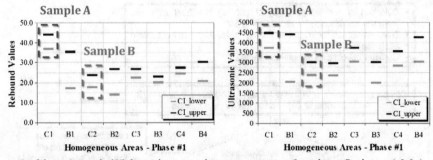

Confidence intervals (CI) for each primary homogeneous area. C = column, B = beam, 1-2-3-4 are the number of floor.

Fig. 4.7 Analysis of the confidence intervals for each elementary sample

Rebound and velocity	B1	C2	B2	C3	B3	C4	B4
C1	NO	NO	NO	NO	NO	NO	NO
B1		OK	OK	OK	OK	OK	OK
C2			OK	NO	OK	NO	OK
B2				NO	OK	OK	OK
C3					NO	OK	OK
B3						NO	NO
C4							OK

Fig. 4.8 Analysis of the confidence intervals, pair-comparison of samples (OK when samples can be assumed as belonging to the same population)

| Investigated Area | Student's t-test (two-tailed, α=0.05) | | | | | | Final Decision |
| | Rebound Method [R] | | | Ultrasonic Method [V] | | | [R]∩[V] |
	t-value	t_c-value	Accepted/Rejected	t-value	t_c-value	Accepted/Rejected	
B1-C2	2.43	2.31	Rejected	1.96	2.31	Accepted	Rejected
B1-B2	2.31	2.78	Accepted	2.00	2.78	Accepted	Accepted
B1-C3	0.95	2.26	Accepted	-0.56	2.26	Accepted	Accepted
B1-B3	2.87	2.45	Rejected	2.26	2.45	Accepted	Rejected
B1-C4	0.26	2.45	Accepted	0.04	2.45	Accepted	Accepted
B1-B4	0.40	2.20	Accepted	-1.37	2.20	Accepted	Accepted
C2-B2	0.14	2.31	Accepted	0.25	2.31	Accepted	Accepted
C2-B3	-0.47	2.23	Accepted	0.93	2.23	Accepted	Accepted
C2-B4	-2.28	2.31	Accepted	-4.24	2.31	Rejected	Rejected
B2-B3	-0.78	2.45	Accepted	0.60	2.45	Accepted	Accepted
B2-C4	-4.03	2.20	Rejected	-1.82	2.20	Accepted	Rejected
B2-B4	-2.74	2.78	Accepted	-6.25	2.78	Rejected	Rejected
C3-C4	-1.34	2.12	Accepted	0.80	2.12	Accepted	Accepted
C3-B4	-0.54	2.26	Accepted	-0.96	2.26	Accepted	Accepted
C4-B4	0.43	2.20	Accepted	-1.41	2.20	Accepted	Accepted

Fig. 4.9 Results of t-student test

Group of Floor Elements	ANOVA TEST'S RESULTS (one-tailed, α=0.05)						Final Decision [R]∩[V]
	Rebound Method [R]			Ultrasonic Method [V]			
	F value	F_c value	Accepted/Rejected	F value	F_c value	Accepted/Rejected	
B1, C3, B4, C4	0.83	3.10	Accepted	0.94	3.10	Accepted	Accepted
B1, C2, B2, B3	4.15	3.34	Rejected	3.17	3.34	Accepted	Rejected
C2, B2, B3	0.21	3.88	Accepted	0.62	3.88	Accepted	Accepted

Fig. 4.10 Final decision and constitution of the groups

If hypothesis H0 is verified then "Accepted" in Fig. 4.9 means that the two samples are comparable. So the two samples can be considered as part of the same population and then analysed as a single homogeneous area.

This step indicates the samples that can be grouped together and those that cannot. Additionally, in order to confirm the clustering provided by applying in pairs the Student's t-test, or to replace it in the case of more than two samples, the analysis of variance (ANOVA) could be performed.[9] ANOVA can be used to verify if the mean values of several groups of data can be considered equal or not, and therefore it generalizes the t-test to more than two groups allowing the comparison of three or more mean values regarding statistical significance. On this basis, several possibilities for larger groups are tested, and final results are given in Fig. 4.10 after having applied ANOVA.

At the end of the process, three TR are identified (namely C1; B1 + C3 + C4 + B4; C2 + B2 + B3) that are assumed to belong to different populations. One can come back to Fig. 4.7 to check that these three TR correspond to progressively decreasing material properties. The three TR contain respectively 8, 25 and 21 TL. Their mean rebound value is respectively 40, 25 and 21, with a COV between 8 and 11%. Their mean UPV is respectively 4110, 3220 and 2620 m/s, with a COV between 10 and 13%.

4.2.2.3 Identification of Homogeneous Regions by the Analysis of Confidence Intervals

Description of the Procedure

We describe here two approaches to identify homogeneous regions that should be used depending on the parameters that are selected to characterize concrete strength in a given region. The first approach can be used when concrete strength is only characterized by its mean value while the second approach can be used when both the mean and the standard deviation are used. The first approach consists in analysing the variation of the ratio between the sample mean and the true mean of concrete strength. The second approach identifies homogeneous regions by combining the first approach with the confidence interval of the coefficient of variation of strength. Both approaches are defined with respect to the three knowledge levels (KL1, KL2

---

[9]Breysse [2].

and KL3) proposed in EC8.The theoretical background of the two approaches is firstly described, then the two approaches are applied to a real dataset (§4.2.2.3.6).

Basis for the First Approach—The Confidence Interval of the Mean Value

The two-sided confidence interval of the true mean value μ of a given distribution for the case where the variance of the distribution is unknown and for a confidence level α is given by:

$$\overline{X} - t_{\alpha/2,n-1} \times \frac{sd}{\sqrt{n}} \leq \mu \leq \overline{X} + t_{1-\alpha/2,n-1} \times \frac{sd}{\sqrt{n}} \tag{4.1}$$

where $\overline{X}$ is the sample mean, sd is the sample standard deviation, $t_{j,n-1}$ is the jth percentage point of the t distribution with $n - 1$ degrees of freedom, and n is the sample size. By dividing all the terms by $\overline{X}$ and rearranging the expression, one obtains:

$$\overline{1} - t_{\alpha/2,n-1} \times \frac{COV}{\sqrt{n}} \leq \frac{\mu}{\overline{X}} \leq \overline{1} + t_{1-\alpha/2,n-1} \times \frac{COV}{\sqrt{n}} \tag{4.2}$$

where COV is the sample coefficient of variation. To account for the fact that the underlying population may have a finite number of elements, Eq. 4.2 can be modified as follows to include a finite population correction factor:

$$1 - t_{\alpha/2,n-1} \times \frac{COV}{\sqrt{n}} \sqrt{\frac{N-n}{N-1}} \leq \frac{\mu}{\overline{X}} \leq 1 + t_{1-\alpha/2,n-1} \times \frac{COV}{\sqrt{n}} \sqrt{\frac{N-n}{N-1}} \tag{4.3}$$

where N is the number of elements in the population. This expression can be further rearranged to obtain the following form:

$$\frac{1}{1 - t_{\alpha/2,n-1} \times \frac{COV}{\sqrt{n}} \sqrt{\frac{N-n}{N-1}}} \geq \frac{\overline{X}}{\mu} \geq \frac{1}{1 + t_{1-\alpha/2,n-1} \times \frac{COV}{\sqrt{n}} \sqrt{\frac{N-n}{N-1}}} \tag{4.4}$$

Redefining the terms of the expression for the concrete strength, one obtains:

$$\frac{1}{1 - t_{\alpha/2,n-1} \times \frac{COV_{f_c}}{\sqrt{n}} \sqrt{\frac{N-n}{N-1}}} \geq \frac{\overline{X}_{f_c}}{\mu_{f_c}} \geq \frac{1}{1 + t_{1-\alpha/2,n-1} \times \frac{COV_{f_c}}{\sqrt{n}} \sqrt{\frac{N-n}{N-1}}} \tag{4.5}$$

where COV_{f_c} is the sample coefficient of variation of the concrete strength. Since the value of COV_{f_c} is initially unknown, it can be replaced by an approximation defined in terms of the sample coefficient of variation of available rebound number (R) data (COV_R) or of the sample standard deviation of available UPV data (s_{UPV})

similar to the following proposal:

$$COV_{f_c} \approx \begin{cases} k_1 \times sd_{UPV} \\ k_2 \times sd_R \end{cases} \tag{4.6}$$

These empirical relations are coming from an important database, all the details can be found in the study from Pereira and Romão.[10]

Basis for the Second Approach—The Confidence Interval of the Coefficient of Variation

Even though a reliable and simple non-parametric confidence interval for the true value of the coefficient of variation (COV) of a given distribution is unavailable, research has shown that a confidence interval with an adequate performance for different types of distributions can be obtained based on the confidence interval for the true variance of a normal distribution.[11]

The following expression defines a confidence interval for the true variance σ^2 that includes finite population correction factors and considers a confidence level of α[12]:

$$\left(\frac{n-1}{N-1} + \frac{N-n}{N-1} \times \frac{1}{F_{1-\alpha/2,n-1,N-n}} \right) \times sd^2$$

$$\leq \sigma^2 \leq \left(\frac{n-1}{N-1} + \frac{N-n}{N-1} \times \frac{1}{F_{\alpha/2,n-1,N-n}} \right) \times sd^2 \tag{4.7}$$

where $F_{j,n-1,N-n}$ is the jth percentage point of the F distribution with $n-1$ and $N-n$ degrees of freedom. Dividing the interval by μ^2 yields:

$$\left(\frac{n-1}{N-1} + \frac{N-n}{N-1} \times \frac{1}{F_{1-\alpha/2,n-1,N-n}} \right) \times \frac{sd^2}{\mu^2}$$

$$\leq \left(\frac{\sigma}{\mu} \right)^2 \leq \left(\frac{n-1}{N-1} + \frac{N-n}{N-1} \times \frac{1}{F_{\alpha/2,n-1,N-n}} \right) \times \frac{sd^2}{\mu^2} \tag{4.8}$$

Since μ is not known, it can be replaced in the upper and lower bounds by its unbiased estimate $\overline{X}$ to obtain:

---

[10]Pereira and Romão [7].

[11]Gulhar et al. [8].

[12]O'Neill [9].

$$\left(\frac{n-1}{N-1} + \frac{N-n}{N-1} \times \frac{1}{F_{1-\alpha/2, n-1, N-n}} \right) \times \left(\frac{sd}{\bar{\bar{X}}} \right)^2$$

$$\leq \left(\frac{\sigma}{\mu} \right)^2 \leq \left(\frac{n-1}{N-1} + \frac{N-n}{N-1} \times \frac{1}{F_{\alpha/2, n-1, N-n}} \right) \times \left(\frac{sd}{\bar{\bar{X}}} \right)^2 \qquad (4.9)$$

This expression can then be written in terms of the sample coefficient of variation COV and the true coefficient of variation COV_{true}:

$$\sqrt{\frac{n-1}{N-1} + \frac{N-n}{N-1} \times \frac{1}{F_{1-\alpha/2, n-1, N-n}}} \times COV$$

$$\leq COV_{true} \leq \sqrt{\frac{n-1}{N-1} + \frac{N-n}{N-1} \times \frac{1}{F_{\alpha/2, n-1, N-n}}} \times COV \qquad (4.10)$$

Considering that only the upper bound of this interval is relevant for the present analysis leads to:

$$COV_{true} \leq \sqrt{\frac{n-1}{N-1} + \frac{N-n}{N-1} \times \frac{1}{F_{\alpha, n-1, N-n}}} \times COV \qquad (4.11)$$

Redefining the terms of the expression for the concrete strength, one obtains:

$$COV_{true, f_c} \leq \sqrt{\frac{n-1}{N-1} + \frac{N-n}{N-1} \times \frac{1}{F_{\alpha, n-1, N-n}}} \times COV_{f_c}$$

$$\Leftrightarrow COV_{true, f_c} \leq k_\alpha \times COV_{f_c} \qquad (4.12)$$

As before, COV_{f_c} can be replaced by the approximation established by $COV_{f_c} \approx \begin{cases} k_1 \times sd_{UPV} \\ k_2 \times sd_R \end{cases}$ Eq. 4.6. This expression insures that an estimate of the coefficient of variation COV_{f_c} multiplied by factor k_α bounds the true value COV_{true, f_c}. However, if the case of interest is to insure that a certain admissible value of the coefficient of variation COV_{adm, f_c} is not exceeded, Eq. 4.12 needs to be complemented as:

$$COV_{true, f_c} \leq \sqrt{\frac{n-1}{N-1} + \frac{N-n}{N-1} \times \frac{1}{F_{\alpha, n-1, N-n}}} \times COV_{f_c} \leq COV_{adm, f_c}$$

$$\Leftrightarrow COV_{true, f_c} \leq k_\alpha \times COV_{f_c} \leq COV_{adm, f_c} \qquad (4.13)$$

For the purpose of identifying a homogeneous region (i.e. with an upper bound for the coefficient of variation), only the following part of Eq. 4.13 is relevant since

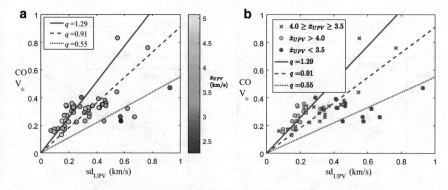

Fig. 4.11 Empirical correlation trend between the variability of concrete strength and the variability of UPV data: **a** disaggregation of the data according to the mean value of the corresponding dataset and **b** representation of the proposed multi-linear correlation model

it also insures a safe-side estimate of the coefficient of variation (Eq. 4.14):

$$\sqrt{\frac{n-1}{N-1} + \frac{N-n}{N-1} \times \frac{1}{F_{\alpha,n-1,N-n}}} \times COV_{f_c}$$
$$\leq COV_{adm,f_c} \Leftrightarrow k_\alpha \times COV_{f_c} \leq COV_{adm,f_c} \qquad (4.14)$$

Definition of Parameters k_1 and k_2

By analysing the empirical correlation trend between the variability of the concrete strength and the variability of UPV for 50 datasets, the relations presented in Fig. 4.11 were obtained.[13] As can be seen, an estimate of COV_{f_c} can be obtained by using the Eq. 4.15:

$$COV_{f_c} = \begin{cases} 1.29 \times sd_{UPV} & \overline{x}_{UPV} > 4km/s \\ 0.91 \times sd_{UPV} & 4km/s \geq \overline{x}_{UPV} \geq 3.5km/s \\ 0.55 \times sd_{UPV} & \overline{x}_{UPV} < 3.5km/s \end{cases} \qquad (4.15)$$

By analysing the empirical correlation trend between the variability of the concrete strength and the variability of R for 68 datasets, the relations presented in Fig. 4.12 were obtained. As can be seen, an estimate of COV_{f_c} can be obtained by using the following relation:

$$COV_{f_c} = 1.94 \times COV_R \qquad (4.16)$$

---

[13]Breysse [2].

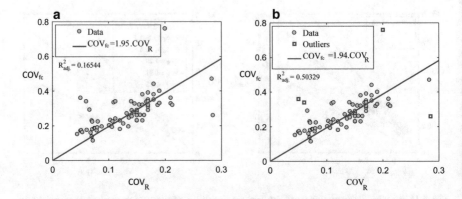

Fig. 4.12 Empirical correlation trend between the variability of the concrete strength and the variability of R data: all the 68 datasets (**a**); when discarding 4 outlying datasets (**b**)

Implementation Details of the Proposed Approaches

By considering Eq. 4.5 combined with Eq. 4.15 or Eq. 4.16, it can be seen that bounding values for the relation between the sample estimate of the mean concrete strength and its true value can be defined based solely on the sample size considered for the sample estimate and on a predefined confidence level α. A similar reasoning can then be applied to verify if the safe-side estimate of the coefficient of variation of concrete strength defined by Eq. 4.14 combined with Eq. 4.15 or Eq. 4.16 complies with a certain admissible coefficient of variation COV_{adm,f_c}.

The first approach can be used to identify homogeneous regions by:

(1) Combining Eq. 4.5 with Eq. 4.15 or Eq. 4.16
(2) Associating specific confidence levels with certain knowledge levels
(3) Comparing the admissible bounding values of Eq. 4.5 for a given confidence/knowledge level with the bounding values determined from the sample data obtained from a certain group of test results.

A certain group of elements can be considered as homogeneous region if the limits of Eq. 4.5 are met for the selected knowledge level.

With respect to the confidence levels, it is suggested that α values of 5, 10 and 15% are associated to the KL1, KL2 and KL3 knowledge levels, respectively. The reasoning behind this proposal is that a smaller value of α leads to a wider confidence interval, thus reflecting the larger uncertainty in the estimate of the mean value that is associated to a lower level of knowledge. In terms of admissible bounding values for each case of Eq. 4.5, the following values are therefore suggested:

$$1.15 \geq \frac{\overline{X}_{f_c}}{\mu_{f_c}} \geq 0.87 \text{ for KL1}$$
$$1.10 \geq \frac{\overline{X}_{f_c}}{\mu_{f_c}} \geq 0.91 \text{ for KL2} \qquad (4.17)$$
$$1.05 \geq \frac{\overline{X}_{f_c}}{\mu_{f_c}} \geq 0.95 \text{ for KL3}$$

Table 4.3 Quantiles of the upper bound of Eq. 4.5 for different α values

	Quantiles				
	0.50	0.75	0.80	0.85	0.90
KL1 ($\alpha = 5\%$)	1.05	1.11	1.12	1.15	1.20
KL2 ($\alpha = 10\%$)	1.05	1.09	1.10	1.12	1.16
KL3 ($\alpha = 15\%$)	1.04	1.08	1.09	1.11	1.14

To demonstrate to which quantiles of the distribution of $\overline{X}_{f_c}/\mu_{f_c}$ these limit values correspond to, a numerical study was carried out to simulate the upper bound of Eq. 4.5. The simulations considered cases with sample sizes n between 5 and 100, assuming that N is 100, and considering values of cv_{f_c} ranging between 10 and 50%. For these conditions, the quantiles presented in Table 4.3 were obtained for the simulated upper bound data considering α values of 5%, 10% and 15% and are comparable to the values of 1.05 (KL1), 1.10 (KL2) and 1.15 (KL3) that are suggested.

With respect to the second approach, it can be used to identify homogeneous regions by:

(1) Applying the first approach
(2) Combining Eq. 4.14 with Eq. 4.15 or Eq. 4.16
(3) Selecting an admissible value of the coefficient of variation COV_{adm,f_c}
4) Associating specific confidence levels with certain knowledge levels
5) Comparing COV_{adm,f_c} with the left hand side of Eq. 4.14 (which depends on COV_{f_c}) for a given confidence/knowledge level.

A certain group of elements can be considered to be part of a single homogeneous region by complying with Eq. 4.5 or Eq. 4.14 for the selected knowledge level. It is noted that Step 5 can be carried out by evaluating factor k_α of Eq. 4.14 for the specific case under analysis or, as a simplification, by using pre-defined values of k_α. With respect to the confidence levels, unlike for the previous case, it is now suggested that α values of 15, 10 and 5% are associated to the KL1, KL2 and KL3 knowledge levels, respectively. The reasoning behind this proposal is that a smaller value of α leads to a wider confidence interval (i.e. a larger value of k_α), thus reflecting a larger constraint in complying with Eq. 4.14, which is associated to a higher level of knowledge. Simulations were also carried out to determine the quantiles of k_α and establish safe-side estimates for this parameter. The values obtained from the simulations are presented in Table 4.4.

Based on the results of these simulations, the following admissible values of k_α for each case of Eq. 4.14 are suggested:

$$1.15 \times COV_{f_c} \leq COV_{adm,f_c} \quad \text{for KL1}$$
$$1.20 \times COV_{f_c} \leq COV_{adm,f_c} \quad \text{for KL2}$$
$$1.25 \times COV_{f_c} \leq COV_{adm,f_c} \quad \text{for KL3} \tag{4.18}$$

Table 4.4 Quantiles of k_α for different α values

	Quantiles				
	0.50	0.75	0.80	0.85	0.90
KL1 ($\alpha = 15\%$)	1.08	1.15	1.17	1.20	1.26
KL2 ($\alpha = 10\%$)	1.10	1.19	1.21	1.26	1.33
KL3 ($\alpha = 5\%$)	1.14	1.25	1.29	1.35	1.45

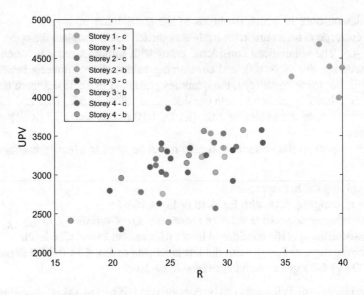

Fig. 4.13 R-UPV datasets provided to define homogeneous zones

Application to Identify Homogenous Regions in the Case-Study

The methodology is here applied to the datasets of rebound (R) and velocity (UPV) test results taken from beams and columns of the four-storey building that has been described in the previous section (§4.2.2.1, Fig. 4.5). The methodology was first applied for the case where only the mean value of concrete strength needs to be characterised and a second application was then performed for the case where both the mean and the variability of concrete strength need to be considered.

The datasets are represented in Fig. 4.13 separated by storeys and type of element. A visual analysis of the datasets shows that some of the columns of the first storey exhibit R-UPV pairs that are distant from the remaining ones. They correspond to the C1 sample that had been identified previously as TR.

To analyse the possible homogeneous regions that may be found in the datasets, the two proposed approaches were applied. In terms of the first approach, the values of the upper bounds defined by Eq. 4.5 combined with Eqs. 4.15 and 4.16 were determined for all the knowledge levels and for the following combinations of elements:

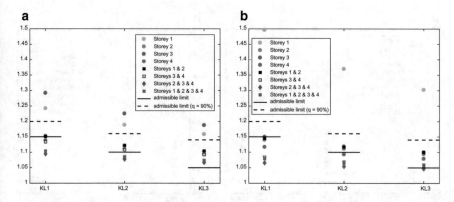

Fig. 4.14 Results of the first approach for different combinations of elements according to the proposed methodology: R datasets (**a**); UPV datasets (**b**)

- Beams and columns storey by storey
- Beams and columns of the 1st and 2nd storeys and beams and columns of the 3rd and 4th storeys
- Beams and columns of the 2nd, 3rd and 4th storeys
- Beams and columns of all the storeys.

Datasets considering only the beams or only the columns of a given storey were not analysed since the size of the beam datasets would be too small to obtain a reliable estimate of the sample variability. It must be noted that it was also chosen here to consider regions containing all elements (i.e. beams and columns) of a given storey (therefore beams and columns of a same storey cannot belong to two different TR). This prior assumption was not done in §4.2.2.3.2 and the conclusions regarding TR identification may thus be different.

The upper bounds results obtained for these combinations of elements for all knowledge levels are presented in Fig. 4.14 (only the upper bound of Eq. 4.5 was analysed herein). As can be seen from the results, several options can be considered, depending on the selected knowledge level. Grouping all the elements of the four storeys appears to be possible (for KL1 and KL2). If one considers both R and UPV datasets, the best grouping option appears to be the one involving the elements of the 2nd, 3rd and 4th storeys. However, the zone involving only the elements of the 1st storey does not comply with the admissible limits. Still if one considers admissible limits consistent with larger quantiles, other grouping options can be considered, as can be seen from Fig. 4.14 where the scenario in which the limits of Eq. 4.5 based on the 90% quantile is also plotted.

When applying the second approach, where both the mean and the variability of concrete strength need to be analysed, an admissible value of COV_{adm, f_c} needs to be defined first. In the current case, a value of 0.25 was selected. The results of the second approach are presented for the case where Eq. 4.14 is evaluated for the specific case under analysis and for the case where the approximations defined by

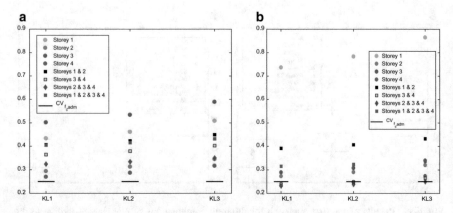

Fig. 4.15 Results of the second approach for different combinations of elements considering Eq. 4.14: R datasets (**a**); UPV datasets (**b**)

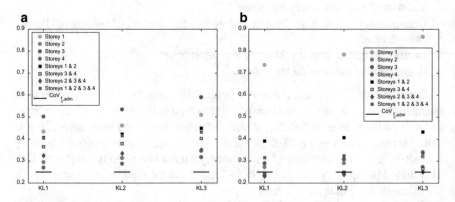

Fig. 4.16 Results of the second approach for different combinations of elements considering the approximation of k_a defined by Eq. 4.18: R datasets (**a**); UPV datasets (**b**)

Eq. 4.18 are used. Results obtained for the part of the second approach that deals with the admissibility of the coefficient of variation are presented in Figs. 4.15 and 4.16. Figure 4.15 presents the results obtained with Eq. 4.14 while Fig. 4.16 presents the results obtained with Eq. 4.18.

As can be seen, the results of Figs. 4.15 and 4.16 yield similar conclusions which indicates that the approximations considered in Eq. 4.18 are adequate. In both analyses, no group of R dataset complies with COV_{adm,f_c} and only a few groups of UPV datasets comply with COV_{adm,f_c}. **These analyses also confirm the difficulty in identifying a homogenous region within these datasets**, as was seen from the results obtained by the first approach. Given the heterogeneity of the data considered in this example (see Fig. 4.13), it is noted that a realistic value for COV_{adm,f_c} needs to be set case-by-case. Establishing a constant value for COV_{adm,f_c} may invalidate the applicability of the second approach in cases where the intrinsic variability of the data (i.e. without considering the sampling variability) is already larger than COV_{adm, f_c}.

4.2.3 Conclusions

We have proposed different methods to discriminate test regions on an investigated domain. In a first part we have worked on synthetic data representing three tanks. A set of data was available for these tanks involving rebound hammer, UPV and true compressive strength values. It was proposed to identify the test regions by means of NDT test results only by combining rebound and UPV. The proposed methodology uses the k-means clustering method. The conclusions of this study are:

- Tank 1 can be considered as a unique TR, which was coherent regarding the simulations,
- Tank 2 can be divided into two TR, which is in good agreement with the simulations,
- For Tank 3, the method was not able to identify a small area of low strength.

In a second part, the determination of TR is performed on the same set of real data by using two methods. The set of data is obtained from a four-storey building with beams and columns on which rebound number and UPV measurement were performed. For these two methods, structural considerations were applied for grouping elements, for instance by considering differently beams and columns according to casting conditions and mechanical solicitations. This can affect the conclusions. Consequently, even if each beam or column can contain several TL, the identification of TR only concerns the elements. So the possibility to have several TR in a same element is not considered.

The first method is based on a statistical analysis of the data and a determination of confidence intervals which allows the discrimination of test regions. Then a combination of T-student test and ANOVA is used for validation. The method determined three TR, with one or several elements, namely [C1], [B1 + C3 + C4 + B4], [C2 + B2 + B3] containing respectively 8, 25 and 21 TL.

For the second method, the first approach has shown that the grouping of elements in TR depends on the considered KL. A TR could be identified by grouping the elements of the 2nd, 3rd and 4th storeys but, on the other hand, a group made of the elements of the first floor cannot be considered as a homogeneous region. By the second approach where both mean and variability are considered it was not possible to identify different TR.

4.3   Choice of Conversion Models

4.3.1   Description of the Process

The flowchart presented in Fig. 4.17 describes the process that can be followed to define, identify and check the relevant conversion model(s) corresponding to the data available on an investigation domain. Figure 4.18 illustrates what type of data is available at this stage.

As explained in 1.5.1 of Guidelines, a conversion model can be identified and applied:

– either with a specific model for each test region
– or with a same model covering several or all test regions.

The choice between these options is based on:

– the consideration of prior knowledge about the structure and the concrete,
– information provided by NDT test results and their statistical distribution,

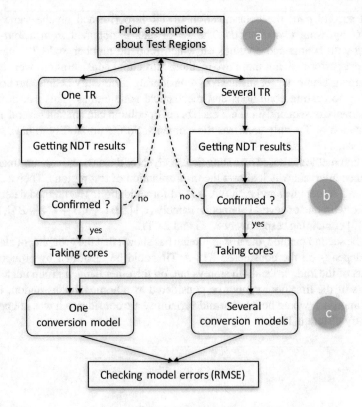

Fig. 4.17 Flowchart for the identification of test regions and conversion models

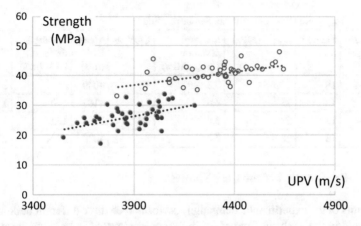

Fig. 4.18 Illustration of the UPV-strength correlation when two options can be considered (i.e. 2 TR, with a specific model in each TR against 1 TR with a common model)

– the number of cores that will be used for identifying the conversion model(s), since the uncertainty of strength estimates decreases when the number of cores increases whereas each model requires a minimum number of cores.

The first stage ("a" of flowchart of Fig. 4.17) consists in making prior assumptions based on prior knowledge about the structure and the concrete. This leads to consider either a unique Test Region or several Test Regions (see §1.5.1 of the guidelines).

The second stage ("b") is to get NDT results, whose analysis will confirm or not the prior assumptions.

Whatever the choice (one or several test regions), conversion models have to be identified at stage "c". A conversion model can be specific for each test region or can cover several test regions. A relevant number of cores must be taken in each test region (see §1.6).

Once the conversion model(s) is/are identified (see §1.5.5), it is necessary to finally check the predictive model error (see §1.5.6). It is important to note that revising the choice (from several TR to one TR) at last stage "c" does not imply any additional data. Thus the comparison between model errors with the two options can lead to revise easily this last choice.

4.3.2 Example of Application on Three Case Studies

The purpose is here to show on three case studies how the methodology illustrated on Fig. 4.17 can be followed and how the efficiency of the different options can be tested.

Table 4.5 $f_{c,mean}$ and UPV data of Punta Perotti building

Part of the building	Sample Size	$f_{c,mean}$ (MPa)	Standard deviation $f_{c,mean}$ (MPa)	COV %	Mean UPV (m/s)	Standard deviation UPV (m/s)	COV %
Floor −1	18	33.5	5.3	16	4049	85	2
Floor 1	18	33.1	3.6	11	4107	140	3
Floor 2	18	28.1	3.8	14	3942	123	3
Floor 3	18	28.9	2.2	8	4029	99	2

4.3.2.1 Presentation of the Case Studies

The results of the experimental campaigns performed on three different buildings are considered in the analysis. Two of the buildings are located in Italy, the third one is in Algeria. Punta Perotti (PP) building lays in Bari, in Apulia region, and the in situ experimental tests were performed by ENEA.[14] The second Italian building is a public building in Basilicata (BB) and the experimental tests were performed by University of Basilicata. On the third building, Algerian building (AB), the experimental campaign was the object of a Ph.D. thesis at University of Chlef.[15]

4.3.2.2 Identification of the TR

The analysis of homogeneity of data is a preliminary step toward the definition of the correlations between direct and indirect test results, in the present study between compressive strength ($f_{c,core}$) and ultrasonic pulse velocity (UPV) data. The analysis has the scope of identifying the Test Regions (TR), namely parts of the building whose data belong to the same population.[16] The analysis started from dividing the dataset into groups on assumptions based on physical considerations. Then a statistical analysis, namely ANOVA, was performed to compare the groups and to check if they belong to the same population.

4.3.2.3 Punta Perotti (PP) Building

UPV-$f_{c,mean}$ pairs of data measured on columns of floors -1, 1, 2 and 3 were considered. Data for each floor are reported in Table 4.5. The basement and the first floor have similar compressive strength mean values, as well as Floor 2 and Floor 3. The coefficients of variation of all the floors are quite similar between each other and they varies between 8% (floor 3) and 16% (basement). As also known from the literature, the COVs of UPV data are considerably lower than that of f_c, being equal to 2–3%.

---

[14]Luprano et al. [10] and Pfister [11].

[15]Ali-Benyahia [12].

[16]EN 13791:2007 [13].

All data of the same floor are supposed to belong to a same TR: this assumption is consistent with the fact that each floor in the construction of RC buildings is typically related to a different concrete casting phase. An ANOVA analysis was then performed to check if several TRs (i.e. floors) could be grouped to form a unique test region. The ANOVA analysis is based on the hypothesis of normality of data within each group and of equality of variances between groups. The Shapiro–Wilk test was performed on $f_{c,mean}$ data of each group and it stated that, at 95% level of confidence, data were normally distributed in all groups. The Leven's test, used to check if the variances of the groups were equal, was not satisfied, thus it was not possible to perform the ANOVA test. The equivalent non-parametric test, Kruskal–Wallis, was used to verify if the groups belonged to the same population. At 95% level of confidence the null hypothesis (considering that the groups belong to the same population) was not satisfied. The test didn't evidence which groups were different from the others: comparing the values of the ranks, floor −1 and 1 were very similar as well as floors 2 and 3. Thus, the ANOVA test was performed on floors −1 and 1: the two hypotheses (normality of the groups and equality of variances) were satisfied and at 95% level of confidence, the two groups come from the same population. The same result was obtained for floors 2 and 3.

Thus, the four prior test regions could be changed into two test regions: TR1 made of 36 data of floors −1 and 1; TR2 made of 36 data of floors 2 and 3 (Fig. 4.19a).

In real practice, it would be useful to perform the homogeneity analysis on NDT measurement in order to identify test regions before coring. Statistical analysis performed on UPV data may not give the same results as the analysis on $f_{c,i}$. For this database, it was not possible to perform the ANOVA analysis on UPV data, since the variances of the groups were significantly different. Thus, theKruskal-Wallis test was used, giving the same results than those obtained from $f_{c,mean}$ data. The frequency distribution of UPV data of TR1 and TR2 is reported in Fig. 4.19b.

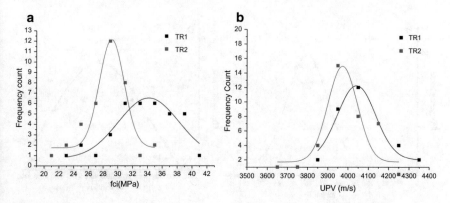

Fig. 4.19 Frequency distribution of $f_{c,i}$ (**a**) and UPV (**b**) data belonging to TR1 and TR2 of PP database

4.3.2.4 Basilicata Building *(BB)*

Seven structurally independent parts of the hospital were assumed as TRs and they were considered for the homogeneity analysis. The mean value, the standard deviation and the coefficient of variation of UPV and $f_{c,mean}$ data of each test region are reported in Table 4.6. The values of $f_{c,mean}$ vary from 25.3 MPa for E2 block to 39.5 MPa for EF3 block. Most of the blocks have a coefficient of variation higher than 15% which evidences a high variability of f_c within each group, higher than that between the groups (13%). The COV of UPV data within each group is quite high, reflecting the f_c dispersion of results.

$f_{c,i}$ data belonging to the same part of the building were considered as a group for ANOVA analysis. Both the Shapiro–Wilk test for normality and the Leven's test for the variance equality were satisfied for $f_{c,mean}$ data, thus the ANOVA analysis was performed. At 95% level of confidence, the means of the seven groups were

Table 4.6 $f_{c,mean}$ and UPV data of Basilicata Building (BB)

Part of the building	Sample size	$f_{c,mean}$ (MPa)	Standard deviation $f_{c,mean}$ (MPa)	COV %	Mean UPV (m/s)	Standard deviation UPV (m/s)	COV %
E1	3	38.2	7.8	20.3	4886	163	3
E2	3	25.3	2.1	8.3	4588	228	5
EF1	8	34.7	7.2	20.8	4518	328	7
EF2A	5	35.5	5.5	15.4	4775	104	2
EF2B	5	37.3	4.5	12.1	4606	197	4
EF3	8	39.5	12.1	30.5	4629	199	4
F2	4	37.5	4.2	11.3	4605	182	4

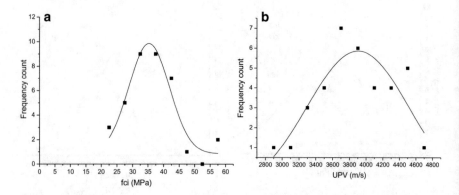

Fig. 4.20 Frequency distribution of $f_{c,mean}$ (**a**) and UPV (**b**) data of BB database

not significantly different, thus a unique test region (BB) of 36 pairs of data was considered (Fig. 4.20a). The same result was obtained with the ANOVA analysis on UPV data (Fig. 4.20b).

4.3.2.5 Algerian Building (AB)

Experimental data of columns and beams of the first floor of block1 of the Algerian building were considered as two TRs for the homogeneity analysis (Table 4.7). A high variability of both UPV and $f_{c,i}$ data can be observed within each group: the COV values were equal to 11–13% and 35% respectively for UPV and $f_{c,i}$ data. After verifying the hypotheses (Shapiro–Wilk test and Leven's test) the ANOVA analysis was performed on $f_{c,i}$ data of the two groups. At 95% level of confidence, the means of the two groups were not significantly different, thus the two groups can be considered as a unique test region (AB of 36 data) (Fig. 4.21a). The same result was obtained performing the analysis on UPV data (Fig. 4.21b).

Table 4.7 UPV and $f_{c,mean}$ data of the Algerian building

Part of the building	Sample size	$f_{c,mean}$ (MPa)	Standard deviation $f_{c,mean}$ (MPa)	COV %	Mean UPV (m/s)	Standard deviation UPV (m/s)	COV %
Block1 floor1 beams	17	19.3	6.7	35	3955	508	13
Block1 floor1 columns	19	18.7	6.6	35	3803	420	11

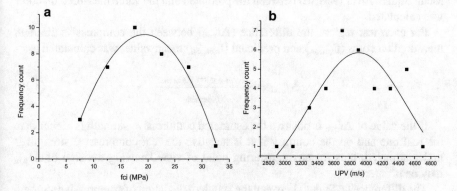

Fig. 4.21 Frequency count of fc,i (**a**) and UPV (**b**) data of Algerian building

4.3.2.6   Quality of Compressive Strength Prediction: Comparison Between One Specific Model Per TR and a Unique Model for All TR

It is well known that the quality of a model depends on several factors, such as the number of NDT-DT results used to construct the model, the compressive strength range of variability and the precision of the tests (§1.5).

In this section, two main questions are addressed. The first one is: is it more convenient to use a specific model ($Model_i$) for each test region (TR) or to use a unique model obtained by grouping all the data of the different test regions ($Model_{All}$)?

$Model_{All}$ is built from a higher number of NDT-DT data than $Model_i$, which reduces the statistical uncertainty at the conversion model identification stage. However, $Model_{All}$ is generally less accurate than the specific models since it covers a wider range of compressive strength values.

The second question is: is there a minimum number of cores below which it is more convenient to use a model for all TR instead of a specific model for each TR? In fact, when the number of cores must be low for cost or safety reasons, it could be better to use a unique model instead of several models obtained from a very low number of data pairs.

The quality of compressive strength prediction within each TR was evaluated by dividing the TR database in two parts: the "Calibration Database" (CD) and the "Assessment Database" (AD). The first group of data was used to fit the Model ($Model_i$), by means of the linear regression analysis, while the second one was used to test it. NDT-DT data of CD were chosen from the entire database (CD + AD) following the principles of conditional coring.[17] Starting from NDT data of AD, compressive strength was estimated by using $Model_i$ equation, and a mean of estimated compressive strength ($f_{c,est,cal}$) was calculated and compared with the value obtained from results of the tests on cores ($f_{c,I}$ of AD dataset). Furthermore, the Root Mean Square Error (RMSE) between the predicted and the value measured on cores was calculated.

For each test region, the difference ($\Delta f_{c,m}$) between the compressive strength measured on cores ($f_{c,mean}$) and predicted ($f_{c,est,cal}$) mean value, was calculated as:

$$\Delta f_{c,m} = \frac{f_{c,est,cal} - f_{c,mean}}{f_{c,mean}} \tag{4.19}$$

If the value of $\Delta f_{c,m}$ is positive the estimated compressive strength is higher than the real one and on the contrary if it is negative the real compressive strength is underestimated: thus for an engineering point of view knowing the sign of $\Delta f_{c,m}$ may be of interest.

The difference ($\Delta St.dev$) between the standard deviation of compressive strength measured on cores (St. Dev.) and the estimated one (St.Dev.est) is:

---

[17]Vasanelli et al. [14].

$$\Delta St.dev = \frac{St.Dev. - St.Dev.est}{St.Dev.} \tag{4.20}$$

The same process is followed with the two options of Fig. 4.17 (several specific models, i.e. one for each TR, or one unique model for all TR) and different populations which have all the same size and combine the four TR identified at §4.3.2.2: TR1 and TR2 (from PP), BB and AB.

The process followed to calculate the performance when a specific model (Model$_i$) is used on each TR is illustrated on Fig. 4.22. Figure 4.23 explains what is done following the alternate option, with a common model Model$_{All}$ identified on the composite population grouping the two TR. In both cases, the models are used to

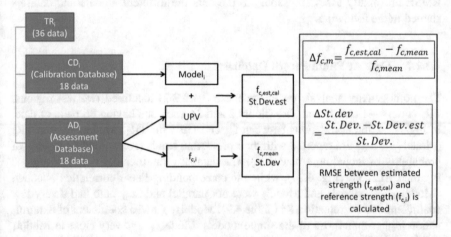

Fig. 4.22 Procedure followed to evaluate the performance of a specific model for each TR (Model$_i$)

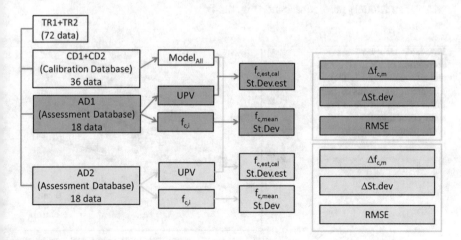

Fig. 4.23 Procedure followed to evaluate the performance of a unique model for all TRs (Model$_{All}$)

Table 4.8 Definition of tested datasets and models

		Specific models	Unique model
Case A	Punta Perotti	PP-TR$_1$ Model$_1$ PP-TR$_2$ Model$_2$	PP-TR$_1$ and PP-TR$_2$ Model$_{All\_A}$
Case B	Punta Perotti + Algerian Building	PP-TR$_2$ Model$_2$ AB Model$_{AB}$	PP-TR$_2$ and AB Model$_{All\_B}$
Case C	Algerian Building + Basilicata Building	AB Model$_{AB}$ BB Model$_{BB}$	AB and BB Model$_{All\_C}$

assess strength on the AD database and the three performances $\Delta f_{c,m}$, ΔSt.dev. and RMSE are finally assessed. Table 4.8 presents the different models and datasets studied in the following.

4.3.2.7 Case A: Punta Perotti Database

The homogeneity analysis on PP database (§4.3.2.2) identified two test regions: TR$_1$ (floors −1 and 1) and TR$_2$ (floors 2 and 3), each one having 36 pairs of data. In Fig. 4.24, the correlations between $f_{c,I}$ and UPV for TR$_1$ (Model$_{1\_36}$) and TR$_2$ (Model$_{2\_36}$) data are reported with their prediction bands (corresponding to 95% of probability of including a new data). The database that includes both test regions is called ALL TR (Fig. 4.24), while the corresponding linear correlation is called "Model$_{All\_A}$". Model$_1$ and Model$_2$ were not parallel and they both had a very low coefficient of determination R^2 (Table 4.9). Model$_{Al\_A}$ had a coefficient of determination higher than those of the single models. Model$_{All\_A}$ is very close to Model$_1$ and Model$_2$. The prediction band of the Model$_{All\_A}$ is very close to Model$_1$ in the interval between 3900 and 4300 m/s while, outside from this interval, it is slightly wider than Model$_2$ prediction band (Fig. 4.24).

Fig. 4.24 UPV-$f_{c,I}$ correlation of TR1, TR2 data and all the data (ALL TR). The prediction bands of each linear correlation are of the same color than the corresponding database

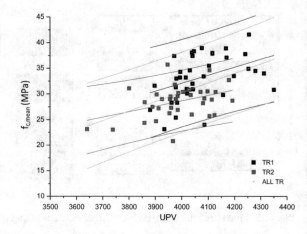

Table 4.9 Linear correlation parameters obtained for TR1, TR2 and All_A databases

	Calibration dataset			Full dataset		
	Slope	Intercept	R^2	Slope	Intercept	R^2
$Model_1$	0.016	−31.96	0.24	0.016	−30.36	0.15
$Model_2$	0.012	−18.96	0.36	0.011	−13.70	0.14
$Model_{All\_A}$	0.017	−38.87	0.39	0.018	−43.10	0.25

Left part: calibration on half set (18 pairs of data for $Model_i$ and 36 pairs of data for $Model_{All}$); right part: model identified on the full set (36 pairs of data for $Model_i$ and 72 pairs of data for Model $_{All}$)

$Model_1$ and $Model_2$ were determined fitting 18 UPV-$f_{c,i}$ data of CD dataset of TR_1 (CD_1) and TR_2 (CD_2) respectively. The same procedure was performed on $Model_{All\_A}$, from the composite ($CD_1 + CD_2$) dataset, with 36 UPV-$f_{c,I}$ pairs. The models (Table 4.9) identified on the calibration dataset are very close, and they are similar to those obtained from all the data. The explanation is the use of the conditional coring procedure that ensures a correct coverage of the property distribution.

Table 4.10 provides the performances of the models (specific or unique) when they are applied on the ADs. The relative error on the mean strength is below 5% for AD_1 and not more than 13% for AD_2, and it is slightly larger with a unique model. The true standard deviation is not captured by the models, whatever the options, with relative errors always over 50%. The mean error on local strength (RMSE) is about 4 MPa in all cases, and slightly larger with a unique model. One can relate this RMSE value to the width of the prediction bands on Fig. 4.24.

The same comparison between the two options (specific models vs. unique model) has also been tested by further reducing the dataset size. In particular, CD1 and CD2 of 10 (Case A1) and 4 (Case A2) data pairs were considered selecting data based on conditional coring. $Model_{1\_10}$ and $Model_{2\_10}$ were calculated in Case A1, while $Model_{1\_4}$ and $Model_{2\_4}$ were calculated in Case A2. $Model_{All\_A}$ was fitted using 20 and 8 pairs of data for Case A1 ($Model_{All\_A\_20}$) and Case A2 ($Model_{All\_A\_8}$), respectively. Table 4.11 summarizes the identified regression models whereas Table 4.12 provides the results of the cross-validation procedure.

Reducing the number of cores from 18 to 10 in calibration datasets, did not alter slopes and intercepts of Models 1 and 2 significantly (Table 4.11). Model 2 remained almost unchanged varying the number of cores used for calibration from 18 to 10 and 4, while Model 1 of case A2 differed in both slope and intercept from the same model determined using 18 and 10 cores (Table 4.11). $Model_{All}$ of case A1 and A2 showed significant variations from the same model of case A (Table 4.11).

The results of the test (Table 4.12) showed that in case A1, $Model_i$ and $Model_{All\_A}$ gave comparable results for both AD1 and AD2 datasets. In Case A2, when only four pairs of data were used for calibration, $\Delta f_{c,m}$ of $Model_1$ and $Model_2$ were both under 5%, while $\Delta f_{c,m}$ of $Model_{All\_A}$ approached 10%. RMSE values of $Model_1$ and $Model_{All\_A}$ were comparable in both case A1 and A2; when AD_2 dataset is considered, $Model_2$ gave lower values of RMSE than $Model_{All\_A}$ in both the case A1

Table 4.10 Results of cross validation procedure calculated on AD_1 and AD_2 databases (18 UPV-$f_{c,i}$ data)

	Data for validation	Models	$f_{c,m}$	$f_{c,est,cal}$	$\Delta f_{c,m}$ (%)	St. Dev.	St.Dev. est	ΔSt. Dev. (%)	RMSE
Specific model	AD_1	$Model_1$	33.07	32.37	−2.1	4.27	1.18	−72	4.01
	AD_2	$Model_2$	27.41	29.58	+7.9	2.74	0.93	−66	3.46
Unique model	AD_1	$Model_{All\_A}$	33.07	31.52	−4.7	4.27	1.29	−70	4.24
	AD_2		27.41	30.93	+12.8	2.74	1.34	−51	4.50

Table 4.11 Linear correlation parameters obtained for TR1, TR2 and All_A databases

	Calibration dataset				Calibration dataset		
	Slope	Intercept	R^2		Slope	Intercept	R^2
$Model_{1\_10}$	0.016	−32.15	0.22	$Model_{1\_4}$	0.0048	12.46	0.041
$Model_{2\_10}$	0.012	−16.72	0.30	$Model_{2\_4}$	0.011	−14.41	0.63
$Model_{All\_A\_20}$	0.015	−28.77	0.34	$Model_{All\_A\_8}$	0.012	−16.59	0.34

Left part: calibration on 10 pair-set for $Model_i$ and 20 pair-set for $Model_{All}$; right part: model identified on 4 pair-set for $Model_i$ and 8 pair-set for $Model_{All}$

and A2. The standard deviation of f_{cm} was strongly underestimated using 10 and 4 cores for calibration, as in case A.

From PP data analysis we can conclude that when an high number of cores can be sampled it is preferable using one model for each test region, but when the number of cores has to be reduce it is preferable, from a statistical point of view, to consider a unique model. In fact even if the results of $Model_{All}$ were worse than those of $Model_i$ in Case A2, they were anyway acceptable giving good values of RMSE and a $\Delta f_{c,m}$ around 10%.

4.3.2.8   Case B: Punta Perotti (TR$_2$)—Algerian Building Database

The databases considered in this case study come from two different buildings (PP and AB) which have nothing in common. They are however considered as two different test regions of a same hypothetical building. First of all, an homogeneity analysis of data was performed, resembling the procedure used for data of Punta Perotti in order to check if, from a statistically point of view, the two databases can be considered as two distinct test regions. f_c data of both TRs (PP and AB) can be drawn from a normally distributed population, but the variances of the two groups were statistically different, thus a non-parametric test (Mann–Whitney test) was used. The test confirmed that at 95% level of confidence the two databases were statistically non-homogeneous. Considering the frequency distribution of $f_{c,I}$ data of the two groups, it can be observed that the range of variation of f_c within AB is wider than that of PP, and the mean values of the two TR differs of about 10 MPa (Fig. 4.25a).

Repeating the same homogeneity analysis for UPV data, the Mann–Whitney test stated that at 95% level of confidence the two distribution were not different, in contrast with $f_{c,m}$ results. Looking at the frequency distribution of UPV data within the two groups, the range of UPV variation in AB is wider than that of PP, but the mean values of the two groups were very similar (Fig. 4.25b). This aspect has to be taken into consideration in the planning of the diagnostic project, when the number of cores has to be established since it would be a sign of different characteristics of the materials employed in the building.

Table 4.12 Results of the cross validation procedure on AD_1 and AD_2 in case A_1 (CD_i of 8 UPV-$f_{c,m}$ data) and in case A_2 (CD_i of 4 UPV-$f_{c,m}$ data)

Data for validation	Case/models	$f_{c,m}$	$f_{c,est,cal}$	$\Delta f_{c,m}$ (%)	St. Dev.	St. Dev. est	ΔSt. Dev. (%)	RMSE
Specific model								
	AD_1	$A1Model_{1\_10}$	33.07	31.37	**−5.1**	4.27	1.16	**−73**	4.30
		$A2Model_{1\_4}$		31.88	**−3.6**		0.36	**−92**	4.23
	AD_2	$A1Model_{2\_10}$	27.41	30.21	**+10.2**	2.74	0.90	**−67**	3.88
		$A2Model_{2\_4}$		28.52	**+4.0**		0.83	**−70**	2.90
Unique model								
	AD_1	$A1Model_{All\_A\_20}$	33.07	31.12	**−5.9**	4.27	1.10	**−74**	4.41
		$A2Model_{All\_A\_8}$		30.34	**−8.3**		0.86	**−80**	4.82
	AD_2	$A1Model_{All\_A\_20}$	27.41	30.62	**+11.7**	2.74	1.14	**−58**	4.22
		$A2Model_{All\_A\_8}$		29.95	**+9.3**		0.90	**−67**	3.69

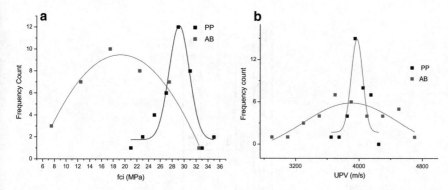

Fig. 4.25 Frequency distribution of $f_{c,i}$ (**a**) and UPV (**b**) data belonging to PP and AB databases

Model$_{PP\_36}$ and Model$_{AB\_36}$ were determined fitting all the data of PP_TR2 database and AB database, respectively Model$_{All\_B\_72}$was then calculated considering all the data of PP_TR2 and AB databases (right part of Table 4.13). In contrast with Case A, for which the linear correlations of TR$_1$ and TR$_2$ data were very close and not parallel, in this case the linear correlation of PP and AB data have a similar slope but a very different intercept. Because of its wider range of properties, the coefficient of determination of Model$_{AB\_36}$ was higher than Model$_{PP\_36}$ and Model$_{All\_B\_72}$. The prediction band of Model$_{All\_B\_72}$ was sensibly wider than those of the models of single test regions were.

Following the same logic, Model$_{PP}$ and Model$_{AB}$ were determined fitting 18 UPV-$f_{c,i}$ data of CD belonging to PP (CD$_{PP}$) and AB (CD$_{AB}$) databases, respectively. Model$_{All\_B}$ was obtained considering the merged dataset (CD$_{PP}$ + CD$_{AB}$). As for Case A, the models with the full data sets are very close to those obtained considering half datasets, with higher coefficients of determination.

Table 4.14 provides the performances of the models (specific or unique) when they are applied on the AD databases. The difference between the results of single models (Model$_{PP}$ and Model$_{AB}$) and those obtained from a unique model (Model$_{All\_B}$) are more significant than for Case A. This is easy to understand by looking at Fig. 4.26, as

Table 4.13 Linear correlation parameters obtained for AB, PP and All_B databases

	Calibration dataset			Full dataset		
	Slope	Intercept	R^2	Slope	Intercept	R^2
Model$_{AB}$	0.012	−26.49	0.88	0.012	−28.71	0.75
Model$_{PP}$ [a]	0.012	−18.96	0.36	0.011	−13.70	0.14
Model$_{All\_B}$	0.014	−32.34	0.57	0.014	−31.44	0.46

Left part: calibration on half set (18 pairs of data for Model$_i$ and 36 pairs of data for Model $_{All}$); right part: model identified on the full set (36 pairs of data for Model$_i$ and 72 pairs of data for Model $_{All}$)
[a]Model$_{PP}$ is similar to Model$_2$ of Table 4.9

Table 4.14 Results of cross validation procedure calculated on AD_1 and AD_2 databases (18 UPV-$f_{c,i}$ data)

	Data for validation	Models	$f_{c,m}$	$f_{c,est,cal}$	$\Delta f_{c,m}$ (%)	St. Dev.	St. Dev. est	ΔSt. Dev. (%)	RMSE
Specific model	AD_{AB}	$Model_1$	20.44	20.68	**1.2**	5.98	3.86	**35**	3.79
	AD_{PP} [a]	$Model_2$	27.41	29.58	**7.9**	2.74	0.93	**66**	3.46
Unique model	AD_{AB}	$Model_{All\_B}$	20.44	25.22	**23.4**	5.98	4.71	**21**	6.05
	AD_{PP}		27.41	25.43	**7.2**	2.74	1.11	**59**	3.38

[a]$Model_{PP}$ is similar to $Model_2$ of Table 4.10

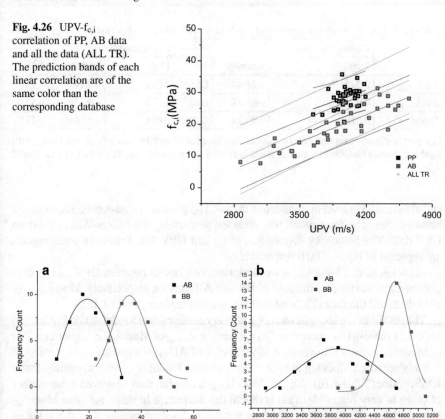

Fig. 4.26 UPV-$f_{c,i}$ correlation of PP, AB data and all the data (ALL TR). The prediction bands of each linear correlation are of the same color than the corresponding database

Fig. 4.27 Frequency distribution of $f_{c,i}$ (**a**) and UPV (**b**) data of AB and BB test regions

the unique model is not capable of properly describing the two TR: it underestimates the strength in PP and overestimates it in AB. The relative error on Model$_{All\_B}$ amounts 23.4%. Another consequence is the much larger RMSE value on the AB set, for the same reason. Regarding the standard deviation, the results confirm that the true standard deviation cannot be captured by the regression models. The unique model in this case did not work better than single models even with the higher number of cores, thus we have not considered the case B1 and B2 with 10 and 4 cores for calibration.

4.3.2.9 Case C: Basilicata Building—Algerian Building

As for Case B, we consider here a hypothetical structure, which groups two TR respectively corresponding to the Basilicata building and Algerian building datasets. The homogeneity analysis of the two databases was performed. The ANOVA analysis

Table 4.15 Linear correlation parameters obtained for AB, BB and All_C databases

	Calibration dataset)			Full dataset		
	Slope	Intercept	R^2	Slope	Intercept	R^2
Model$_{AB}$	0.012	−26.49	0.88	0.012	−28.71	0.75
Model$_{BB}$	0.023	−68.87	0.48	0.020	−56.57	0.32
Model$_{All\_C}$	0.018	−49.24	0.77	0.018	−51.01	0.74

Left part: calibration on half set (18 pairs of data for Modeli and 36 pairs of data for Model All); right part: model identified on the full set(36 pairs of data for Modeli and 72 pairs of data for Model All)

stated that at 95% level of confidence the two $f_{c,m}$ groups can be considered to be not homogenous. The same result was obtained performing the Mann–Whitney test on UPV data. The frequency distribution of $f_{c,i}$ and UPV data for the two test regions are reported in Fig. 4.27a, b respectively.

Model$_{BB\_36}$ and Model$_{AB\_36}$are the linear correlations between UPV and $f_{c,i}$ data obtained considering all the data of BB and AB dataset, respectively. Model$_{All\_C\_72}$ considered all the data (72 points) of the two test regions (Table 4.15).

The results of the analysis on AD$_{BB}$ are very similar when Model$_{BB}$ and Model$_{All\_C}$ are used (Table 4.16), except for $f_{c,m}$ prediction: $\Delta f_{c,m}$ of Model$_{BB}$ is higher than that of Model$_{All\_C}$, even if lower than 10%. In case of AD$_{AB}$, the quality of estimation of f_c is higher when Model$_{AB}$ is used compared with Model$_{All\_C}$, with a sensibly lower RMSE value (Table 4.16). $\Delta f_{c,m}$ of Model$_{All\_C}$ are higher than 15% even if the number of cores is very high (36). This is due to the difference in slope between Model$_{AB}$ and Model$_{ALL\_C}$: the UPVs of AD$_{BB}$ dataset had values mostly over the value of intersection between Model$_{AB}$ and Model$_{ALL\_C}$ (about 3500 m/s in Fig. 4.28). This caused an overestimation of the f_{cm} predicted by Model$_{ALL\_C}$ compared to the real value, which is closer to that estimated by Model$_{AB}$. For the same reason, in case of AD$_{AB}$, the RMSE value of the unique model was higher than that of Model$_{AB}$. The RMSE of Model$_{BB}$ and Model$_{ALL\_C}$ are comparable.

Case C1 and Case C2 considered 10 and 4UPV-$f_{c,i}$ data, respectively, to fit the models for each test regions. Model$_{All\_C\_20}$ and Model$_{All\_C\_8}$ were fitted using 20 and 8 data pairs in case C1 and Case C2, respectively. Models of Case C1 were quite similar to the corresponding models of Case C as well as Model$_{AB\_4}$ and Model$_{All\_C\_8}$. Model$_{BB\_4}$ was different from Model$_{BB}$ and Model$_{BB\_10}$ and this reflected on the quality of f_c estimation (Table 4.17).

While the results obtained for Case C1 remained similar to those of Case C, they changed in Case C2 (Table 4.18). In Case C2 it is better to use Model$_{All\_C}$ than Model$_{BB}$ for f_c prediction and also for f_c local estimation (lower RMSE value). Model$_{AB}$ performs better in Case C2 than Model$_{All\_C}$.

Table 4.16 Results of cross validation procedure calculated on AD_{BB} and AD_{AB} databases (18 UPV-$f_{c,i}$ data)

	Data for validation	Models	$f_{c,m}$	$f_{c,est,cal}$	$\Delta f_{c,m}$ (%)	St. Dev.	St. Dev. est	ΔSt. Dev. (%)	RMSE
Specific model	AD_{AB}	Model$_{AB}$	20.44	20.68	**1.2**	5.98	3.86	**35**	3.79
	AD_{BB}	Model$_{BB}$	36.16	39.42	**9.0**	6.93	3.55	**49**	6.95
Unique model	AD_{AB}	Model$_{All\_C}$	20.44	23.91	**17.0**	5.98	5.99	**0.2**	5.25
	AD_{BB}		36.16	36.92	**2.1**	6.93	2.83	**59**	6.15

Fig. 4.28 UPV-$f_{c,m}$ correlation of BB, AB data and all the data (ALL TR). The prediction bands of each linear correlation are of the same color than the corresponding database

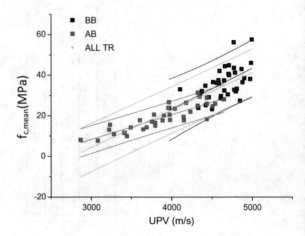

Table 4.17 Linear correlation parameters obtained for BB and AB databases in Case C1 and C2

		Slope	Intercept	R^2
Case C1	Model$_{BB\_10}$	0.022	−63.37	0.58
	Model$_{AB\_10}$	0.012	−29.10	0.95
	Model$_{All\_C\_20}$	0.019	−51.47	0.80
Case C2	Model$_{BB\_4}$	0.032	−105.06	0.91
	Model$_{AB\_4}$	0.012	−25.69	0.98
	Model$_{All\_C\_8}$	0.020	−53.03	0.81

4.3.3   Conclusions

Three different real database with a high variability of both UPV and $f_{c,i}$ data were analyzed: the COV values were ranging between 3–13% and 15–35% respectively for UPV and $f_{c,i}$ data. Different homogeneity analysis were performed to identify homogeneous test regions.

The one specific model for each TR *vs* a unique model for all TR alternative was tested in three different cases: case A with data from the same real structures, cases B and C where data from different sources were merged as if they were TR of a hypothetical structure. These three cases correspond to different features in the (UPV, f_c) diagram as shown by Figs. 4.24, 4.26 and 4.28. All comparisons were carried out by identifying the models on a first subset of data and by assessing their performances on a complementary subset.

Main conclusions are that:

– a unique model is not a relevant option in cases similar to case B (Fig. 4.26) where the two specific models give a significantly different strength for an identical UPV value. Different UPV distribution with similar UPV$_m$ may be an alarm of the presence of different materials in the same building.

Table 4.18 Results of the cross validation procedure on AD_{BB} and AD_{AB} in case C1 and in case C2

Data for validation		Models	$f_{c,m}$	$f_{c,est,cal}$	$\Delta f_{c,m}$ (%)	St. Dev.	St. Dev. est	ΔSt. Dev. (%)	RMSE
Case C1	AD_{BB}	Model$_{BB}$	36.16	39.74	**9.9**	6.93	3.38	**51**	7.09
		Model$_{All\_C}$	36.16	37.05	**2.5**	6.93	2.90	**58**	6.17
	AD_{AB}	Model$_{AB}$	20.44	20.06	**1.9**	5.98	4.03	**33**	3.77
		Model$_{All\_C}$	20.44	23.68	**15.9**	5.98	6.15	**28**	5.14
Case C2	AD_{BB}	Model$_{BB}$	36.16	44.19	**22.2**	6.93	4.90	**29**	10.27
		Model$_{All\_C}$	36.16	39.25	**8.5**	6.93	3.03	**56**	6.85
	AD_{AB}	Model$_{AB}$	20.44	21.07	**3.1**	5.98	3.83	**36**	3.84
		Model$_{All\_C}$	20.44	25.31	**23.8**	5.98	6.42	**7.4**	6.37

– in other configurations (i.e. that of Fig. 4.24 of overlapping (UPV, f_c) clouds and that of Fig. 4.28 where the two clouds follow a same overall "tendency"), the two approaches are almost equally efficient regarding the estimation of the mean strength or the local error RMSE. However, the specific models generally lead, with this large number of cores, to better estimates (smaller errors). In addition, it is confirmed that, bases on these approaches, no model is able to capture the true value of the strength standard deviation.

Reducing the number of cores, increases the statistical risk and global conclusions are not easy to draw because of statistical fluctuations. Using a unique model would reduce this risk by keeping a larger number of cores.

Acknowledgements The 5th author would like to acknowledge the financial support by Base Funding - UIDB/04708/2020 of CONSTRUCT - Instituto de I&D em Estruturas e Construções, funded by national funds through FCT/MCTES (PIDDAC).

References

1. EN 13791:2007: Assessment of In situ Compressive Strength in Structures and Precast Concrete Components. CEN, Brussels (2007)
2. Breysse, D.: Nondestructive evaluation of concrete strength: an historical review and a new perspective by combining NDT methods. Constr. Build. Mater. **33**, 139–163 (2012)
3. Wu, J.: Advances in K-means Clustering: A Data Mining Thinking, 1st edn. Springer, Berlin, Heidelberg (2012)
4. Javadi, S., Hashemy, S.M., Mohammadi, K., Howard, K.W.F., Neshat, A.: Classification of aquifer vulnerability using K-means cluster analysis. J. Hydrol. **549**, 27–37 (2017)
5. Masi, A., Chiauzzi, L., Manfredi, V.: Criteria for identifying concrete homogeneous areas for the estimation of in-situ strength in RC buildings. Constr. Build. Mater. **121**, 576–587 (2016)
6. EN 1998-3:2005: Eurocode 8: Design of Structures for Earthquake Resistance—Part 3: Assessment and Retrofitting of Buildings. European Committee for Standardization (CEN), Brussels, Belgium (2005)
7. Pereira, N., Romão, X.: Assessing concrete strength variability in existing structures based on the results of NDTs. Constr. Build. Mater. **173**, 786–800 (2018)
8. Gulhar, M., Kibria, B.M.G., Albatineh, A.N., Ahmed, N.U.: A comparison of some confidence intervals for estimating the population coefficient of variation: a simulation study. SORT (Barc) **36**(1), 45–68 (2012)
9. O'Neill, B.: Some useful moment results in sampling problems. Am. Stat. **68**(4), 282–296 (2014)
10. Luprano, V.A.M., Pfister, V., Tundo, A., Ciniglio, G., Tati, A.: Prove Non Distruttive e prove Distruttive su Calcestruzzo: campagna prove di Punta Perotti (BA) e correlazione delle misure, ENEA Technical Report RT/2009/18/FIM (2009)
11. Pfister, V., Tundo, A., Luprano, V.A.M.: Evaluation of concrete strength by means of ultrasonic waves: a method for the selection of coring position. Constr. Build. Mater. **61**, 278–284 (2014). https://doi.org/10.1016/j.conbuildmat.2014.03.017https://doi.org/10.1016/j.conbuildmat.2014.03.017
12. Ali-Benyahia, K.: Contrôle de la qualité du béton par les essais non destructifs, Ph. D. Report, Univ. Hassiba Ben Bouali of Chlef, Algeria (2017). (in French)
13. EN 13791:2007: Assessment of In-situ Compressive Strength in Structures or in Precast Concrete Products. European Committee for Standardization (CEN), Brussels, Belgium (2007)
14. Vasanelli, E., Colangiuli, D., Calia, A., Luprano, V.A.M.: Estimating in situ concrete strength combining direct and indirect measures via cross validation procedure. Constr. Build. Mater. **151**, 916–924 (2017)

Chapter 5
Identification and Processing of Outliers

Xavier Romão and Emilia Vasanelli

Abstract When analyzing real data sets, observations different from the majority of the data are sometimes found. These observations are usually called outliers and can be defined as individual data values that are numerically distant from the rest of the sample, thus masking its probability distribution. Outliers require special attention because they can have a significant impact in the concrete strength estimation process and because they may signal the presence of a different concrete population that deserves a separate assessment. The two-step process involved in an outlier analysis (outlier identification and outlier handling) is presented, discussing several statistical methodologies that are available for its implementation. To illustrate the application of an outlier analysis, examples involving univariate and multivariate datasets are presented. Several statistical methodologies are implemented for outlier identification, while outlier handling is illustrated by using robust statistics, i.e. outlier accommodation approaches that reduce the effect of existing outliers on the outcomes of statistical analyses of the data.

The identification, analysis and processing of possible outliers among the series of test results is required before the stage of identification of the conversion model (i.e. T7 and T8 in the flowchart at Fig. 1.9 of Chap. 1). The detailed guidelines in this same chapter have pointed out (Sect. 1.5.3) why and how a careful attention must be paid to possible outliers.

This chapter details what methods can be used and illustrates how they work on a series of examples.

X. Romão (✉)
CONSTRUCT-LESE, Faculty of Engineering, University of Porto, Porto, Portugal
e-mail: xnr@fe.up.pt

E. Vasanelli
Institute of Heritage Science - National Research Council (ISPC-CNR), Lecce, Italy

D. Breysse and J.-P. Balayssac (eds.), *Non-Destructive In Situ Strength Assessment of Concrete*, RILEM State-of-the-Art Reports 32,
https://doi.org/10.1007/978-3-030-64900-5_5

5.1 Context and Principles

The non-destructive concrete strength assessment methodology is based on the analysis and processing of experimental datasets. These datasets are mostly series of test results, either destructive (core strength) or non-destructive (NDT test results). The data processing consists in establishing statistical distributions and correlations.

However, when analyzing real data sets, it is often found that some observations are different from the majority of the data. Such observations are usually called outliers and can be defined as individual data values that are numerically distant from the rest of the sample, thus masking its probability distribution. In most situations, outliers are seen to be unusually large or small data values when compared with the remaining data values in a data set. The main sources of outlying observations may be:

- A measurement or recording error.
- Sampling from a different population than that of the rest of the data (also called contamination and occurs when the distribution of the outlying data is different from that of the remaining data due to a physical or technical reason).
- An incorrect assumption regarding the distribution of the data (i.e. some data does not fit the distribution assumption).
- Rare observations (it is possible that a given observed datum represents a rare event of the underlying population, e.g. if the distribution of the population is heavy-tailed).

Outliers always deserve a careful attention, either because they may reveal specific intrinsic properties of the data previously undetected, or because they are a sign of unexpected problems in the data. Two possible situations may occur:

- Situation A: only a series of test results Tr_i is considered (univariate data set), which typically corresponds to results obtained from test locations where only one NDT is used.
- Situation B: a series of test results for two parameters is analyzed (bivariate data set) which corresponds to results obtained from test locations where two NDTs are used.

An outlier appears either because it departs from the general distribution of the data or because it departs from the general correlation that can be identified for the global data set. In Situation A, one analyzes if the statistical distribution of test results shows something unexpected. In Situation B, one analyzes if there are inconsistencies between test results obtained from two NDTs when they are considered together.

An outlier analysis can usually be seen as a two-step process:

- in the first step, several methods can be used to **identify potential candidate outliers** of the data. As far as possible, it is recommended to analyze the NDT measurements while on the field, in order to be able to perform additional measurements if suspicious data are detected.
- in the second step, the candidate outliers **need to be handled**. After identifying the presence of candidate outliers and considering that they are not due to errors

that can be eliminated by repeating measurements, the second step of the outlier analysis process needs to be carried out, **to account for their effect** in subsequent statistical analyses of the data. This second step depends on how much it is possible to know about the source(s) of the outlying data. Among the several approaches that can be followed, the use of outlier accommodation procedures is recommended.

To illustrate the two steps of outlier analysis, one case of Situation A and one case of Situation B are analyzed in the following.

5.2   Outlier Identification

5.2.1   General Considerations and Methods

5.2.1.1   General Considerations

Outliers can occur in both univariate and multivariate data. For the first type of data, the identification of outliers is usually based on the analysis of the statistical distribution of the data and on how it departs from the expected characteristics. For multivariate data, one might think that a similar type of analysis could be conducted for the corresponding univariate data sets (i.e. that multivariate outliers could be detected by analyzing the corresponding univariate data sets of the multivariate sample). However, this is not true since a multivariate outlier may not be identified as a univariate outlier in the corresponding one-dimensional data sets and observations identified as outliers in those one-dimensional data sets may not be true multivariate outliers.

Figures 5.1 and 5.2 illustrate two different cases. The first case (Fig. 5.1a) corresponds to a situation where the scatter plot of a bivariate data set of rebound hammer (RH) and ultrasonic pulse velocity (UPV) test results enables the identification of

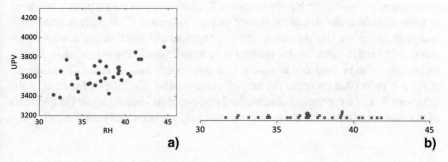

Fig. 5.1 Scatter plot of the RH-UPV data (**a**) and dot plot of the RH data (**b**) with one outlier that is only visible when looking at the bivariate relation

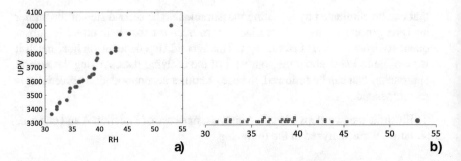

Fig. 5.2 Scatter plot of the RH-UPV data (**a**) and dot plot of the RH data (**b**) with one outlier that is less visible when looking at the bivariate relation and easier to identify from the univariate data set

an outlying data (the red data point in the plot). However, when looking at the RH data only (Fig. 5.1b), the corresponding value of this outlying data (i.e. the red data point) does not stand out since it is totally mixed within the data set. The second case (Fig. 5.2a) corresponds to a situation where the visual analysis of the scatter plot is not more informative than the univariate RH data analysis (Fig. 5.2b). Given these two scenarios, multivariate outlier identification is often carried out by analyzing the correlation between the one-dimensional variables and on how it departs from the expected characteristics. Still, this type of analysis is not without a certain amount of subjectivity. In other cases, the outlier identification process can also be performed by analyzing the distance between data points and groups of data points. However, this type of approach usually requires larger samples of data in order to provide more decisive results.

5.2.1.2 Visual Screening

Outlier identification **should usually start by a visual screening** of the data using different types of plots able to display the variability of the data and reveal the existence of extreme/outlying observations. Typical non-parametric plots considered in these situations are dot plots, scatter plots, histograms, CDF plots or stem-and-leaf plots. Boxplots (for univariate data) or bagplots (for multivariate data) are also commonly used but assume that the data is symmetric or close-to-symmetric. Nevertheless, they can be modified to account for asymmetric data. In terms of parametric plots, p-p plots (that compare the empirical cumulative distribution function of the data with that of a theoretical distribution) and q-q plots (that compare the quantiles of the data with those of a theoretical distribution) are commonly considered.

5.2.1.3 Formal Statistical Tests and Their Limits

If the distribution of the data is known, outlier identification can also be carried out **using formal statistical tests**. Unfortunately, **such type of tests is not available for all types of distributions**. A relatively large number of tests were specifically developed for the case where data is expected to follow a normal distribution. If data follows an exponential, type I extreme-value, Weibull or gamma distribution, several specific tests are also available. Some formal statistical tests are able to identify single outliers while others target the identification of multiple outliers. For the case of normal or close-to-normal univariate data, some of the most popular outlier identification statistical tests include the Grubbs test, the Dixon test, the Tiejen-Moore test, the z-score test and its variants and the generalized extreme studentized deviate many-outlier procedure. Several of these tests have been discussed and reviewed extensively [1–4] and some of them are also suggested by standards such as ISO 16269-4 [5], ASTM D7915 [6] and ASTM E178 [7].

Formal statistical tests can be powerful in identifying outlying values when the data follows the statistical distribution assumptions but they are also susceptible to masking or swamping problems. Masking can occur when the test assumes a number of outliers lower than the real one, for example when testing for a single outlier when in fact there are two (or more) outliers. These additional outliers may influence the value of the test statistic so that no data are declared as outliers. On the contrary, swamping can occur when the test assumes a number of outliers larger than the real one. In this case, when testing, for example, for two outliers when in fact there is only one, both data may be declared outliers. Due to the possibility of masking and swamping effects, the use of formal statistical tests should normally be complemented by the visual analysis of the data.

Even though formal statistical tests are applicable in many situations, it is found that most real data distributions may be unknown or may not follow the specific distribution assumed by those tests. To overcome this situation data transformation techniques are often used to obtain new data compatible with the distribution assumptions of the test. Given that many tests were developed for the case where data is expected to follow a normal distribution, non-normal data is often transformed to close-to-normal data using a Box-Cox transformation.

5.2.1.4 Multivariate Outliers

Some of the more popular approaches for multivariate outlier detection involve distance-based methods. Among those, methods based on the Mahalanobis distance are often used. The theory behind methods using the Mahalanobis distance assumes that the data are generated from an elliptical distribution, among which the multivariate normal is the most popular one. To classify the observations into regular points and outliers, Mahalanobis distances are determined and compared with a cutoff value based on the χ^2 distribution which represents the distribution of these distances [8]. However, values larger than this cutoff value are not necessarily outliers, they could still belong to the data distribution. The Mahalanobis distance incorporates both the variability of each marginal direction and the correlation structure of the data. Usual estimators needed from the data are location and dispersion measures that are usually defined by the sample means and sample covariance matrix. However, these estimators are sensitive to outliers and alternative robust approaches have been developed to estimate them. Among these methods, the minimum covariance determinant estimator [9] is one of the most well-known.

Projection-based (or depth-based) methods are another type of approach often used to identify multivariate outliers. These methods are based on computational geometry using the notion of depth of one data point among a data set [10]. Depth can be seen as the relative location of an observation with respect to an upper or lower edge layer of the data set. The data are organized in convex hull layers and outliers are data on outer layers (i.e. with shallow depth values). One method of identifying outliers based on projection or depth measures of the data can be carried out using the Stahel-Donoho outlyingness [11]. However, this measure is only adequate if the data exhibits elliptical symmetry. To overcome this limitation, an adjusted outlyingness measure has also been proposed that accounts for skewed data [12].

Cluster analysis is also a technique that can be used for multivariate outlier identification. It searches for homogeneous groups in the data, which afterward may be analyzed separately. One of the most popular methods for cluster analysis is k-means clustering [13]. The method aims to partition n observations into k clusters in which each observation belongs to the cluster with the nearest mean, serving as a prototype of the cluster. However, this method is not robust as it uses group averages and to overcome this problem several robust approaches have been also proposed (e.g. see the Partitioning around Medoids method [14]). Since these approaches are especially suited to handle large data sets, they may not be particularly adequate for the scope covered by the RILEM TC 249-ISC Recommendations.

5.2.2 Identification of Candidate Outliers for Univariate Data Sets (Situation A)

The implementation of several methods for outliers identification in univariate data sets is illustrated using the RH and UPV test results synthesized in Table 5.1. These data correspond to a 100 m long and 5 m high concrete circular wall, which was constructed in successive batches of about 25 m-length and 0.60 m-height (see Chap. 7). This represents a scenario involving four different concrete batches along the length and 9 batches in elevation. The data is obtained by considering 30 test locations involving the following test results: 10 test locations with both RH and UPV test results, 10 test locations with only RH test results, 10 test locations with only UPV test results.

The outlier identification methods that are considered to analyze univariate data sets involve both visual and analytical methods/statistical tests. Visual methods are

Table 5.1 NDT test results (RH and UPV) and their location along the wall (where x and z are the horizontal coordinate and the vertical coordinate, respectively)

z	x	RH	UPV	z	x	RH	UPV
(m)	(m)		(m/s)	(m)	(m)		(m/s)
0.3	0	35.5	4335	2.1	80	39.3	
0.9	10	39.5	4425	2.1	90	37.0	
0.9	20	39.0		2.7	0		4336
0.9	30		4253	2.7	10		4333
0.9	50	36.6		2.7	20		4306
0.9	70	36.3		2.7	30		4211
1.5	20	35.9	4415	2.7	40	37.9	4322
2.1	0	36.0		2.7	50		4120
2.1	10	37.1		2.7	60		4314
2.1	20	37.7		2.7	70		4479
2.1	30	37.9	4322	2.7	80		4355
2.1	40	37.4		2.7	90		4320
2.1	50	32.0	4291	3.3	60	39.4	4489
2.1	60	34.1		3.9	80	34.6	4281
2.1	70	37.0	4257	4.5	90	34.0	4272

based on special plots of the univariate data enabling a visual understanding of several statistical characteristics of the data (namely its skewness) while allowing also to flag candidate outliers. The visual methods that are considered are the histogram, the dot plot, the boxplot and the adjusted boxplot. Other types of plots could also be used such as CDF plots or steam-and-leaf displays. In terms of analytical methods and statistical tests, most available approaches assume the data follows some known statistical distribution. Most of the well-known methods assume that this statistical distribution is the normal distribution. Given that, in many cases, the statistical distribution followed by the data is unknown or not defined by a known theoretical model, identifying outliers using analytical methods needs to be carried out carefully. As such, only two methods are considered herein. One is a robust z-score and the other is the generalized extreme studentized deviate (GESD) many-outlier procedure suggested by standards ASTM D7915 [6] and ISO 16269-4 [5]. Although the GESD procedure was originally defined for data that follows a normal distribution, the procedure can still be applied by transforming the non-normal data to close-to-normal data using a Box-Cox transformation.

Figure 5.3 shows the histograms of the RH and UPV samples of size 20 under analysis. As can be seen, both data sets are asymmetric. This fact will make the outlier identification process more difficult since one or more extreme values of a data set can simply be the result of being in the presence of a heavy-tailed distribution. However, when determining the sample skewness using Pearson's coefficient, skewness values of -0.542 and 0.009 are obtained for the RH and UPV samples, respectively. The

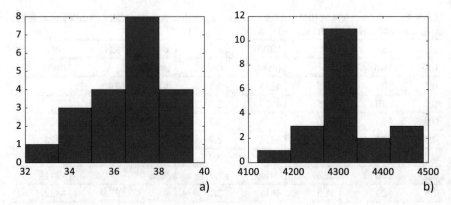

Fig. 5.3 Histograms of the RH data (**a**) and of the UPV data (**b**)

fact that the skewness value of the UPV data set is very close to zero (the value for a symmetric distribution) indicates that the apparent asymmetry of the UPV histogram is not caught by this skewness estimator. By using a more robust estimator of skewness such as the medcouple [15], robust skewness values of −0.115 and − 0.080 are obtained for the RH and UPV samples, respectively. By analyzing these results, it can be seen that, for the RH data, the non-robust skewness estimate provides an adequate reading (the medcouple value is now much closer to zero but its range is between −1 and 1). For the UPV data, since the robust estimate of skewness is now negative (as opposed to the non-robust estimate), this indicates the possibility of candidate outliers in the right tail.

Figure 5.4 shows the dot plots of the RH and UPV samples and their analysis indicates that several test results could be identified as outliers. **For the case of the RH data, the lowest value (RH = 32) may be flagged as a candidate outlier. For the case of the UPV data, the analysis is more complex: the lowest value (UPV = 4120 m/s) may be flagged as a candidate outlier and the top two or even top four values** also (which is consistent with previous analysis of the skewness estimates). This means that, for UPV, five measurements could be seen as candidate outliers. However, since this amount of data represents 25% of the sample, it deserves further attention.

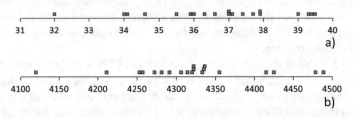

Fig. 5.4 Dot plots of the RH data (**a**) and of the UPV data (**b**)

To provide a more detailed visual interpretation of the data sets, Fig. 5.5 shows the boxplots of the RH and UPV samples, where the top and bottom fences of the plot are defined by 1.5 times the interquartile range (IQR) below the lower quartile and 1.5 times the IQR above the upper quartile of the data, respectively. **These plots confirm that several test results could be identified as outliers, namely the lowest value of the RH data and the lowest and top two values of the UPV data.** Even though boxplots are recommended by ISO 16269-4 for outlier identification, several data results can be erroneously identified as outliers when the data are skewed. To overcome this limitation of the original boxplot, an adjusted boxplot [16] is also considered herein which accounts for a robust measure of skewness (the medcouple) when defining the fences of the boxplot. Figure 5.6 shows the adjusted boxplots of the RH and UPV samples and the results that are obtained provide further interpretations. **For the case of the RH data, the adjusted boxplot indicates that the lowest value is not an outlier. On the other hand, for the case of the UPV data, the adjusted boxplot indicates that the lowest value and the top four values are candidate**

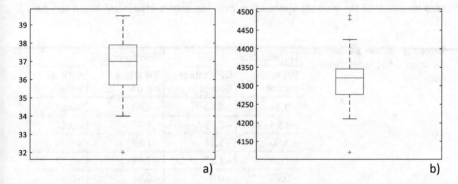

Fig. 5.5 Boxplots of the RH data (**a**) and of the UPV data (**b**)

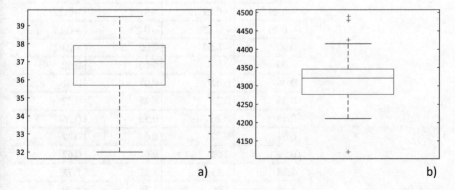

Fig. 5.6 Adjusted boxplots of the RH data (**a**) and of the UPV data (**b**)

outliers. As referred before, since this amount of data represents 25% of the UPV sample, any decision about these values needs to be made carefully.

To obtain further insights on the data, additional numerical approaches are also used to complement the visual results obtained so far. Robust z-scores [17] are determined as an alternative to the classical z-score approach. The classical approach flags candidate outliers as those observations that are away from the mean of the data by ±2 or 3 times the standard deviation of the data. The robust version of the z-score replaces the mean and the standard deviation by robust estimators (i.e. less affected by potential outliers), namely the median and the median of all absolute deviations from the median (MAD), respectively. Table 5.2 presents the robust z-scores that were obtained for the RH and UPV data (the 2 columns on the left). By considering that outliers are observations that are away from the median by more than ±2.5 times the dispersion estimator, the results obtained correspond to the bold values in Table 5.2 and are those also obtained from the boxplot. Other proposals[1] however suggest the value of 2.5 should be replaced by 3.5. In this case, no outliers are found for the RH data and only the lowest value of the UPV data is flagged as an outlier. It is also known that the z-score approaches provide higher efficiency when the data is

Table 5.2 Robust z-scores of the original and transformed data	Original data		Transformed data	
	RH robust z-score	UPV robust z-score	RH robust z-score	UPV robust z-score
	−3.21	**−3.66**	**−2.58**	**−3.67**
	−1.93	−2.01	−1.73	−2.01
	−1.86	−1.24	−1.68	−1.24
	−1.54	−1.17	−1.43	−1.17
	−0.96	−0.89	−0.94	−0.89
	−0.71	−0.73	−0.70	−0.73
	−0.64	−0.55	−0.64	−0.55
	−0.45	−0.27	−0.46	−0.27
	−0.26	−0.13	−0.27	−0.13
	0.00	−0.02	0.00	−0.02
	0.00	0.02	0.00	0.02
	0.06	0.02	0.07	0.02
	0.26	0.22	0.28	0.22
	0.45	0.26	0.50	0.26
	0.58	0.27	0.64	0.27
	0.58	0.62	0.64	0.62
	1.28	1.71	1.52	1.71
	1.48	1.90	1.78	1.89
	1.54	**2.88**	1.87	**2.88**
	1.61	**3.06**	1.96	**3.06**

symmetric and close to normal. Since both the RH and UPV data are asymmetric, an alternative outlier analysis could be carried out after transforming the data to close-to-normal data using a Box-Cox transformation. By applying this type of transformation, Box-Cox parameters of 5.13 and 0.91 are obtained for the RH and UPV data sets, respectively. It should be noted that for the UPV data, the transformation parameter is close to 1.0, the value corresponding to no transformation of the data. By applying the robust z-scores to the transformed data, the results presented in the 2 columns on the right of Table 5.2 were obtained. As can be seen, analyzing the transformed data does not change the conclusions made for the original data.

Finally, the GESD procedure recommended by ISO 16269-4 for outlier identification is also applied to the transformed data since the procedure is only applicable when the data is normal or approximately normal. **The results obtained for the RH and the UPV data indicate that no outlier was identified by the GESD procedure**.

In conclusion, it can be seen that different outlier identification procedures may lead to different numbers of outliers being flagged in a given data set. Given the statistical characteristics of the considered data sets, it is accepted that one value of the RH data could be considered an outlier, as well as one or more values of the UPV data. With respect to the latter, the lowest value of UPV was flagged as a potential outlier by all procedures except the GESD procedure.

5.2.3   Identification of Candidate Outliers for Bivariate Data Sets (Situation B)

The analysis is illustrated using the ten RH-UPV test results obtained at the same locations by the investigator (Table 5.1). As for the univariate case, there a number of techniques to identify outliers in bivariate data sets, both visual and analytical. However, most outlier detection rules for multivariate data are based on the assumption of elliptical symmetry of the underlying distribution. Furthermore, the complexity of the procedures involved increases significantly and their reliability is highly dependent on the sample size. Since the RH-UPV data set only involves 10 pairs of test results, most methods are not expected to be able to identify outliers correctly. To illustrate the data under analysis, Fig. 5.7a presents a scatter plot of the bivariate RH-UPV data set. As can be seen, the positive correlation trend between the two parameters is not very clear.

To provide a more detailed visual interpretation of the data, Fig. 5.7b presents a bagplot [18] of the data, a bivariate generalization of the univariate boxplot. The main components of the bagplot are three and can be interpreted as those of the boxplot. The darker inner part is the "bag" and contains at most 50% of the data. The outer limit of the lighter area is the fence and is formed by inflating the bag by a certain factor (usually 3). Data outside the fence are flagged as outliers. The lighter area between the bag and the fence is the loop. The point marked with a cross symbol (+) marks the center of the plot and corresponds to the point with the highest possible

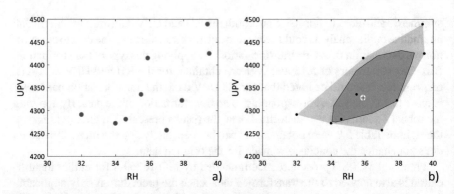

Fig. 5.7 Scatter plot (**a**) and bagplot (**b**) of the bivariate RH-UPV data

Tukey depth [18]. Like the univariate boxplot, the bagplot also provides a visual interpretation of several characteristics of the data: its location (the depth median), spread (the size of the bag), correlation (the orientation of the bag), skewness (the shape of the bag and the loop), and tails (the points near the boundary of the loop and the outliers). As can be seen, for the present case, the bagplot does not flag any data as an outlier.

In terms of analytical methods, the Stahel and Donoho outlyingness and the skewness-adjusted outlyingness [12] were analyzed. Figure 5.8a presents the values of these measures for each data pair (two pairs have a zero value of both measures) along with the corresponding cutoff limits above which a given data pair would have been flagged as an outlier. As can be seen, no data was identified as a potential outlier by both approaches. While the first measure assumes the distribution of the data is approximately symmetric, the second measure does not. Even though both approaches lead to the same outcome, it is noted that these results might be conditioned by the small sample size of the data. The Mahalanobis distances were also analyzed to identify candidate outliers and different approaches were used to determine the estimators needed to determine such distances. Aside from the classical approach in which the Mahalanobis distances are determined based on the classical mean and covariance matrix of the data, alternative robust procedures were also used to obtain these estimators: the Minimum Covariance Determinant (FAST-MCD) method [9], the Orthogonalized Gnanadesikan-Kettenring (OGK) method [19] and the Olive-Hawkins (OH) method [20]. To illustrate the results that were obtained, Figs. 5.8b–d present the values of these robust distances against the corresponding classical Mahalanobis distances for each data pair, along with the corresponding cutoff limits above which a given data pair is flagged as an outlier. As can be seen, both the classical approach and the approach based on the FAST-MCD method do not flag any outlier. On the other hand, the approach based on the OGK method flags one outlier and the approach based on the OH method flags three outliers. The data pairs flagged as outliers are the three upper pairs identified in the scatter plot of Fig. 5.9 (the single outlier identified by the distances based on the OGK method corresponds to the data pair with the highest UPV value).

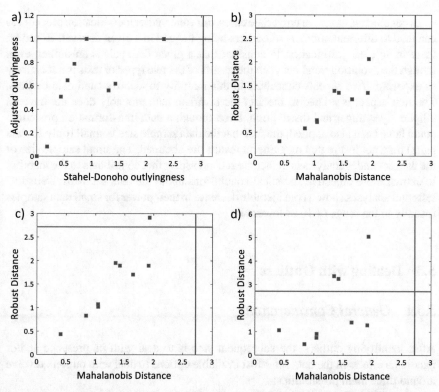

Fig. 5.8 Adjusted outlyingness versus Stahel-Donoho outlyingness (**a**), Mahalanobis distances versus robust distances using the FAST-MCD method (**b**), Mahalanobis distances versus robust distances using the OGK method (**c**) and Mahalanobis distances versus robust distances using the OH method (**d**) of the bivariate RH-UPV data

Fig. 5.9 Scatter plot of the bivariate RH-UPV data set with outliers identified in red

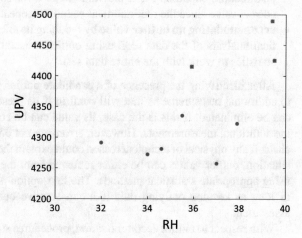

In conclusion, as for the univariate cases, different outlier identification procedures can lead to different numbers of outliers being flagged in a given data set. However, there is no clear justification to consider that a given data pair is an outlier, even though the evolution trend between the results of the two types of tests is not entirely as expected. Two aspects contribute to this inability to identify candidate outliers. The first aspect is related to the fact the bivariate data probably does not have an elliptically symmetrical distribution. Even though a data transformation procedure could have been also applied, the fact that the data sample size is small (only 10 data pairs) may not insure that meaningful results are obtained. The small sample size of the data set under analysis is in fact the second aspect that contributes to the inability to extract more objective statistical conclusions about the data set. Most statistical tests and analyses show a considerable decrease in their power for small data samples (usually in the range of 10 or lower).

5.3   Dealing with Outliers

5.3.1   *General Considerations*

After identifying outliers, the subsequent step is to deal with its presence and/or handle its effects in the data set. Most available options to deal with outliers revolve around three main possibilities:

- **correcting an outlier value**, if an outlier has been identified as an error and if it is possible to replace that observation by a correct value,
- **removing an outlier value** from the data set. This action requires an explicit justification, explaining that the value is an error and cannot be replaced by a correct value since the measurement cannot be repeated,
- **accommodating an outlier value** by reducing its effect on the subsequent statistical analysis of the data (e.g. using outlier-resistant techniques such as robust statistics to work with the entire data set).

After identifying the presence of a candidate outlier, the ideal way is to proceed to additional measurements that will confirm if its presence is due to an error that can be eliminated. If this is the case, its value can be corrected by using data from the additional measurements. However, errors are not often identified and one must check if any physical or technical reason could explain the origin of the outlier. In this situation, outlier values can be either removed from the data set or accommodated using appropriate statistical methods. The first option is not recommended herein, as long as keeping outlying data is a conservative option regarding the strength assessment.

With respect to outlier accommodation, procedures such as those based on robust statistics should be followed to reduce the effect of outliers. The fundamentals of robust statistics are comprehensively addressed in a number of reference books [21–25]. Still, some of the more important concepts are briefly reviewed and illustrated for the case where estimating the central value of a sample (e.g. a mean value) is required. These concepts can then be extended for the purpose of estimating other parameters of the data (e.g. a dispersion value) or for regression analysis between two sets of data.

In the presence of a "well behaved" data sample, i.e. a sample without outliers, the best estimate of the central value is expected to be the sample mean. The *efficiency* of an estimator is judged by its variance which can be obtained after applying the estimator to several samples of a given size drawn from the reference population. The most efficient estimator is that which exhibits the smallest variance. However, outlying observations are likely to occur in most practical situations, since a real data sample seldom follows an exact theoretical distribution model. Having decided that these outlying observations are not errors, there is no compelling reason to exclude them from the analysis. In this case, the adequate characterization of the central value of the underlying distribution must be carried out with methods having adequate *resistance* properties. The *resistance* of an estimator refers to its sensitivity to misbehaviour of the data. The *breakdown point* and the *influence function* are two widely used measures of the resistance of an estimator. The *breakdown point* is the smallest percentage of the sample observations that can be replaced by arbitrarily small or large values before the estimator no longer provides reliable information. For the case of the mean, the breakdown point is known to be 0, while for the median it is 50%. The *influence function* measures the sensitivity of an estimator to different values of the observations and it may be used to describe the effect of outliers on the estimator. To illustrate this concept, Fig. 5.10 presents the influence functions of several estimators. From Fig. 5.10, it can be observed that, for the mean, the absolute value of the influence function increases as the distance between a certain observation and the central value of the data (zero in the case represented) increases, meaning

Fig. 5.10 Influence functions of different estimators

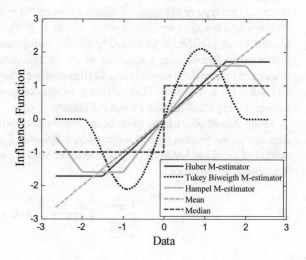

that very large (or very small) observations will have a significant contribution to the estimate. On the other hand, for other (robust) estimators the influence function is seen to have a bounding value, beyond which the influence of the observations remains constant or decreases. For such estimators, the influence of very large (or very small) values is clearly reduced.

The *resistance* properties of robust estimators can also be seen as indirect indicators of their *sufficiency*. A sufficient estimator for a certain parameter θ of the data is one that captures all the information about θ contained in the sample. Assuming that all the data in the sample is important to estimate θ, a given estimator is seen to be sufficient if it makes use of the whole sample to estimate θ. Since robust estimators reduce the importance of certain parts of the sample (either by trimming or by giving less weight to extreme values of the data), a loss in *sufficiency* is a characteristic of their definition. However, robust estimators are "nearly as" sufficient for a given parametric model as the classical ones. Still, when using robust estimators to deal with samples with potential outlying values, i.e. when not all the data in the sample can be considered to be important to estimate θ, the gains in the accuracy of the inference and in the efficiency outweighs the loss in sufficiency. Robust estimation is therefore a good strategy representing a compromise between excluding outlying observations entirely from the analysis and including the whole sample in the analysis, treating all the data equally.

5.3.2 Outlier Accommodation Techniques for Univariate Data Sets (Situation A)

As referred in the previous section, the recommended second step of the outlier analysis process is outlier accommodation. This process involves the use of outlier-resistant statistical techniques known as robust statistics to extract information from the data. This type of approach eliminates the need to make the decision of removing or keeping a certain candidate outlier, which is particularly important when their identification is difficult to make. Since for univariate data sets one is typically interested in determining a measure of the central value and of the scale (i.e. the dispersion) of the data, several robust estimators are considered herein to analyze the RH and UPV test results. Their values are then compared to those obtained by the corresponding standard non-robust estimators.

Many robust procedures have been proposed to estimate the central value of a data set. In the present case, a series of M-estimators [26] are selected based on the results of previous research [27]. These estimators T_n are defined implicitly as the solution of the equation:

$$\frac{1}{n} \sum_{i=1}^{n} \psi \left(\frac{x_i - T_n}{S_n} \right) = 0$$

Table 5.3 Robust estimates of the central and scale values compared with the mean and the standard deviation for the RH and UPV data

Central value			Scale value		
Estimator	RH data	UPV data	Estimator	RH data	UPV data
T_{Hub}	36.79	4321.31	S_t	1.81	76.82
$T_{Ham,1}$	36.78	4321.65	τ_s	1.94	80.77
$T_{Ham,2}$	36.74	4322.99	Standard deviation	1.99	86.13
T_{uk}	36.76	4321.58			
$T_{\log}$	36.78	4320.35			
Mean	36.71	4321.80			

for a real weight function ψ and a sample x of size n. The denominator S_n is a robust scale estimate such as the MAD (median of all absolute deviations from the median). The estimators selected herein are defined for different ψ functions and consist of T_{Hub} (involving the Huber ψ function [26] and a tuning constant of 2), $T_{Ham,1}$ and $T_{Ham,1}$ (involving the Hampel ψ function [21] and tuning parameters of (2.1, 4.0, 8.2) and (2.5, 4.5, 9.5), respectively), T_{Tuk} (involving Tukey's biweight ψ function [21] and a tuning constant of 8) and $T_{\log}$ (involving the logistic ψ function [26]). The central value estimates for the RH and UPV data obtained by these estimators are presented in Table 5.3 along with the corresponding mean value of each data set. As can be seen, **the robust estimates and the mean value are all very close which leads to conclude that the data sets do not possess values that may create a bias in the central estimate**. Furthermore, it can also be seen that any of the robust estimators could be used without any preliminary outlier analysis (which is actually one of the objectives of using these estimators).

For the case of estimating the scale of the data, several robust procedures have also been proposed. In the present case, two estimators are selected based on the results of previous research [28]. The selected scale estimators are the t-estimator of scale S_t based on the t-distribution [29] with a tuning constant of 4.5, and the truncated standard deviation defined by the tau-scale τ_s [30]. The scale value estimates for the RH and UPV data obtained by these estimators are presented in Table 5.3 along with the corresponding standard deviation of each data set. As can be seen, the robust estimates and the standard deviation are all very close which leads to **conclude that the data sets do not possess values that may create a bias in the scale estimate**. Even though the UPV scale estimates appear to be less close, they represent coefficients of variation ranging between 1.8 and 2%. Therefore, any of the robust estimators could also be used without any preliminary outlier analysis to estimate the scale of the data.

Table 5.4 Robust estimates of the covariance matrix and correlation matrix compared with the corresponding non-robust estimates for the RH and UPV data

Estimator	Covariance matrix	Correlation matrix
FAST-MCD	$\begin{bmatrix} 9.43 & 123.87 \\ 123.87 & 6352.93 \end{bmatrix}$	$\begin{bmatrix} 1 & 0.51 \\ 0.51 & 1 \end{bmatrix}$
OGK	$\begin{bmatrix} 4.74 & 62.26 \\ 62.26 & 3193.36 \end{bmatrix}$	$\begin{bmatrix} 1 & 0.51 \\ 0.51 & 1 \end{bmatrix}$
OH	$\begin{bmatrix} 8.11 & 37.06 \\ 37.06 & 1453.64 \end{bmatrix}$	$\begin{bmatrix} 1 & 0.34 \\ 0.34 & 1 \end{bmatrix}$
COV	$\begin{bmatrix} 5.88 & 117.66 \\ 117.66 & 5901.21 \end{bmatrix}$	$\begin{bmatrix} 1 & 0.63 \\ 0.63 & 1 \end{bmatrix}$

5.3.3 Outlier Accommodation Techniques for Bivariate Data Sets (Situation B)

For the case of bivariate datasets, the key aspects that are usually required are either an estimate of the correlation between the two data sets or a regression function between the two variables. Even though a regression function may not be necessary for the present case, a simple analysis is nonetheless carried out herein to illustrate the performance of robust regression methods.

To obtain robust estimates of the correlation between RH and UPV, the FAST-MCD, the OGK and the OH methods are considered herein and the robust covariance estimates are compared with the non-robust covariance matrix (COV). The robust covariance estimates for the RH-UPV data are presented in Table 5.4 along with the non-robust covariance matrix. Even though these covariance estimates may seem to be very different, when they are transformed into correlation matrices (see Table 5.4), it can be seen that only the estimate based on the OH method is significantly different (it is recalled that the OH-based distances identified three outliers in the data, contrary to the other methods). Therefore, based on the closeness of most of the covariance matrices **it is found that the data sets do not possess values that may create a bias in the covariance estimate**.

As referred, to illustrate the performance of robust regression in this case, Fig. 5.11a presents the linear regression analysis of the RH-UPV data using an ordinary least-squares regression and two robust regressions, one using a logistic influence function and another using the Andrews influence function [22]. As can be seen, there is little to gain from using robust regression in this case. The reason for this outcome is connected to the reduced size of the RH-UPV data sample but also to the fact that the sample does not exhibit a clear linear trend. To complement this analysis, the performance of an alternative linear regression was also examined. Based on the results obtained from the ordinary least-squares regression, the alternative regression discards data that has a residual value (computed using the ordinary least-squares regression information) higher than 1.5 times the standard error of the

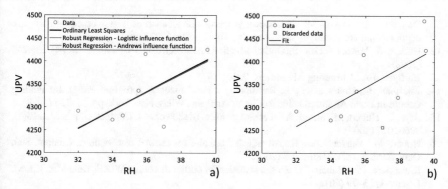

Fig. 5.11 Ordinary least squares and robust regressions (**a**) and alternative linear regression (**b**) for the bivariate RH-UPV data set

residuals. This alternative regression is presented in Fig. 11b and it can be seen that only one data point was discarded according to the referred condition. Although this new regression is not significantly different than the one obtained by ordinary least-squares regression, some advantages can be seen when analyzing the adjusted coefficients of determination. For ordinary least-squares regression, a value of 32% was obtained, while the alternative linear regression yielded a value of 46% instead.

Acknowledgements The first author would like to acknowledge the financial support by Base Funding—UIDB/04708/2020 of CONSTRUCT—Instituto de I&D em Estruturas e Construções, funded by national funds through FCT/MCTES (PIDDAC).

References

1. Iglewicz, B., Hoaglin, D.: How to Detect and Handle Outliers. ASQC Quality Press, Milwaukee (1993)
2. Barnett, V., Lewis, T.: Outliers in Statistical Data. Wiley, New York (1994)
3. Davies, L., Gather, U.: The identification of multiple outliers. J. Am. Stat. Assoc. **88**(423), 782–792 (1993)
4. Thode, H.C.: Testing for Normality. Marcel Dekker, New York (2002)
5. ISO 16269-4:2010: Statistical interpretation of data—Part 4: Detection and treatment of outliers (2010)
6. ASTM D7915.14: Standard Practice for Application of Generalized Extreme Studentized Deviate (GESD) Technique to Simultaneously Identify Multiple Outliers in a Data Set. ASTM International, West Conshohocken (2014)
7. ASTM E178-16: Standard practice for dealing with outlying observations, ASTM International, West Conshohocken (2016)
8. Aggarwal, C.C.: Outlier Analysis. Springer Publishing Company, Inc (2017)
9. Rousseeuw, P.J., Van Driessen, K.: A fast algorithm for the minimum covariance determinant estimator. Technometrics **41**(3), 212–223 (1999)
10. Tukey, J.W.: Mathematics and the picturing of data. In: Proceeding of the International Congress of Mathematicians. Vancouver, Canada, 21.29 Aug 1974, vol. 2, pp. 523–531

11. Donoho, D.L., Gasko, M.: Breakdown properties of location estimates based on halfspace depth and projected outlyingness. Ann Stat **20**(4), 1803–1827 (1992)
12. Hubert, M., Van der Veeken, S.: Outlier detection for skewed data. J. Chemom. **22**(3–4), 235.246 (2008)
13. Hartigan, J.A.: Clustering Algorithms. Wiley, 1975
14. Kaufman, L., Rousseeuw, P.J.: Partitioning around medoids (program PAM), In: Finding Groups in Data: An Introduction to Cluster Analysis. Wiley, New York, pp. 68–125 (1990)
15. Brys, G., Hubert, M., Struyf, A.: A robust measure of skewness. J. Comput. Graph. Stat. **13**(4), 996–1017 (2004)
16. Hubert, M., Vandervieren, E.: An adjusted boxplot for skewed distributions. Comput. Stat. Data. Anal. **52**(12), 5186–5201 (2008)
17. Rousseeuw, P.J., Hubert, M.: Robust statistics for outlier detection. WIREs Data Min. Knowl. Discov. **1**, 73–79 (2011)
18. Rousseeuw, P.J., Ruts, I., Tukey, J.W.: The bagplot: a bivariate boxplot. Am. Stat. **53**(4), 382–387 (1999)
19. Pison, G., Van Aelst, S., Willems, G.: Small sample corrections for LTS and MCD. Metrika **55**(1.2), 111.123 (2002)
20. Olive, D.J.: A resistant estimator of multivariate location and dispersion. Comput. Stat. Data Anal. **46**(1), 93–102 (2004)
21. Hoaglin, D.C., Mosteller, F., Tukey, J.W. (eds.) Understanding Robust and Exploratory Data Analysis. Wiley, New York (1983)
22. Hampel, F.R., Ronchetti, E.M., Rousseeuw, P.J., Stahel, W.A.: Robust Statistics: the Approach Based on Influence Functions. Wiley (1986)
23. Staudte, R.G., Sheather, S.J.: Robust estimation and testing. Wiley, New York (1990)
24. Wilcox, R.: Introduction to Robust Estimation and Hypothesis Testing, 2nd edn. Academic press, Cambridge (2005)
25. Maronna, R.A., Martin, D.R., Yohai, V.J.: Robust Statistics—Theory and Methods. Wiley, New York (2006)
26. Rousseeuw, P.J., Verboven, S.: Robust estimation in very small samples. Comput. Stat. Data Anal. **40**(4), 741.758 (2002)
27. Romão, X., Delgado, R., Costa, A.: Statistical characterization of structural demand under earthquake loading. Part 1: Robust estimation of the central value of the data. J. Earthq. Eng. **16**(5), 686–718 (2012)
28. Romão, X., Delgado, R., Costa, A.: Statistical characterization of structural demand under earthquake loading. Part 2: Robust estimation of the dispersion of the data. J. Earthq. Eng. **16**(6) (2012)
29. Randal, J.A.: A reinvestigation of robust scale estimation in finite samples. Comput. Stat. Data Anal. **52**(11), 5014–5021 (2008)
30. Yohai, V.J., Zamar, R.H.: High breakdown-point estimates of regression by means of the minimization of an efficient scale. J. Am. Stat. Assoc. **83**(402), 406–413 (1988)

Part II
Applications

Chapter 6
How Investigators Can Assess Concrete Strength with On-site Non-destructive Tests and Lessons to Draw from a Benchmark

Denys Breysse, Jean-Paul Balayssac, Samuele Biondi, Adorjan Borosnyoi, Elena Candigliota, Leonardo Chiauzzi, Vincent Garnier, Michael Grantham, Oguz Gunes, Vincenza Anna Maria Luprano, Angelo Masi, Valerio Pfister, Zoubir Mehdi Sbartaï, and Katalin Szilagyi

Abstract A benchmark is carried out in order to compare how 13 experts define and can carry out an NDT investigation program and derive strength values from NDT measurements. The benchmark is based on simulations, which reproduces a synthetic data set corresponding to a grid of twenty 3m-high columns defining the structure of

---

[1]The interested reader will find relevant information in other chapters, and we have chosen not to repeat this information here, for the sake of clarity.

D. Breysse · Z. M. Sbartaï
University Bordeaux, I2M-UMR CNRS 5295, Talence, France

J.-P. Balayssac (✉)
LMDC, Université de Toulouse, INSA/UPS Génie Civil, Toulouse, France
e-mail: jean-paul.balayssac@insa-toulouse.fr

S. Biondi
Engineering and Geology Department, inGeo, Gabriele d'Annunzio, University of Chieti-Pescara, Pescara, Italy

A. Borosnyoi
Budapest University of Technology and Economics, Presently Hilti, Budapest, Hungary

E. Candigliota · V. A. M. Luprano · V. Pfister
ENEA - Italian National Agency for New Technologies, Energy and Sustainable Economic Development, Bologna, Italy

L. Chiauzzi · A. Masi
School of Engineering, University of Basilicata, Potenza, Italy

V. Garnier
Laboratory of Mechanic and Acoustic, L.M.A - AMU/CNRS/ECM - UMR7031, Aix Marseille University, Marseille, France

M. Grantham
Consultant - Sandberg LLP, London, UK

O. Gunes
Instanbul Technical University, Istanbul, Turkey

© RILEM 2021

D. Breysse and J.-P. Balayssac (eds.), *Non-Destructive In Situ Strength Assessment of Concrete*, RILEM State-of-the-Art Reports 32,
https://doi.org/10.1007/978-3-030-64900-5_6

183

a building made up of beams and columns. The experts must assess the mean and the standard deviation of compressive strength. Three levels of assessment are considered corresponding to different quantities of test results (destructive or non destructive) available for the experts. The comparison of the various strategies used by the experts and the analysis of results enables the identification of the most influential parameters that define an investigation approach and influence its efficiency and accuracy. A special emphasis is placed on the magnitude of the measurement error. A model of the investigation strategy is also proposed.

6.1 Introduction

The idea of carrying out an international benchmark for comparing the efficiencies of various investigation strategies emerged during the meetings of RILEM TC 249-ISC Committee. It was justified by several considerations:

- we first needed to have a representative view over the possible options that can be chosen by investigators, regarding the ND techniques used, the number of test locations for ND tests and core taking, the choice of the conversion model or the method chosen for calibrating this model and many other issues,
- despite the many publications that have been devoted to the issue of ND concrete strength assessment, it was impossible to find any comparison between alternative investigation options, each publication being focused on one case-study, with a single investigation strategy,
- the statements about the influence of many factors mostly remain an open issue, as the experts are lacking any tangible database against which the options can be tested,
- some innovations have been promoted recently in the scientific literature[1] and seem to be promising, but they were still waiting for a broader validation.

The RILEM TC 249-ISC Committee therefore organized an international benchmark between experts, with the purpose of getting new information that could be shared and validated on a broader basis, and further be used for improving the technical Recommendations. It must be underlined that the RILEM TC 249-ISC Recommendations still had to be written at this time, and that some of the specific tasks that are now included in the Recommendations emerged from the analysis of this benchmark. Some results of the benchmark were published in 2017 [1]. This chapter will describe what has been done in more detail and show how it contributed to the elaboration of the RILEM TC 249-ISC Recommendations. The reader may find a double interest in this chapter as (a) it illustrates what an expert can do, and with what results, (b) it tries to draw some more generic conclusions about the added-value

K. Szilagyi
Budapest University of Technology, Presently Concrete Consultant, Budapest, Hungary

of some options/variants in the investigation strategy. This chapter will be complemented by the next one which details the results of a second benchmark, with more specific objectives.

6.1.1 Original Idea of Benchmark and Methodology

The first purpose of the benchmark is to offer a common dataset for comparing various strategies that can be used for assessing strength on existing structures. Of course, the benchmark is carried out as a blind test and each expert contribution has to remain anonymous. A letter (or a number) is assigned to each participant and all results are analyzed on an anonymous basis.

The benchmark is not carried out on a real structure, but on a synthetic structure, i.e. on a synthetic data set. This means that all data are generated within the computer but following rules that make them similar in many patterns to real data. Three reasons justify this fundamental choice:

- getting synthetic data is by far much less expensive that getting real data from a real structure,
- generating the structure in silico enables to generate both "true values" (i.e. material properties which correspond to the reference strength) and "test results" that may differ from true values because of test result uncertainties. This point is crucial. The RILEM TC 249-ISC committee wanted to address the model error issue, which requires to quantify the distance between the estimated value and the true one for the same material property. In silico simulation makes this possible, whereas it remains impossible with a real structure in which the "true values" always remain unknown,
- thanks to computer simulations, the conclusions that will be drawn at the end of the benchmark can be tested in a different case, by simply changing some input parameters in the simulations (these can be parameters describing the structure and the material as well as parameters defining the investigation program). This was done after the conclusion of the benchmark in order to check some of the conclusions that had been drawn [2].

As for the similarity between real and synthetic data, it will be better discussed at Sect. 1.2 that using synthetic data works provided that there is a physically sound relationship between (generated) "true values" and related "test results". The synthetic data set was generated once and was common for all participants, but it remained unknown for them until the last stage of result analysis that was carried out openly. All benchmark participants had to follow the same game rules, that is:

(a) they had to define what investigation program they wanted to follow,
(b) they therefore received the data (synthetic "test results") corresponding to this investigation program,

(c) they analyzed the test results, and by applying their own methodology, they derived the assessed properties. These assessment results were communicated only to the benchmark organizer.

The stage of shared analysis consisted in comparing on one hand the various investigation programs and the ways of processing the test results and on the other hand the final assessments. Finally, these assessments could also be compared with the true (reference) synthetic values that were made public only at this stage. These discussions helped us to identify: (a) what can be the alternatives for getting data, (b) what can be the alternative for analyzing data, (c) the benefits and drawbacks of some options and to draw some more general conclusion in order to feed the RILEM Recommendations.

6.1.2 Synthetic Simulations for Assessing Strategies

The two main advantages of the synthetic simulations are that they can reproduce the main patterns of concrete strength variations in a real structure as well as the NDT measurement process, and that they can be used to assess the efficiency of a given evaluation strategy. In the following, an evaluation strategy will be defined as, a given amount of resources being known, the choice of the type, number and precision of tests (both destructive and non-destructive) and the use of any method/model for deriving strength estimates from the test results. This includes the model identification at the calibration stage.

The ability of synthetic simulations to offer a reliable copy of reality is justified by the fact that:

- strength variation patterns in a structure are governed by a number of factors that are widely documented in the literature and can be reproduced in the computer. Among these factors, one can cite: natural variability of concrete, variability induced by the casting process, variability due to carbonation or other phenomena (mechanical damage, humidity …);
- NDT measurements give access to physical properties (UPV velocity, rebound number, pull-out force …) that are all correlated with strength but also affected by (a) measurement errors due to the device, the operator or the environment, (b) additional factors that can modify the reading—and possibly the concrete strength—like moisture content, carbonation, chloride content, damage …
- relationships between strength, influent factors and NDT properties have widely been documented in the literature and reference relationships can be used for simulations.

In addition, unlike what happens in reality, it is possible with synthetic simulations to estimate the precision of the strength assessment. This is explained in Fig. 6.1, where two domains are defined: the upper part of the flowchart refers to the synthetic structure while the lower part refers to the strength assessment process.

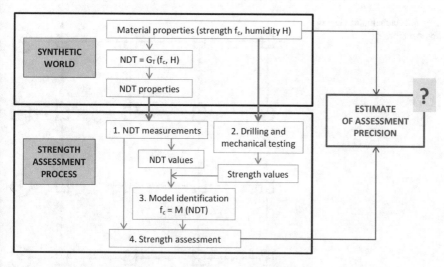

Fig. 6.1 Illustration of the use of synthetic simulation for strength assessment

The synthetic world contains the relationships between strength, influent factors and NDT properties. When a simulation is carried out, a data set is generated according to the rules governing this world. These data/properties are accessible through tests, either with NDT (1) or drilling and coring (2). Thus, ND test results and core strength values (including possible measurement errors) are made available and strength estimates can be done (3), either by identifying a specific model or by using a yet existing model (4). At the end of the process, it is also possible to compare the estimated strength values with the "real" ones (i.e. synthetic strength generated in the synthetic world), and to quantify how far estimates are from truth. Of course, this is never possible with real structures and is a major advantage when one wants to address the issue of model errors. The simulation process can be used either once, for generating a synthetic data set and "play the assessment game" or it can be repeated many times, in order to carry out statistical analysis and provide additional information (see Chap. 8).

6.2   Presentation of the Benchmark: Case Study and Rules to Be Applied

The information given in this section was given to all participants before the beginning of the benchmark.

Fig. 6.2 Schematic view of
the structure

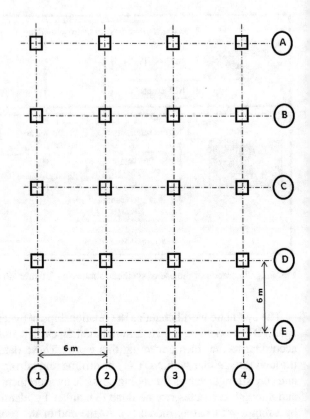

6.2.1 Case Study

The synthetic structure is a one-level structure, made of a column-beam frame in
which only columns are considered (Fig. 6.2). The building shape is rectangular,
with 20 columns in five series (from A to E) of four columns (from 1 to 4), with
a regular spacing of 6 m and a height of 3 m. The cross-section of the columns
is 35×35 cm^2. The building is un-heated in a temperate climate. From visual
inspection, a vertical gradient of properties may be expected in the columns, and a
humidity gradient may be suspected between the extreme areas of the structure. The
carbonation effect is neglected.

It is expected from all benchmark participants to:

- define three "levels of investigation"[2], respectively corresponding to increasing
  amounts of resources that are mobilized for getting data,

---

[2]The three "levels of investigation" can be compared with the three "knowledge levels" defined by
Eurocode-8. They were a first version of the Estimation Quality Level (EQL) concept as it is now
defined in the TC 249-ISC RILEM Recommendations (see Sect. 1.1 and Table 1.1). However, the
EQL concept is attached to the assessment results whereas the "levels of investigation" are attached
to the resources that enable to reach the objective.

- receive the test results they asked and paid for and
- to answer few questions.

The test results obtained with various destructive and non-destructive tests are all affected by some measurement uncertainties (like it happens with any on-site test).

Each participant can play three times, corresponding to the three KL (from KL1 - poor knowledge, to KL3—extended knowledge). He thus receives three data sets, the second and third ones being sent as soon as he has carried out the analysis at the previous stage and sent his/her assessment results.

6.2.2 Resources and Cost

All investigation costs are defined in Cost Units (CU). Each type of test has a cost which depends on the test type and on the precision of the test (higher precision increases the cost but reduces the uncertainty of measurements—see Tables 6.1 and 6.2. Thus, each participant must choose:

- What ND technique(s) is (are) used (single or in combination)
- At what locations (pile, elevation, side) the tests are carried out
- What is the precision of the technique he uses

He must combine the number, type and precision of techniques in order to respect the resource he has for each knowledge level. The available techniques are (the type of device is fixed):

Table 6.1 Increase of cost and reduction of uncertainties, valid for all techniques (the uncertainty for poor precision test results is twice that with medium precision test results)

	High precision	Medium precision	Poor precision
Cost (CU)	X1.40	1	X0.7
Magnitude of test result uncertainty	0.5	1	2

Table 6.2 Unit costs of all tests

		High precision	Medium precision	Poor precision
Drilling and compression test	One core	14	10	7
Pull-out test	One test	5.6	4	2.8
US velocity	One test result, average of 2 readings	2.8	2	1.4
Rebound	One test result, average of 10 readings	1.4	1	0.7

- coring and mechanical tests (the cost covers both drilling, conservation and mechanical compression tests), which provide a compressive strength in MPa,
- NDT1: ultrasonic tests (direct from one side of the column to the opposite side), which provide a velocity in m/s,
- NDT2: rebound hammer tests (can be performed on any of the four sides of the column), which provide a rebound number,
- SD1: pull out tests (Capo), which provide a load in kN.

6.2.3 Resources Available

The total amount of resource available varies with the KL value. It is respectively 20 CU for KL1, 40 CU for KL2 and 60 for KL3. Each participant is fully free to share this resource at his/her convenience between on-site NDT, cores and NDT on cores. Any combination of tests which respects the total amount is possible.

6.2.4 Cores and Conservation

The location of cores can be defined either:

- Option P: the location being Predefined before beginning the investigation or
- Option C: Conditional, i.e. that it is defined only after a first series of ND tests, on the basis of the analysis of ND test results. In this case, the participant must explicit how the location is chosen (what column, what elevation).

In case one takes some cores, he can choose the conservation method (how to keep the cores from the time of drilling to that of mechanical tests. Three possibilities exist, with no impact on cost:

- C1: cores are kept at 20 °C in the air in the laboratory until mechanical tests (duration about 28 days),
- C2: cores are immersed in water during 28 days at 20 °C and superficially dried before testing,
- C3: cores are sealed in a tight envelope, in order to maintain their internal humidity and kept at 20 °C until testing.

It is also possible to perform non-destructive tests (NDT1 and NDT2) on cores just before the destructive compression tests. The cost and precision of ND test is the same than for on-site tests.

6.2.5   What is Expected from Each Benchmark Participant

For each data set (and thus for each KL), two quantitative values are expected:

- The estimate of the average value of strength on the whole structure
- The estimate of the minimum local strength that could be reached on the whole structure

Of course, it is also expected that each participant explains how (method, assumptions, model …) the estimated strengths have been derived from the test result data set.

6.3   Generation of Synthetic Data

6.3.1   How Simulation Works

More information is given here to the reader about what is done inside the "SYNTHETIC WORLD" box of Fig. 6.1. This information was not available to benchmark participants before the benchmark.

It is assumed that the same concrete mix is used for all columns but that, due to the casting process, three levels of variability are considered:

- A batch-to-batch variability, corresponding to a difference between various series of columns,
- A within-batch variability, corresponding to differences between columns of the same batch,
- A within column variability, with a degree of compaction that can vary according to the elevation.

Only one influent factor is considered here, that is moisture content. It is considered as varying between columns depending on their (X, Y) location, and constant along elevation.

In a given batch, the concrete strength f_c is generated by assuming a Gaussian distribution $N(f_{cm}, s(f_c))$.

Synthetic values for the velocity V (ultrasonic pulse velocity technique), the rebound number R (rebound hammer technique) and pull-put force F are produced using relationships established after an in-depth literature review, containing both original data sets and already existing empirical laws describing observed relationships. In these relationships, concrete strength and humidity are the two leading variables, from which NDT properties are derived.

Like it is the case in real world, measurement errors are added to the synthetic values of V, R and F whose magnitudes depend on the precision of measurements. These errors are assumed to have Gaussian distribution $N(0, s)$, where s denotes the standard deviation, which depends on the type of NDT ($s(R)$, $s(V)$ and $s(F)$

Table 6.3 Values of standard deviation of ND test results for different precisions of tests

Test results precision level Q	s(V), m/s	s(R)	s(F), kN
High	50	1	0.5
Medium	100	2	1
Poor	200	4	2

for rebound, UPV and pull-out tests respectively). In Table 6.3 the values of standard deviations for different precision levels are given. It must be pointed here that choosing a lower precision test result (i.e. pay less for the result) leads to a larger variability of the reading, whose consequences can be prejudicial at further stages. In practice, at each test location, a random standard error is added to the theoretical value and the measured value is the result:

$$NDT_m = NDT_{true}(f_c, \ H) + \varepsilon_{NDT}(Q))$$

where NDT_m corresponds to the measured NDT value (i.e. test result), NDT_{true} denotes the relationships assumed between leading parameters (strength, humidity) and NDT parameters, and where ε_{NDT} is the test result error, which is Gaussian distributed and whose magnitude depends on the precision Q and on the standard deviation s, according to Table 6.3.

With twenty ($= 4 \times 5$) columns of 3 m-height and possible test locations with a 10 cm-spacing, the number of possible test locations is $NT = 620 = 4 \times 5 \times 31$. The computer thus generates a series of NT 4-uplets (V, R, F, f_c). For deriving strength values from the simulated strength value, it is assumed that coring does not induce any bias or noise, and that the change in strength only depends on conservation conditions, since the compression test can be carried out in three conditions: saturated, as on site, air-dried. In all cases, the "reference strength" will remain available for comparison with the estimated strength provided.

6.3.2 Analysis of Synthetic Data

Before going further, it must be noted that the synthetic data set is identical for all participant BUT that the data each participant gets can be not identical, for two reasons:

(a) different choices regarding the investigation program: type of techniques, location of tests, number and precision of tests
(b) the effect of chance: even with the same type, number and precision of tests, because of random test result errors, two participants that get values at two very close points can get values which are different.

Thus, it will be possible to establish some statements about the efficiency of the strategies from the analysis of assessments, but it will not be possible to compare

Fig. 6.3 Cumulative distribution of strength in the whole structure (620 test locations)

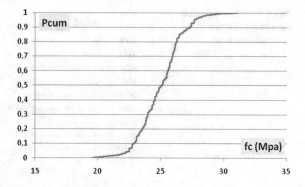

Fig. 6.4 Rebound—strength correlation (synthetic reference values)

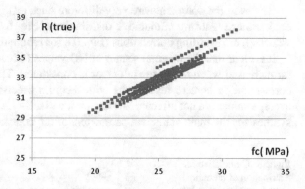

directly two strategies. The data that are processed are only the result of A SINGLE SIMULATION (one building, one data set) whereas, to be effective, a comparison must consider the variability of the assessment, when the same strategy is repeated many times.[3]

The following series of figures illustrates what has been simulated in the benchmark and some important issues that have to be considered for assessment. Figure 6.3 provides the cumulative distribution of strength, while Figs. 6.4 and 6.5 respectively illustrate the existing correlations between strength and rebound values (Fig. 6.4) and between rebound and velocity values (Fig. 6.5). The scatter of properties (resulting from the variability at three scales) is clearly visible on Fig. 6.4. Figures 6.4 and 6.5 highlight that the strength-rebound and velocity-rebound relationships are not unique. The reason is that an underlying influent factor (here humidity) is variable, which prevents to establish a unique relationship. However, on each figure the set of points shows that a relationship could be established at the scale of each column.

---

[3]This limitation is not inherent to synthetic simulations and the same limitation exists with real on-site studies. If two strategies had to be compared, because the effect of chance, they would have to be repeated several times in similar situations and their statistical fluctuations would have to be analyzed (for instance, as an indicator of robustness of the methodology).

Fig. 6.5 UPV—rebound correlation (synthetic reference values)

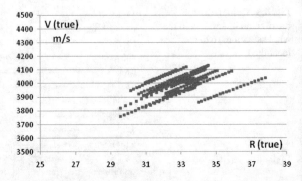

However, the comments are very different once test result uncertainty is considered, either regarding cumulative distributions (Fig. 6.6 for rebound, Fig. 6.7 for velocity), or regarding correlations (Fig. 6.8 for rebound, Fig. 6.9 for velocity).

It is clear that the original scatter of ND values (already visible in Fig. 6.5) is by far increased when the test result error is considered. The cumulative distributions, corresponding here to test results with medium precision, show that, even if the average values are not affected, the apparent variability is highly increased. Table 6.4 synthesizes the main statistical parameters of the concrete properties. The standard

Fig. 6.6 Cumulative distribution of rebound: synthetic reference values and medium precision test results

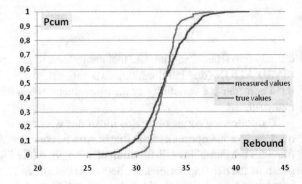

Fig. 6.7 Cumulative distribution of ultrasonic velocities: synthetic reference values and medium precision test results

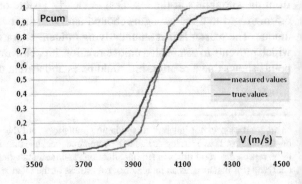

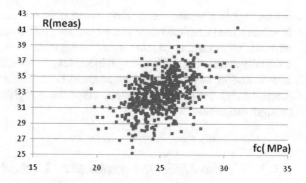

Fig. 6.8 Rebound—strength correlation (medium precision test results)

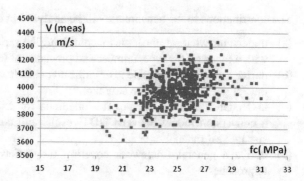

Fig. 6.9 UPV—rebound correlation (medium precision test results)

Table 6.4 Statistical parameters for strength, rebound and velocity (reference and measured values, with medium precision test results)

	Strength (MPa)	Rebound value		Velocity (m/s)	
		Reference	Measured	Reference	Measured
Average	25.0	32.9	32.8	3990	3985
Median	25.1	32.9	32.8	3994	3979
Standard deviation	1.7	1.2	2.3	63	116
5% percentile	22.5	31.2	28.9	3875	3788

deviation of NDT results is about twice that of synthetic values, and the lower 5%-percentile is significantly lowered.

Figures 6.8 and 6.9 can be compared with Figs. 6.4 and 6.5, the only difference being that true ND reference properties are replaced with NDT results. The first point is that while the correlations were very strong for reference values, they are much less visible with test results. If a positive correlation still can be identified, the width of the cloud of points means that this correlation is much less reliable to directly estimate the strength value from the NDT results. The second very important point is that, due to a weaker correlation, when only few cores are taken (i.e. only few fc values), the statistical relationship between NDT results and strength can be very unstable, even leading in some cases to negative correlation (i.e. NDT parameter unusually decreasing while strength increases).[4] This effect of test result uncertainty is well known in practice and highlighted here. This explains why the test result precision and the sampling strategy can have a high impact on the efficiency of strength estimation.

6.3.3 How to Model the Assessment Methodology

The contributions to the benchmark will be analyzed in order to:

(a) understand what was the logic for combining the various types of measurements, with a focus on the type, number and precision of test results,
(b) derive a general model able to describe all possible strategies that can be carried out for getting data for the structure,
(c) understand what are all possible methods for using/establishing/calibrating a conversion model between NDT results and concrete strength and derive strength estimate,
(d) derive a general framework covering the variety of possibilities for such approaches.

Finally, the benchmark contributions will be analyzed regarding the strength estimates they have provided, while pointing the fact that it will not be possible to derive any general conclusion at this stage (i.e. from the analysis of a single realization of a Monte-Carlo process).

6.4 Feedback from the Benchmark Contributions—Defining Strategies

6.4.1 Participants

A total of thirteen TC members were involved in this benchmark, either alone or grouped by two. Some of them proposed variants in their approaches. It finally resulted in 39 answers that can be shared into 11, 11 and 10 contributions (and 14, 13, 12 variants) for KL1, KL2 and KL3 respectively. The three knowledge levels (KL1, KL2, KL2) respectively corresponded to increasing resources (20, 40 and 60

---

[4]This fact is not inherent to synthetic simulations and can be found identically with real on-site studies. It has been identified as a main source of uncertainty at the conversion model identification stage, as it induces the trade-off between model parameters (Sect. 12.4).

Table 6.5 Number of variants for each type of approach

	KL1	KL2	KL3
Without cores	4	1	0
Cores + V	3	1	2
Cores + R	3	5	3
Cores + combined	4	6	7
Total	14	13	12
NDT on cores	2	5	4
Conditional coring	2	2	2

cost units), progressively offering a larger variety of possibilities in the investigation strategy.

For keeping anonymity, the contributions were coded with letters, from A to K, according to the increasing percentage of NDTs in cost. Numbers are used after the letter in the code for describing the variants. Table 6.5 synthesizes how participants have chosen to combine destructive and non-destructive tests at the three KL. The last two lines indicate when NDTs were also performed on cores and when core location has been specified from a previous NDT series of test results (i.e. conditional coring).

Because of the limited amount of resources, assessment without core is common at KL1 (4 cases out of 14) but becomes marginal for KL2. The percentage of resources devoted to NDTs varies a lot according to the participants and KL, between 25 and 100% (four participants have chosen to take no cores at KL1 and to assess concrete only from NDTs). In average, cores and NDTs weigh about half of the resources whatever KL. When increasing the level of knowledge, the possibility of combining two ND techniques is more frequent and becomes dominant. Figure 6.10 synthesizes for all contributions the relative weight of NDTs in terms of cost. Table 6.6 provides an additional information for the 12 variants at KL3 level. For each contribution, it points how the 60 cost units are shared among cores and NDTs, either on site or on cores. The second column gives what percentage of cost comes from cores and a color code indicates the precision of test results, from poor (orange) to high (green). It must be pointed that, the total cost being fixed, choosing a higher precision technique reduces the possible number of tests, inducing a necessary trade-off between quantity and precision of data. This is illustrated by the fact that the lower part of Table 6.6 corresponds to strategies where the cost of cores is the larger, then reducing the amount available for NDT, thus tending to decrease the precision of test results, in order to maintain the total cost.

6.4.2 Analysis and Modelling of the Investigation Strategies

After having analyzed the feedback from all participants, it is possible to derive a simplified model describing what could be called an "investigation strategy".

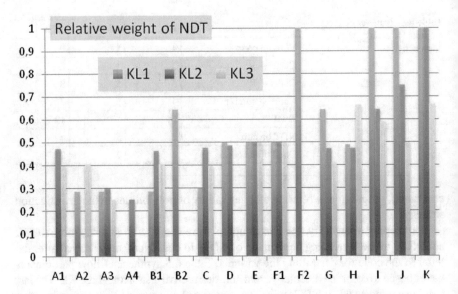

Fig. 6.10 Relative weight (in cost) of NDT for all participants and all knowledge levels

Table 6.6 Distribution of costs at KL3 for the 12 contributions (color indicates the test result precision: orange = poor, yellow = medium, green = high)

	% cores	US	R	pull out	core fixed	core dependent	US on core	R on core
K	33	20	4	16	20			
H	34	28	11,6			20		
I	41	8,8	12,2	13,6	24			
J	50		29		30			1
F1	50	14	10		30		4	2
E	50	12	12		30		5,6	
C	58		25		35			
G	58	15,4	7		35		1,4	0,7
A1	60	15,4	7,7		35			
B1	60	23,8				35		
A2	60	23,8			35			
A3	75		14		42			

The total amount of resources S being defined, this model describes a step-by-step approach, as shown in the flowchart of Fig. 6.11:

- defining [1] what part of the available resources is devoted to destructive tests (cores—DT) and to NDT. If α is the part of resources for NDT, one gets αS for NDT and $(1 - \alpha) \times S$ for DT;
- if some cores have to be taken, define ([a] vs. [b]) if the location of cores is defined after an analysis of NDT results or not (i.e. conditional coring or predefined coring);

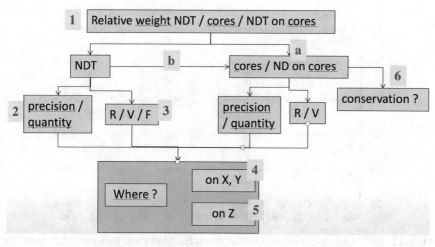

Fig. 6.11 Model for defining an investigation strategy

- if some NDT measurements have to be planned, either on the structure or on the drilled cores, one needs to know how the available resource αS is used. The choice is on three factors:
- choosing the type of tests: rebound R, ultrasonic velocities V or pull-out F,
- choosing the precision of tests[5]: poor / medium / high,
- calculating the number of tests.

If one denotes respectively CU_i and n_i the unit cost of the test of type T_i (i $= 1$ for R, i $= 2$ for UPV and i $= 3$ for F), the cost constraint writes:

$$\sum (n_i CU_i) \leq \alpha S$$

This choice can be seen as a 2-step decision: (a) choosing the part of resources dedicated to each type of NDT, (b) choosing the precision level, which leads to the number of tests. This choice corresponds to [2] and [3] in Fig. 6.12 (respectively [2′] and [3′] for NDT on cores, where pull out is meaningless).

Of course, this decision could follow another scheme, like choosing a minimal number of tests, which thus would imply constraints on type and precision.[6]

---

[5]If one considers real situations, the choice of test precision is usually not explicit. However, it is implicit if one considers that a better precision comes, for instance, to follow standards or guidelines, or to work with expert practitioners ... In both cases, a precision increase induces a higher cost and a smaller number of tests. In all cases (real structure or synthetic benchmark), the precision is negatively correlated with the test result uncertainty.

[6]In real practice, many other factors play a role, like the prior information already available or collected through visual inspection, time constraints, technical constraints... Our purpose here is not to model the process in its overall complexity.

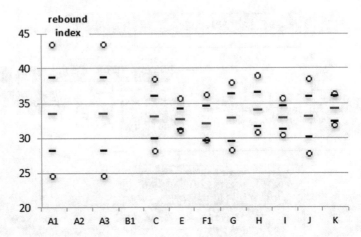

Fig. 6.12 Statistical properties of rebound test results for the different strategies (mean = red bar, mean ± standard deviation = black bars, minimum and maximum = circles)

- Choosing the specific location (X, Y, Z) for each test ([4] and [5]). This choice can correspond to different scenarios, which are not discussed at this step. One can make the difference between a more or less systematic spatial distribution or some kind of random distribution.
- If cores are to be taken, the last decision regards the conservation conditions [6], and especially the target moisture condition at the time of compression tests.

It is easy to understand that this series of decisions quickly leads to a combinatory set of solutions and that the number of possible alternatives may be very large, especially when the amount of resources increases. Regarding the first two steps, with the simplifying assumption that the precision level of the two NDT techniques is identical, one has a very large number of possibilities at KL1 (resource available = 20 CU), as illustrated on and Table 6.8, in which the strategies corresponding to some contributions have been pointed. For instance, if we consider a medium precision of cores (Table 6.7), the choice is between taking no core, 1 cores or 2 cores. The last option prevents any NDT as each core costs 10 CU. One can take the example of Strategy F2 which shares the resources (20 CU) between 1 core (10 CU), 3 UPV tests (6 CU) and 4 rebound tests (4 CU). Table 6.9 indicates how many combinations can exist with poor precision cores, with for instance 9 possibilities for cores and velocity tests, 9 possibilities for cores and rebound tests and 50 possibilities with cores and a mix of velocity and rebound tests. Table 6.8 looks like Table 6.7 but considers poor precision of cores. Thus, taking 2 cores lets some extra resources for carrying out some NDTs. The large number of possible combinations can be even larger if the precision level is different for the two ND techniques. It also increases when the amount of available resources increases, i.e. at KL2 and KL3. Any systematical analysis of the strategy would therefore be based on numerical simulations in order to process the huge amount of data.

Table 6.7 All possibilities offered for tests, medium precision of cores (KL1)

	Ncores	NUPV			Nrebound		
		P	M	H	P	M	H
UPV only	0	14	10	7			
R only	0				28	20	14
combined	0	13	9	6	2	2	2
	0	12	8	5	4	4	4
	0	11	7	4	6	6	6
	0	10	6	3	8	8	8
	0	9	5	2	10	10	10
	0	8	4	1	12	12	12
	0	7	3		14	14	
	0	6	2		16	16	
	0	5	1		18	18	
	0	4			20		
	0	3			22		
	0	2			24		
	0	1			26		
UPV only	1	7	5	3			
R only	1				14	10	7
combined	1	6	4	2	2	2	3
	1	5	3	1	4	4	5
	1	4	2		6	6	
	1	3	1		8	8	
	1	2			10		
	1	1			12		
UPV only	2	0	0	0			
R only	2				0	0	0
combined	2	0	0	0	0	0	0
	2	0	0	0	0	0	0
	2	0	0	0	0	0	0
	2	0	0	0	0	0	0
	2	0	0	0	0	0	0
	2	0	0	0	0	0	0
	2	0	0	0	0	0	0
	2	0	0	0	0	0	0

Callouts: J, K, F1, D, F2, E

Table 6.8 Number of possibilities of combinations—with poor precision of cores

	Cores + V	Cores + R	Cores + R + V
0 core	3	3	28
1 core	3	3	16
2 cores	3	3	6

Table 6.9 All possibilities offered for tests, poor precision of cores (KL1)

	Ncores	NUPV			Nrebound		
		P	M	H	P	M	H
UPV only	0	14	10	7			
R only	0				28	20	14
	0	13	9	6	2	2	2
	0	12	8	5	4	4	4
	0	11	7	4	6	6	6
	0	10	6	3	8	8	8
	0	9	5	2	10	10	10
	0	8	4	1	12	12	12
combined	0	7	3		14	14	
	0	6	2		16	16	
	0	5	1		18	18	
	0	4			20		
	0	3			22		
	0	2	B1		24		
	0	1			26		
UPV only	1	9	6	4			
R only	1				18	13	9
	1	8	5	3	2	3	3
	1	7	4	2	4	5	5
	1	6	3	1	6	7	7
combined	1	5	2	G	8	9	
	1	4	1		10	11	
	1	3			12		
	1	2			14		
	1	1	B2, A2		16	A1	C
UPV only	2	4	3	2			
R only	2				8	6	4
	2	3	2	1	2	2	2
combined	2	2	1		4	4	
	2	1			6		

6.4.3 Short Note About Precision and Representativeness of Test Results

The three following figures synthesize which type of data (i.e. synthetic test results) the investigators have worked with, for rebound, velocity and core strength respectively. For each investigator, these results are those of KL3 level, with average ± standard deviation and minimum/maximum. The alphabetic denomination of strategies is such that first letters correspond to strategies that give priority to quantity where last letters give priority to precision of test results. Strategies A1, A2, A3, B1 and G correspond to poor precision test results (see Table 6.6). The standard deviation is not provided on Fig. 6.14 as the number of cores varies between only 2 (H, I, K) and 5 (A1, B1, C and G).

Choosing poor precision test results in order to increase the dataset size (either the number of cores or the number of NDT results) clearly induces a larger scatter of the experimental data, as visible on Fig. 6.12 for A1 and A3, on Fig. 6.13 for A1, A2, B1 and G or on Fig. 6.14 for B1. This may be a problem at the stage when these data are used for identifying the conversion model. It can be noted that F1 took three cores which all provided very close strength values. This effect of chance can also

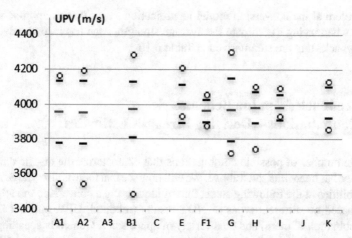

Fig. 6.13 Statistical properties of UPV test results for the different strategies (mean = red bar, mean ± standard deviation = black bars, minimum and maximum = circles)

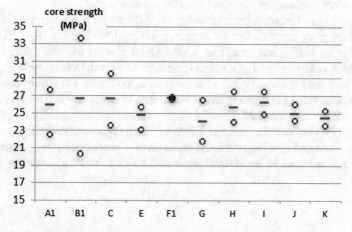

Fig. 6.14 Statistical properties of core strength test results for the different strategies (mean = red bar, minimum and maximum = circles)

Table 6.10 Respective advantages and drawbacks of privileging number against precision

	Advantages	Drawbacks
Large number/poor precision	• Better coverage of the structure (spatial variation) • Location of extrema	• Problems for model fitting (uncertainty, influence of errors) • Overestimation of the strength standard deviation
Small number/high precision	• More accurate test results • Possibility to estimate the standard deviation	• Problems for model identification (low number of calibration data)

be a problem at the conversion model identification stage. If both approaches may be valid when trying to estimate the average strength, each has its own advantages and drawbacks that are summarized at Table 6.10.

6.5 Feedback from the Benchmark Contributions—Deriving Strength Estimates

The large number of possible combinations that characterize the on-site data gathering stage has been illustrated above. We will now see that there is also a large number of possibilities at the following stage, that of identifying a conversion model, which can be based on different types of NDT results (rebound, UPV, pull-out force or any combination of them) and on a variety of approaches. Generally speaking, three types of approaches can be used in order to assess strength values from NDT test results: (a) use a pre-defined model, without any calibration, (b) use a pre-defined model after calibration, (c) build a specific model. The calibration option is only possible when some reference core strengths are available.

Table 6.11 synthesizes what approaches the different participants have chosen. In each cell are given the participant reference, the type of NDT results (R, V, F) and the knowledge level. If a participant has modified his/her approach with the knowledge level (which is the case for B, G, K and J, the approach for the lower levels is given in italics. Since the results provided by models with NDT were not always considered as conclusive, some participants have also chosen to direct assess strength from the core strength results they had.

In the following section the main features of each approach are outlined, pointing out differences and specificities. Full details are not provided, but only some information as it helps us to understand what is done or to point at issues of specific interest.

Table 6.11 Synthesis of approaches chosen by the participants

Only cores	Prior model, without calibration	Prior model and calibration	Specific model
	[F] R and V without combination		
	[J]no core at KL1, R	[J] (KL2 – KL3) cal on 1 or 3 cores, R	
		[C] R	
		[E] R + V	
		[H] R, V, R + V (lacks details)	
			[A] (R, V, R + V) not always conclusive
[G](KL1 and KL2), R	[G](KL1 and KL2), R and V		[G] KL3, R and V
	[K]KL1, V		[K] KL3, V (F and R unconclusive)
[B]KL1, V			[B] KL1 (unconclusive), KL2 and KL3, V
	[I] progressive assessment, using at different levels F, US and R		

6.5.1   Approaches Using a Prior Model Without Calibration

• *Participant F*

This participant mixes V and R tests. For each NDT technique, a prior model is used, which provides local f_c estimate:

$$f_c = 0.0228V - 65.1$$
$$f_c = 1.76\,R - 35.4$$

Thus, the strength can be estimated at each point where a NDT result is available. Three cores are also taken but not used for calibration. The strength assessment combines all estimations, whose values are respectively for the mean and standard deviation at KL3:

• 24.5 and 1.8 MPa if assessed from UPV (9 test results)
• 21.2 and 4.2 MPa if assessed from rebound (12 test results)
• 26.7 and 0.2 MPa if assessed from cores (3 cores)
• 23.1 and 3.7 MPa from all estimates (24 estimates).

Approaches using a calibrated model.

Four participants (C, E, H, J) used this approach, with the same calibration method. The mean value of prior local strength estimates is first calculated from N strength estimates with a prior conversion model.

$$f_{c\,est\,m} = 1/N \sum f_{c\,est\,i}$$

Then a calibration factor k is calculated:

$$k = f_{c\,m}/f_{c\,est\,m}$$

where f_{cm} is the average of strength values obtained on cores. Finally, the model is updated, the k multiplying factor being added to the prior model. Two variants exist, considering respectively that N is the number of NDT locations, or the number of cores.

- *Participant J*

The approach is based on an extensive use of rebound and the use of a prior model which is derived from an in-depth literature review:

$$f_c = 0.013R^{2.2271}$$

No core is taken at KL1, and 1 core (KL2) or 3 cores (KL3) are used for calibrating the multiplying factor at other levels. Having a large number of low cost NDTs, it is possible to obtain a large set of estimated strengths. The estimated mean strength at KL3 is 22.7 MPa and the estimated strength standard deviation is 4.4 MPa (from 29 rebound values). Thanks to the large number of NDT results, which were provided by choosing test locations at three different elevations in columns, it is also possible to analyze possible vertical variations in strength estimates. However, due to the large scatter of test results (Fig. 6.15 with a standard deviation on rebound about 3) no significant trend can be identified.

Fig. 6.15 Variation of rebound index with elevation

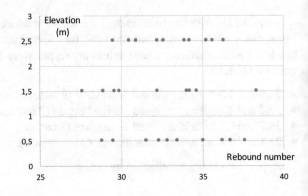

- *Participant C*

The approach is based on the exclusive use of rebound, with a prior model which is calibrated from core strength. Respectively 2, 3 and 5 (poor precision) cores are used at the three knowledge levels. The prior model, derived from Gaede-Schmidt writes:

$$f_c = k(R^2 - 30.5R + 447.8)$$

where k is calibrated from the average values provided by cores. The assessed values from 25 rebound test results for the mean strength and strength standard deviation at KL3 are 26.4 and 5.5 MPa.

- *Participant E*

The approach is based on the use of a combined rebound and UPV test results under the form of a double-power law:

$$f_c = kV^{2.6}R^{1.3}$$

where k is calibrated from the average values provided by cores. Respectively 1, 2 and 3 (medium precision) cores are used at the three knowledge levels. At KL3, the dataset is made of 3 cores, 12 rebound test results and 6 UPV test results, as the last ones were more expensive. Thus, the information at test locations where just a rebound test result is available is used by assuming that the V value at this same location is equal to the mean value of velocity measured on the 6 test locations.

Strength estimates are derived at 12 test locations, which finally lead to an assessed mean strength of 25.6 MPa and a standard deviation equal at 2.3 MPa.

- *Participant H*

The investigation strategy gave priority to V tests (14 tests at KL3) against R tests (10 tests at KL3) and cores (only 2 cores). The location of cores is conditional, i.e. defined after having analyzed the NDT results. The identification approach is based on the use of prior mono-variate (R or V) or a bivariate (R + V) models which are calibrated from core data. In some cases, the identified model is rejected because the exponent value is considered as being unrealistic (typically, being negative). The bi-objective approach is considered, which aims at better capturing the concrete variability. The strength assessment for the mean and standard deviation at KL3 gives respectively 25.8 and 3.0 MPa with the bivariate model.

6.5.2 Approaches Developing a Specific Model

- *Participant A*

This participant has developed three answers in parallel, which respectively gave more weight to V tests or to rebound tests or which had a well-balanced use of these two ND methods. In all cases, the conversion model identification approach consists in fitting a regression model with data provided by cores. Because of regression, a minimum of two cores is required for a mono-variate model and three cores are needed for a bivariate model. Therefore, the bivariate model can be used only at KL2 and KL3. Another consequence is that poor precision NDTs have been privileged, because of cost, thus affecting test result uncertainties. The model shape is not specified and can change depending on the data features (linear, power ...). A new conversion model is calibrated each time new data are considered. At KL3, the combined assessment is based on 11 pairs of V and R values and 5 poor precision cores. The proposed model writes:

$$f_c = 10^{1.86} V^{-0.09} R^{-0.07}$$

The strength properties estimated at KL3 with this model are respectively 26.4 MPa for the mean strength and 1.5 MPa for the strength standard deviation.

- *Participant G*

This approach is based on the combined used of R and V test results with a bi-variate conversion model, that of «nomograms» proposed by former RILEM TC43 (Fig. 6.16). At each KL, for points at which the two NDT results are available together (respectively 5, 7 and 9 at the three KL), a strength is estimated on the nomogram, which is compared to strengths measured on cores (respectively 1, 3 and 5 at the three KL). With increasing KL, the tendency is to increase the relative core/NDT ratio.

Then the estimated strength is based on an «expert» (not explicit) comparison of the two estimates (with NDT or direct). Depending on the results, either core

Fig. 6.16 SonReb nomogram from RILEM [3]

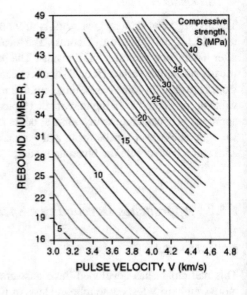

strengths only or the strength estimated with the conversion model are considered as relevant. The estimated strength properties at KL3 are respectively 22.7 MPa for mean strength and 2.0 MPa for the strength standard deviation.

- *Participant K*

This participant has chosen, at the first two KL, to work with NDT results only and to estimate strengths from uncalibrated prior models. For instance, at KL2 level, the test result data set is made of 10 UPV values, 4 rebound values and one has 4 pull-out force values, the latter being assumed as an alternative way of estimating the local strength with a semi-destructive test, without taking cores. Models are used separately for R and V, and the strength estimates obtained at points were both NDTs have been performed are used in a «data fusion process» [4] which considers their overall consistency. The principle of data fusion is to, at the same time, calibrate all the different conversion models (for the different ND used), on all test locations, by maximizing the consistency between the assessment obtained by all techniques independently.[7]

The range of estimated strengths appears to be very far from real values. A reason is that the process maximizes the internal consistency but lacks any external reference like a core strength. It is only at KL3 that two cores are taken, and that the calibration becomes possible. The estimated strength properties at KL3 are respectively 24.9 MPa for mean strength and 1.9 MPa for the strength standard deviation.

- *Participant B*

The approach is based on an extensive use of V test results (respectively 4, 13 and 17 at the three KL), which are calibrated against core strengths (respectively 2, 3 and 5 at the three KL). As for Participant H, it must be noted that the taking of cores is defined only after NDTs, on the basis of the NDT value distribution. The large number of NDTs enables (like for participants J and C) a quite extensive coverage of the structure at KL3. At each level, a power law conversion model is identified. Figure 6.17 provides the NDT results for the 17 test locations at the three KL, the TL corresponding to cores being highlighted with red marks. The test results and identified conversion model identified at KL3 are plotted at Fig. 6.18 for 5 cores. The model writes: $f_c = 3.36 \, V^{2.20}$ but in the range of variation of velocities, the model curve is close to that of a linear regression.

Figure 6.17 illustrates how core location is chosen from the NDT results and shows two interesting points:

- the criteria for core selection, is to take cores at some test locations where V has been measured as «relatively high» and at some other locations where it was minimum. This ensures a good coverage of the presumed strength distribution, while focusing on minimum values. This explains the choice of E4 and D2 at KL1. At KL2, a third core is taken at A3 where the velocity reached a median

---

[7]This method is very specific and it cannot be described into details here. The reader is invited to refer to [4] for more details.

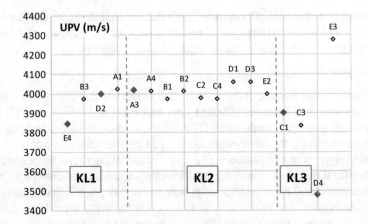

Fig. 6.17 Distribution of NDT results and choice of cores (red marks corresponding to choice of cores)

Fig. 6.18 Power law conversion model at KL3

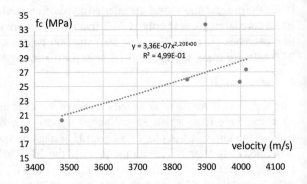

value. At these two KL, the range of velocity test results was quite limited, with values around 4000–4100 m/s.

- at last step (KL3), a strongly different feature emerged, with a much larger scatter of velocity test results. It is not possible to say if the cause is effective variation in the material property, large test result uncertainties or their combination. Two NDT results (D4 and E3) range very much out of the range of the former values. The probability of finding such values is not zero, but remains very small. The last two cores were taken at C1 and D4 in order to consider these NDT results. The regression of Fig. 6.18 shows that the much lower velocity test results of D4 is consistent with the lowest strength of all cores. This illustrates the difficulty which often exists with on-site tests and variations that are not easy to understand.

An original feature of the approach is that the final strength estimate combines information from cores with information derived from NDT results and conversion model. Thus, a weighted average is made between all strength values (direct test results and estimates), by considering that direct values are more reliable. A respective weight of 4 and 1 is given to each estimate. As a consequence, as at KL3, with 5 cores and 17 NDT results, this comes to a contribution which is about equally shared between direct strength measurements and ND estimated strengths. The estimated strength properties at KL3 are respectively 27.6 MPa for mean strength and 3.7 MPa for the strength standard deviation.

6.5.3 Composite Approach

- *Participant I*

This approach is very different from all others, since it combines the three possible NDT (R, V and pull-out) and cores (1 core at KL2 and 2 cores at KL3). Table 6.12 synthesizes the number of test results of each type what are available at the three KL and Fig. 6.19 shows their relative contribution to the total cost of investigation.

The small number of cores prevents the identification of any relevant regression model, but it however provides some estimate of the expected average. Cores are then used for calibration at KL2 and KL3, while pull-out test results are considered as a possible reference (1 pull out at KL1, 3 pull out at KL3) which can provide an alternative «reference strength», by using a prior model, derived from literature:

$$f_c = 1.41F - 2.82$$

Table 6.12 Number of test results available at the three different KL

Level	Cores	Non-destructive tests		
		Rebound	Velocity	Pull-out
KL1	0	8	3	1
KL2	1	10	4	1
KL3	2	11	4	3

Fig. 6.19 Distribution of cost between the various types of tests

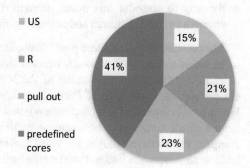

■ US

■ R

■ pull out

■ predefined cores

where F is the pull-out force in kN. Where R and V test results are available, a bi-variate model (same than participant E) is used and calibrated against reference strength value (on cores or, if no core is available on pull-out test results). At test locations where only R is available, a specified linear model $f_c(R)$ is fitted and used for estimating strength. At the end of the process, at KL2 and KL3 levels, one has four types of strength values:

- direct results obtained from cores (respectively 0, 1 and 2 at the three KL),
- strength estimates from pull-out test results (resp. 1, 1, 3),
- strength estimates from R and V combined (resp. 3, 4, 4),
- strength estimates from R only (resp. 5, 6, 7).

The estimated strength properties at KL3 are respectively 25.5 MPa for mean strength and 2.9 MPa for the strength standard deviation. The main feature of this approach is that it uses all available information. It also accounts for the fact that all information has not the same reliability by weighting the information according to what channel it has been obtained from, as was also done by participant B (see above).

6.6 Comparison of Results and Discussion

The brief description of the expert approaches that was followed (independently) by each of the 10 participants has shown a very wide variety. Differences appeared regarding:

- the choice of the ND type of tests used alone or by combining several techniques,
- the choice of the ND test result precision (TRP), that impacts the cost of the test,
- the relative weight between cores (destructive tests) and ND tests, that impacts the number of tests of each type,
- the way of choosing the core locations,
- the type of conversion models that are chosen,
- the way to identify the conversion model parameters,
- the way to consider that direct strength measurements on cores and assessed strength from NDT results and conversion models may have a different reliability.

The combination of all these possible choices (and of a number of variants) results in an undefined number of cases, whose extensive analysis is unfeasible. Another point is that the random character of the investigation and measurement process clearly appeared (the case of Participant B with surprising test results at KL3— Fig. 6.18—is typical of this randomness), that prevents to draw general conclusions from a direct comparison between the participant results. Therefore, we will first compare the assessed properties as estimated by the participants. Then, we will address some specific issues, that deserve further analyses before trying to elaborate some additional statements.

6.6.1   Assessment of Strength Properties at the Various KL

All participants were asked to estimate both the mean concrete strength and the standard deviation of concrete strength. Figures 6.20, 6.21 and 6.22 plot the estimates at the three KL, whereas the values and errors are given in Table 6.13. The last two lines give additional information regarding the distribution of estimates.

All estimates must be compared to the reference strength properties which are respectively 25.0 MPa for mean strength and 1.7 MPa for strength standard deviation in the synthetic structure (these two values where the targets of the benchmark). Regarding the estimation of the mean strength, it appears as being very efficient, even at KL1, as the relative error on mean strength estimates approximately ranges in the [−10%, +10%] interval. It can be noted that passing from KL1 to KL2 and KL3 does not result in significant improvements of the estimate.

Regarding the estimate of the concrete standard deviation, the result is much less conclusive. Figures 6.20, 6.21 and 6.22 show that, whatever the KL, the standard deviation is commonly overestimated by the participants, with mean relative errors that are respectively +40%, +72% and +82% at the three KL. At KL3, 4 participants

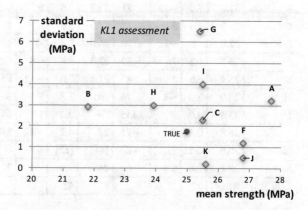

Fig. 6.20 Strength estimates (mean strength and standard deviation) and reference (KL1), E excluded

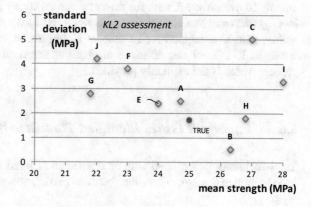

Fig. 6.21 Strength estimates (mean strength and standard deviation) and reference (KL2), K excluded

Fig. 6.22 Strength estimates (mean strength and standard deviation) and reference (KL3)

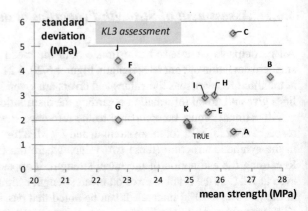

Table 6.13 Strength estimates (mean and sd, in MPa) and relative errors (in %) at all KL

	KL1				KL2				KL3			
	Mean	Sd	er$_m$	er$_{sd}$	mean	sd	er$_m$	er$_{sd}$	Mean	sd	er$_m$	er$_{sd}$
A	27.7	3.2	11	88	24.7	2.5	−1	47	26.4	1.5	6	−12
B	21.8	2.9	−13	71	26.3	0.6	5	−68	27.6	3.7	10	118
C	25.5	2.3	2	35	27.0	5.0	8	194	26.4	5.5	6	223
E	22.0		−12	−100	24.0	2.4	-4	41	25.6	2.3	2	35
F	26.8	1.2	7	−29	23.0	3.8	−8	123	23.1	3.7	−8	118
G	25.4	6.5	2	282	21.8	2.8	−13	65	22.7	2.0	−9	18
H	23.9	3.0	−4	76	26.8	1.8	7	6	25.8	3.0	3	77
I	25.5	4.0	2	135	28	3.3	12	94	25.5	2.9	2	71
J	26.8	0.5	7	−71	22	4.2	−12	147	22.7	4.4	−9	159
K	25.6	0.2	2	−88	[44]	[12]	–	–	24.9	1.9	−1	12
Mean	25.1	2.6	0.4	40	24.8	2.9	−1	72	25.1	3.1	<1	82
CoV (%)	8	73			9	45			7	40		

among 10 even overestimate the concrete variability by a factor larger than 2. All investigation and assessment strategies fail to correctly estimate the concrete variability, with an exception for Participant K, but his/her estimates were completely wrong at KL2 level and estimates are built as the result of a complex numerical process which is not explicitly provided.

6.6.2 Important Issues Identified Thanks to the Benchmark

The benchmark was carried with an open state of mind and important issues have been identified through discussions between participants once all results were shared and submitted to a common critical analysis. The first series of statements deals with the calibration of conversion models and how it is impacted by the uncertainty of test results. These statements can be summarized through three rules:

- Rule 1. It is impossible to reliably estimate concrete strength based on NDT results without at least some cores.[8]
- Rule 2. It is impossible to estimate concrete strength without calibrating a specific conversion model.
- Rule 3. Even after calibration, the conversion model has some uncertainty, which is larger if the precision of NDT results is poor. Quantifying the final model uncertainty would be useful.

6.6.2.1 About the Need of at Least Some Cores

Four participants had no core at KL1 (F, I, J, K) and only K had no core at KL2 (see Fig. 6.10). They therefore had to estimate strength from NDT results only (with pull-out test results for I and K). The case of participant K is interesting as the assessed mean strength at KL2 was fully wrong. All others slightly overestimated the mean strength and had no reliable information about the concrete variability.

6.6.2.2 The Conversion Model Must Be Calibrated

A major utility of core(s) is to enable the identification of a conversion model. This identification is possible through the calibration or a prior model (C, E, H, J) or with a specific model (A, B, G, K), but it always require to have reference strength values. As it is visible from Fig. 6.20, the three participants who used an uncalibrated prior model KL1 (B, G and J) obtained the worst estimations, either regarding mean strength or strength variability.

6.6.2.3 The Conversion Model Precision is Influenced by the Uncertainty of Test Results

In Fig. 6.22, the four participants whose assessment is the closest from the reference values (E, H, I, K) all used medium or high precision test results, which means that they have chosen, because of the limited amount of resources, to have fewer test results, but with a smaller uncertainty. Table 6.14 synthesizes how many tests have been carried out. Due to the different costs of the three types of NDTs (Table 6.2), a synthetic number NT is calculated:

---

[8]It should be noted that current structural codes (e.g. Eurocode 8 - 3 on the seismic assessment of existing structures) prescribes that in-situ concrete strength must be estimated by means of core tests, possibly complemented by ND tests.

Table 6.14 Number of tests for two types of strategies (KL3), N_c is the number of cores

Partic.	N_c	N_{PO}	N_V	N_R	NT	Partic.	N_c	N_{PO}	N_V	N_R	NT
E	3	0	6	12	12	A	5	0	11	11	16.5
H	2	0	14	10	19	B	5	0	17	0	17
I	2	3	4	11	15.5	C	5	0	0	25	12.5
K	2	4	10	4	22	G	5	0	11	10	16
Mean	2.25				17	Mean	5				15

$$NT = 2N_{PO} + N_V + 0.5N_R$$

where N_{PO}, N_V, N_R are respectively the number of tests for pull-out, velocity and rebound tests which are carried out on site (some additional tests may have been carried out on cores).

The difference is clear between the data on the left part of Table 6.14 (corresponding to yellow cells in Table 6.6) and those on the right part (corresponding to orange cells in Table 6.6). Having poor precision test results enables to take twice more cores, for a similar number of NDT results (the mean NT/N_c ratio is respectively 3 for the second set against more than 7 for the first set) but also results in larger test result uncertainty, as it is visible in Figs. 6.12, 6.13 and 6.14. These uncertainties have direct consequences, as it is illustrated in Fig. 6.23 which plots the 11 pairs of NDT results (rebound and velocity) highlighting in red the NDT results corresponding to test locations where cores are taken. The dataset shows no correlation, which is the direct result of the large test result uncertainty. A direct consequence is that the conversion model (that can be built between strength and any of the two types of NDT results) will also have a poor precision.

This consequence has been noted by Participant G who tried to identify specific correlations, like suggested in EN 13791, but with bad results as he obtained a very low value of the determination coefficient ($r^2 = 0.27$) for rebound and even a negative slope for velocity. The expert noted that «the scatter in the data is beyond a tolerable limit and against conventional relationships».

Fig. 6.23 Correlation between velocity and rebound test results at KL3 (Participant A): in blue at test locations without cores, in red at test locations with cores

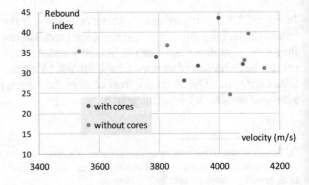

The major role played by TRP of NDT results, as large uncertainties prevent the identification of any reliable conversion model and, therefore, any correct estimation of concrete strength properties. Several participants to the benchmark used some unformal judgement to consider or reject the conversion model they have identified. If expertise can be used at this stage, it would be better to take such decision on a less subjective basis, which is not the case in common practice.

6.7 Conclusions and Contribution of the Benchmark to the Preparation of RILEM TC 249-ISC Recommendations

The benchmark has confirmed the very large number of possibilities offered when an investigator plans to use on-site non-destructive tests to assess concrete strength. The various options have been identified and analyzed. Whereas scientific literature on that topic has been focused on items as the choice of the "best equation" for the conversion model, and the identification of the model parameters, some other priorities have been identified.

The most important issue is about the impact of the test result uncertainty on all further steps of the investigation and assessment process. It appears mandatory to control it and to recommend its estimation through a specific set of test results. **This statement led the TC 249-ISC to the development of the TRP concept** which quantifies the within-test-repeatability in real situations (Sect. 1.5.2) and qualifies the precision of tests. The second most important issue is the need of an objective statement about the precision of the estimated strength as they are delivered at the end of the process. **This led the TC 249-ISC to recommend the systematic estimation of the local error**, $RMSE_{pred}$ (Sect. 1.5.6.4), which quantifies the conversion model uncertainty.

It is because the assessment process is highly influenced by random uncertainties that the TC 249-ISC has also chosen **to introduce the EQL concept** at the very first step of the investigation process. Capturing the concrete strength properties is never sure, but the EQL concept enables to adapt the resources devoted to the task to the assessment precision level that is expected.

Some other points have been tested through the benchmark, like the use of conditional coring for choosing the core location, or the bi-objective method for better estimating the concrete variability, which appeared as a very difficult challenge, with a quasi-general failure of experts for assessing this variability. However, any conclusion based on a single comparison between experts would be premature, because of the random character of the test results.[9] To analyze all these issues, a more

---

[9]Some approaches may have a high potential but not enough robustness (i.e. sensitivity to some random fluctuations). It is therefore dangerous to conclude from a unique comparison (between all experts on a unique case-study), whereas a relevant comparison must be based on a large number of possible situations.

systematic analysis has to be carried out, with the help of synthetic simulations. It is the reason why synthetic simulations are used in the following chapter. They will develop, following a predefined algorithm, a very large number of strategies (choice of techniques, of the number of cores, of models, of methods of calibration, etc.) and draw statistical results. These results will finally be processed in order to derive conclusions which can be applied to a wide application domain.

References

1. Breysse, D., Balayssac, J.P., Biondi, S., Borosnyói, A., Candigliota, E., Chiauzzi, L., Garnier, G., Grantham, M., Gunes, O., Luprano, V.A.M., Masi, A., Pfister, V., Sbartaï, Z.M., Szilágyi, K., Fontan, M.: Non-destructive assessment of in situ concrete strength: comparison of approaches through an international benchmark. Mater. Struct. **50**, 133 (2017). https://doi.org/10.1617/s11 527-017-1009-7https://doi.org/10.1617/s11527-017-1009-7
2. Alwash, M., Breysse, D., Sbartaï, Z.M.: Using Monte-Carlo simulations to evaluate the efficiency of different strategies for nondestructive assessment of concrete strength. Mater. Struct. **50**, 90 (2017). https://doi.org/10.1617/s11527-016.0962-xhttps://doi.org/10.1617/s11527-016.0962-x
3. RILEM TC 43-CND, Facaoaru, I. (Chair): Draft recommendation for in situ concrete strength determination by combined non-destructive methods. Mater. Struct. **26**, 43–49 (1993)
4. Ploix, M.A., Garnier, V., Breysse, D., Moysan, J.: NDE data fusion to improve the evaluation of concrete structures. NDT E. Int. **44**(5), 442–448 (2011)

Chapter 7
How Investigators Can Answer More Complex Questions About Assess Concrete Strength and Lessons to Draw from a Benchmark

Denys Breysse, Xavier Romão, Arlindo Gonçalves, Maitham Alwash, Jean Paul Balayssac, Samuele Biondi, Elena Candigliota, Leonardo Chiauzzi, David Corbett, Vincent Garnier, Michael Grantham, Oguz Gunes, Vincenza Anna Maria Luprano, Angelo Masi, Andrzej Moczko, Valerio Pfister, Katalin Szilagyi, André Valente Monteiro, and Emilia Vasanelli

Abstract This benchmark aims to assess mean compressive strength at several scales and to identify the location and characteristics of possible weak areas in the structure. It concerns synthetic data simulated on a group of four concrete cylindrical structures of identical dimensions with different kinds of strength distribution, based on a real case study. After having received the test results corresponding to their

D. Breysse
University Bordeaux, I2M-UMR CNRS 5295, Talence, France

X. Romão
CONSTRUCT-LESE, Faculty of Engineering, University of Porto, Porto, Portugal

A. Gonçalves
Laboratório Nacional de Engenharia Civil (LNEC), Lisboa, Portugal

M. Alwash
Department of Civil Engineering, University of Babylon, Babylon, Iraq

J. P. Balayssac (✉)
LMDC, Université de Toulouse, INSA/UPS Génie Civil, Toulouse, France
e-mail: jean-paul.balayssac@insa-toulouse.fr

S. Biondi
Engineering and Geology Department, inGeo, Gabriele d'Annunzio, University of Chieti-Pescara, Chieti-Pescara, Italy

E. Candigliota
ENEA - Italian National Agency for New Technologies, Energy and Sustainable Economic Development, Bologna, Italy

L. Chiauzzi · A. Masi
School of Engineering, University of Basilicata, Potenza, Italy

D. Corbett
Proceq SA, Zurich, Switzerland

© RILEM 2021
D. Breysse and J.-P. Balayssac (eds.), *Non-Destructive In Situ Strength Assessment of Concrete*, RILEM State-of-the-Art Reports 32,
https://doi.org/10.1007/978-3-030-64900-5_7

request (non-destructive or destructive), all the experts have to analyze these data and assess the concrete properties and to localize possible weak areas. In addition, they have to define their assessment methodology, i.e. level of investigation, number, type and location of measurements. This study provides information about how the accuracy of the final estimates depend on choices done at the various steps of the assessment process, from the definition of the testing program to the final delivery of strength estimates.

7.1 Introduction

The previous chapter (Chap. 6) explained the objectives and results of a first benchmark that was carried out by RILEM TC ISC-249 members. It focused on the most common approaches selected by experts for the on-site assessment of the concrete strength. After discussing their results, drawing first conclusions and sharing some experience and "good ideas" regarding the assessment strategy, the contributors decided that a second benchmark would offer them the opportunity to analyze some items more into detail.

This chapter details the objectives of the second benchmark and how it was carried out. Since this benchmark is based on the same type of synthetic simulations considered in the first benchmark, the reader is referred to Chap. 6 for details regarding the framework and the interest of synthetic simulations. The principles of the second benchmark are also identical to those detailed previously and all generated data

V. Garnier
Laboratory of Mechanic and Acoustic, L.M.A - AMU/CNRS/ECM - UMR7031, Aix Marseille University, Marseille, France

M. Grantham
Consultant - Sandberg LLP, London, UK

O. Gunes
Istanbul Technical University, Istanbul, Turkey

V. A. M. Luprano · V. Pfister
ENEA - Italian National Agency for New Technologies, Energy and Sustainable Economic Development, Department for Sustainability - Non Destructive Evaluation Laboratory, Brindisi, Italy

A. Moczko
Wrocław University of Science and Technology, Wrocław, Poland

K. Szilagyi
Budapest University of Technology, Presently Concrete Consultant, Budapest, Hungary

A. Valente Monteiro
National Laboratory for Civil Engineering, Lisboa, Portugal

E. Vasanelli
Institute of Heritage Science - National Research Council (ISPC-CNR), Lecce, Italy

remained unknown to all members until the meeting where all results were discussed. The virtual data set (i.e. the synthetic structure) was generated once and is identical for all participants.

This second benchmark targeted a more complex issue than simply assessing the mean value and standard deviation of concrete strength. The targets are:

(a) to assess average strength, if possible at several scales, and
(b) to identify the location and characteristics of possible weak areas in the structure.

All contributors involved in the benchmark had to:

(a) define their assessment methodology, i.e. level of investigation, number, type and location of measurements,
(b) after having received the test results ("virtual measurements") corresponding to their requirement, analyze these data and assess the concrete properties.

The benchmark was designed in order to provide information about how the accuracy of the final estimates depend on choices that are done at the various steps of the assessment process, from the definition of the testing program to the final delivery of strength estimates.

Such results were expected to help the RILEM TC ISC-249 members in identifying key issues and preparing their Recommendations that were presented in Chaps. 1 and 2. The reader will find another interest in reading this chapter since a large variety of expert practices relevant for the investigation and assessment process are illustrated. In this context, the chapter will refer to the investigation process and to its different steps as defined in the flowchart of Fig. 1.9. For the sake of simplicity, it will be simply named "assessment flowchart".

The text of the chapter is organized in order to make the main choices and results explicit and easy to understand. The case study is presented first (Sects. 7.2). Section 7.3 details how synthetic data were generated and explains what are the "right answers" to all questions. Section 7.4 also presents an overview of options regarding the investigation process, as they were chosen by the benchmark contributors. Section 7.5 details the investigation strategies and is divided into four subsections: Sect. 5.1 details the choices made by all contributors and for all knowledge levels whereas Sects. 5.2–5.4 compare and analyze the different options. Section 7.6 discusses the results that were obtained by the contributors regarding each question. Finally, Sect. 7.7 details a specific and successful contribution and draws conclusions regarding the efficiency of the investigation strategies and the identification of good practices. Some additional information is also provided in three Appendices.

7.2 Presentation of the Benchmark: Case Study and Rules to Be Applied

The information presented in this section was provided to all contributors at the beginning of the benchmark.

7.2.1 Case Study

The purpose of this benchmark is to **analyze a group of four concrete cylindrical structures** of identical dimensions, in order to assess the material condition and to locate possible weak areas. The case study is based on the real case of the Stargaard tanks, which were investigated with NDTs in Poland by A. Moczko[1].

The four cylindrical concrete water tanks have all similar dimensions and design characteristics. Each tank is made of a 5-m high and 30-cm thick concrete circular wall. The external perimeter of each wall has 100 m, which corresponds to a radius of 15.9 m. The tanks are about 20-years old, and were cast in place. The casting process involved concrete batches of about 4.5 m³, which approximately corresponds to a wall panel with the following dimensions: thickness 30 cm, height 60 cm, length 25 m. According to the design parameters, all batches have the same mix composition, but there is some variability and it is suspected that a few batches may have been deficient.

At the time of investigation, the tanks are empty and can be accessed from both sides. Owing to the internal and external exposure conditions during the service life, the concrete is very humid and any influence of the moisture variation will be neglected. Due to the age of the structure, carbonation of the few first millimeters on the external side is present. Carbonation on the internal side is neglected, since the tanks had been almost filled with water during the 20 years.

All tests can be carried out up to a height of 2.5 m without specific devices. For larger elevations, some specific installation is required which implies a larger cost and duration for each test. Tests on the internal side of the wall have also a higher cost than those on the external side. It is assumed that the on-site measurements do not need any additional time for the device and/or surface preparation.

---

[1] This case study has been described by Soutsos et al. [1]. The context of the investigation, the general dimensions of the structure and the access conditions for the benchmark were directly derived from the real case study. The material properties were however changed in order to guarantee the objectivity and confidentiality of the benchmark.

7.2.2   The Investigation Strategy

The investigation program is defined to account for three possible knowledge levels (KLs) which correspond to obtaining a progressively refined picture of the structure. Each level corresponds to assigning a given amount of resources to the investigation. It is assumed that the higher the amount, the more refined the picture can get. Three KLs are possible, which correspond to a total investigation time of 25 h for KL1, 50 h for KL2 and 75 h for KL3, respectively, for the four tanks.[2]

The target KL is defined by each benchmark contributor before the investigation. If KL2 is chosen, the contributor can adopt a progressive (step-by-step) approach. In this case, test results corresponding to the lower KL, i.e. KL1, will be given first and the contributor will analyze them before defining its requirements for additional tests in order to reach KL2. If KL3 is chosen, a similar approach is followed, but the test results for a lower KL are now those of KL1 and KL2.

7.2.3   What Can Be Measured

7.2.3.1   Possible Location of Tests

Testing points can be located at any point on a regular 25 cm × 20 cm grid across the tank walls. The testing points are defined by their (x, y) coordinates and the wall side (Exterior/Interior). The geometrical position on each wall is defined from $x = 0$ m for the point facing North, with x increasing when moving anti-clockwise on the external side (and clockwise on the internal side since the corresponding internal point faces South). Thus, possible x locations are 0, 0.25, 0.50, ..., 99.75 m. Regarding the elevation, 0 corresponds to the ground level. The lowest possible position is 0.10, and all upper points are possible with a 0.20 m step from 0.10 to 4.90 m.

It is assumed that the reinforcement grid has been checked previously and that all test locations are such that the measurements are not influenced by rebar (i.e. the test locations are assumed to be at the center of the rebar grid).

7.2.3.2   Possible Type of Tests

- **Non destructive** tests can be:
  - velocity measurements of compression waves (UPV), either direct (through the wall) or semi-direct, i.e. with two sensors (probes) on the same side of the wall and with a 25 cm offset along the same horizontal alignment,
  - rebound hammer measurements (R).

---

[2]The ratio between these three amounts (1/2/3) correspond to a quantitative estimation carried out by S. Biondi on the basis of Eurocode 8.

The value which is provided at any test location is the test result (Sect. 1.3.3.2.4) and corresponds to the mean value of 3 readings for UPV and 9 readings for R.

- **Semi-destructive** tests can be:

    - carbonation test, with coloring technique, at any location after a small drilling. It delivers the carbonation depth, in mm.
    - pull-out test, which provides the value of the pull-out force (PO), in kN. The pull-out test is assumed to also provide a carbonation test result without any additional cost or time.

- **Destructive** tests

It is also possible to core specimens. Each core has a length of 200 mm and a diameter of 100 mm. After coring, all cores are assumed to be kept in an insulating metallic sheet to avoid any change in the moisture content.

There are **two ways for defining the location of cores** (Sect. 1.4.1):

- Predefined cores, i.e. the core locations are set according to a pattern defined before the investigation or
- Conditional cores, i.e. the core locations are set after performing a first series of NDT and on the basis of the analysis of their results (in general using a predefined criterion[3]).

On each core, it is possible to perform:

- dry density measurements,
- velocity (by transmission) and rebound (on lateral sides) tests,
- compressive strength tests on a 100×200 cylinder.

Each core also provides a carbonation test result without any additional cost/time.

7.2.4 Available Resources for the Investigation

According to the assessment flowchart (Fig. 1.9), defining an investigation program involves performing all Tasks described in the data collection stage (from T1 to T6, plus "core sampling and strength measurements"). **Each individual test is defined as "costing" some time, which depends on the type of test and on the accessibility criterion. The maximum time that is available to perform the full on-site investigation is 25 h for KL1, 50 h for KL2 and 75 h for KL3.**

Table 7.1 details the time costs for each type of test, expressed in hours. As mentioned above, the durations depend on the type of test and on the height of the testing location due to site specificities. The durations include the preparation time, namely the handling and preparation of equipment and the surface preparation.

---

[3]Typically, as an example, this criterion can be expressed as "where the NDT result is as close as possible from the average value of all NDT results" or "where NDT results reach extreme values".

Table 7.1 Unit times for each test (in hours)

		Height of the testing location	
Type of test		Low (⇐2.5 m)	High (>2.5 m)
On-site tests			
	R (*)	0.05	0.20
	UPV both sides (**)	0.25	0.40
	UPV one side (**)	0.10	0.25
	PO (***)	0.50	0.65
	Carbonation (****)	0.20	0.35
Sampling cores			
	Cores 100 (***)	0.80	1.0
Laboratory tests (on cores)			
	fc	0.60	0.60
	Density	0.20	0.20
	R	0.05	0.05
	UPV	0.05	0.05

(*) for rebound, duration is that of one test result for one test location (mean value of 9 readings)
(**) for velocity, duration is that of one test result for one test location (mean value of 3 readings)
(***) pull-out test and cores provide a carbonation test result without additional cost
(****) cost of carbonation measurements by drilling

Laboratory measurements always require at least one core on which strength f_c, density, R and UPV can be measured.

Since the TRP level was identified as a major issue in the first benchmark (Sect. 6.6.2.3), it was decided to **fix the same TRP for all tests, in** order to reduce the number of degrees-of-freedom and to make the benchmark analysis easier. Therefore, all test results correspond to a **medium TRP**, which is that found to be compatible with most of the common engineering practice.

7.2.5 What Is Looked for?

Each contributor must answer three questions:

- Q1. Is it possible to consider that the four tanks have the same characteristics? Expected answer is either YES or NO, with explanations regarding differences or similarities.
- Q2. Provide estimates for the average strength and the standard deviation of strength for:

  (a) the four tanks considered as a whole, if you answered YES to Q1
  (b) for each of the four tanks, if you answered NO to Q1.

- Q3. (Question specific to Tank A) Can you identify any areas of weaker strength in Tank A?
  If your answer is positive, provide an estimate of their location, extension and average strength in those areas.

Each contributor is also expected to explain how the estimated values and the conclusions were derived from the test results, by providing relevant details about the method, assumptions and models.

Each investigator is free to distribute the total duration of the investigation program at his convenience between on-site NDT, cores, NDT on cores and laboratory measurements. So, any combination of tests respecting the allowable resources is possible.

Given Q3, more detailed results are expected about Tank A than about the three other tanks. Thus, it may be appropriate to devote more than 25% of the total investigation program to Tank A.

7.3 Generation of Synthetic Data

7.3.1 How the Simulation Process Works

The interest and principles of the synthetic simulations were described in detail in Sect. 7.6.3 in relation with the first benchmark. Therefore, aspects detailed in this section only address features specific to this second benchmark. The data were generated using a synthetic simulation process which can be subdivided into three main stages (Fig. 7.1):

(a) simulation of material properties at different scales (set of four tanks, individual tank, batch, point),
(b) simulation of NDT properties,
(c) simulation of test results.

7.3.2 Simulation of Material Properties

The simulation of material properties accounts for three sources of material variability:

(a) between batch variability, which can be divided into some part of regular variability due to the randomness of the concrete making process and some accidental variability, which may result from many causes and lack of control.
(b) within-batch variability.

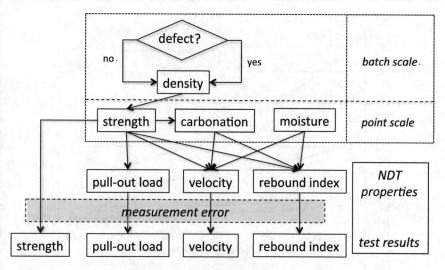

Fig. 7.1 Flowchart of the synthetic generator process

The leading variable is the concrete density (it could have been porosity or directly strength). Since each batch covers a surface of 25 m × 0.6 m, 36 batches are required to cover the surface of each tank.

The first step of the simulation process (Fig. 7.1) is the generation of individual batch density values d according to a prescribed statistical distribution. As long as only regular variability is considered, a Gaussian distribution with a mean value of 2.35 and a coefficient of variation (cv) of 3% is selected. Accidental variability, leading to defective batches, is superimposed in a second stage by simply considering that each batch has a fixed probability p of being defective and, if defective, the previously generated d value is multiplied by a factor 0.95.

Once the batch density is generated, the following step consists in generating strength at a local scale (i.e. a strength value generated at each point of the measurement grid). No (point-to-point) spatial correlation is considered within the batch, and the local strength is generated according to Eq. 7.1.

$$f_c = a\,d + b + \varepsilon \tag{7.1}$$

where a = 60, b = −107 and ε is an error term defined by a random Gaussian variable (white noise), with a zero mean and a 2 MPa standard deviation, which drives the magnitude of the within batch variability. These values correspond to a concrete with a mean strength of 34 MPa for d = 2.35. The batch to batch variability of density with cv(d) = 3% leads to a strength range of [26.1, 42.5 MPa] for +/−2 standard deviations around the mean strength. The moisture content is supposed to be uniform and deterministic, and corresponds to a saturation degree of 80%.

The last step is the generation of the carbonation depth x_c. This parameter is known to depend on time. In our case, the age of the structure (20 years) is a constant. It is

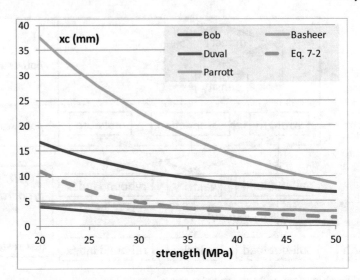

Fig. 7.2 Carbonation depth at 20 years as a function of concrete strength—comparison between the synthetic model of Eq. 7.2 (dotted curve) and four models from the literature

also assumed that the carbonation depth is larger if the concrete is less dense, thus when it has a lower strength. A few authors proposed models allowing to formalize the strength-carbonation development relationship.[4] After analyzing their results, it is assumed that the carbonation depth follows a lognormal distribution, whose average value is:

$$x_{c,\,mean} = k \cdot t^{1/2} / \left(f_c^2 \right) \tag{7.2}$$

This deterministic relation is used to calculate the carbonation depth from the strength value which has been simulated at the previous step.

Figure 7.2 compares strength-carbonation depth (at 20 years) relationship provided by this model and four models from the literature. This figure illustrates the large discrepancies between the different models, both on the value of the average depth as on the strength dependency. The model of Eq. 7.2 corresponds to a kind of "average model" and predicts a carbonation depth of few millimeters for the simulated concrete. Since the local strength varies within a batch, according to Eq. 7.1, the local carbonation depth also varies. Figure 7.3 illustrates the cumulated density of the carbonation depth over a surface corresponding to a single batch of concrete (an area of 25 m × 0.6 m).

---

[4]Bob [2], Parrott [3], Duval [4], Basheer et al. [5].

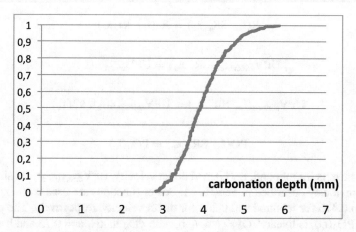

Fig. 7.3 Effect of within batch variability: cumulative distribution function of the carbonation depth value over a surface corresponding to a single concrete batch

7.3.3   Simulation of NDT Properties

As illustrated in the flowchart (Fig. 7.1), NDT properties depend on strength, but can be influenced by physical parameters, namely the moisture of concrete (through the saturation ratio SR) and the carbonation depth x_c. The state of knowledge regarding the influence of carbonation on the NDT properties is not fully established, and the synthetic simulation lays on assumptions that mimic, as best as possible, what can be observed in the "real world". These relationships are assumed to follow the same pattern for all NDT properties, in which the influence of carbonation depth is considered through a multiplying function. The following expressions are therefore assumed to reflect the NDT values (Eqs. 7.3 to 7.6):

$$R(f_c,\ SR,\ x_c) = R_0(f_c,\ SR).\,k1(x_c) \tag{7.3}$$

$$UPV_{direct}(f_c,\ SR,\ x_c) = UPV_{direct,0}(f_c,\ SR).\,k2(x_c) \tag{7.4}$$

$$UPV_{surface}(f_c,\ SR,\ x_c) = UPV_{surface,0}(f_c,\ SR).\,k3(x_c) \tag{7.5}$$

$$PO(f_c,\ SR,\ x_c) = PO_0(f_c,\ SR).\,k4(x_c) \tag{7.6}$$

On the right sides of Eqs. 7.3 to 7.6, the first function describes the dependency for uncarbonated concrete, which is mostly described in the literature, and the second function weights the carbonation depth influences. In practice, the influence of SR on pull out test results and that of carbonation on both UPV_{direct} and pull out test results are neglected. By considering also that the SR value is a constant, the following expressions are obtained:

$$R(f_c, \ SR, \ x_c) = R_0(f_c). \ k1(x_c) \tag{7.7}$$

$$UPV_{direct}(f_c, \ SR, \ x_c) = UPV_{direct,0}(f_c) \tag{7.8}$$

$$UPV_{surface}(f_c, \ SR, \ x_c) = UPV_{surface,0}(f_c). \ k3(x_c) \tag{7.9}$$

$$PO(f_c, \ SR, \ x_c) = PO_0(f_c) \tag{7.10}$$

The models for rebound, $R_0(f_c)$, and direct velocity, $UPV_{direct,0}(f_c)$, are similar to those previously described in Sect. 6.3.1. Power functions are used with exponents equal to 0.476 for rebound and 0.204 for direct velocity, respectively. The pull-out model, $PO_0(f_c)$, is linear: $PO_0 = a \ f_c + b$, with PO_0 in kN, $a = 0.75$ and $b = 2.26$. The surface velocity is known to be slightly lower than the direct velocity.[5] It is here simply assumed that $UPV_{surface, 0} = 0.90 \ UPV_{direct, 0}$

Carbonation effect modelling.

Carbonation influences both rebound (through $k1(x_c)$) and surface velocity (through $k3(x_c)$). For rebound, the correction factor was taken with a value similar to that used in the Chinese JGJ standard.[6] Since this standard only provides tables, the values given in the tables were used to identify a weighting model defined by:

$$k1(x_c) = 10^{-0.0343 \ xc} \tag{7.11}$$

However, this weighting function cannot be lower than 0.6, which is reached for 6.5 mm. For larger carbonation depths, a constant ($= 0.6$) value applies.[7]

For direct velocity, the $k3(x_c)$ model is defined by:

$$k3(x_c) = 1 + a \ x_c \tag{7.12}$$

---

[5]Turgut and Kucuk [6].

[6]JGJ/T23-2011, Technical specification for inspecting of concrete compressive strength by rebound method, 2011 (see also, Proceq, Rebound number corrections with JGJ/T23-2011, online technical documentation).

[7]By combining Eqs. 7.2 and 7.11, one can compute the difference between the R values of two concretes (A and B) that have the same strength, concrete A being uncarbonated and concrete B being carbonated. It appeared (after the beginning of the simulations) that there is some compensation effect: as low strength concrete carbonates more (Fig. 7.2), the k1 correction factor is larger and the R value of the poor strength carbonated concrete can be more or less identical to that of a "good" uncarbonated concrete. A side effect is that the contrast regarding original concrete strength can be masked if the lower strength concrete is more carbonated. Such situations can be found in true life. In the benchmark, this fact had not been anticipated and resulted in some handicap for some strategies which were mostly relying on rebound measurements on the carbonated face (see Sect. 7.6.1 for more details).

which is derived from Pham [7][8], with a $= 0.0025$. This leads to an increase that is less than 2% for $x_c = 6$ mm.

7.3.4   Simulation of Test Results

The effect of measurement uncertainties was considered in a similar way for all test results, by adding a Gaussian error term to the "true" synthetic value. The standard deviation of this random error (TRU) depends on the test type and is summarized in Table 7.2. Another source of measurement error was also added which corresponds to a gross error in the surface velocity measurement on the external face. It is assumed that some surface defect may, with a low probability p, induce a 20% reduction of the velocity. The p value is set as 0.5%.

Once the test results and a conversion model are available, the concrete strength can be estimated. The concrete variability on a large scale (i.e. on a whole tank) also exists and is controlled in the simulation by the density variability. The last line of Table 7.2 provides the standard deviation (SD) of each property (true global variability) for the whole tank. Thus, the two values, TRU and SD can be compared to have some idea about how easy/difficult it is to estimate the "true" material variability when test results are affected by random errors. The larger the SD/TRU ratio, the easier the detection of true material variations. The problem is that true material variations can be masked by the effect of random measurement errors. This is the case in particular for the rebound and the velocity of surface waves where TRU is larger than SD. Even if the influence of random measurement errors can be reduced by repeating the tests, the comparison between TRU and SD remains informative. Thus, core strength, PO, UPV_{direct}, R and $UPV_{surface}$ are ranked following a decreasing efficiency of tests.

7.3.5   What Was Simulated—What Are the Right Answers?

Before analyzing and discussing the efficiency of all strategies carried out by the contributors, we synthesize here what is the result of the simulation process, and what are the "right answers" to the Q1–Q3 questions of Sect. 7.2.5.

---

[8]Pham [7].

Table 7.2 Magnitude of absolute test result uncertainty (TRU) and standard deviation (SD) of the corresponding property

| | Density (g/cm^3) | Cores strength (MPa) | | On site | | | | | On cores | |
		Normal size	Reduced size	R	UPV$_{surface}$ (m/s)	UPV$_{direct}$ (m/s)	PO (kN)	x_c (mm)	R	UPV$_{direct}$ (m/s)
TRU	0.01	1	2	2	100	50	1	1	1.5	50
SD	0.04	3.22		1.7	78	87	2.4	1.01	1.7	87

7.3.5.1 Question 1. Can We Consider that the Four Tanks Have Similar Properties?

Table 7.3 synthesizes the mean value, the standard deviation (sd) and the coefficient of variation (cv) of concrete strength for the four tanks considered individually and the same parameters for the whole data set (four tanks as a whole).

The simulation process was carried out in order to use similar rules and deliver similar properties for all tanks (the cumulative distribution functions of strength values are provided in Fig. 7.4). Therefore, the expected answer to Question 1 is YES.

The only difference between the tanks can come from the process which randomly generates accidental variability (a2), see Sect. 7.3.2, and defective batches with a 5% probability. Each tank is made of 36 batches, and the number of defective batches can vary from 0 to 2 or 3, as a result of chance. In fact, in this specific case, Tanks A, C and D have 2 defective batches while Tank B has only one. The consequence of this small difference can be seen on the left tail of the cumulative distribution functions of the strength (Fig. 7.4), where it can be seen that the left tail of Tank B is slightly shifted to the right. This also explains the slightly larger mean and lower standard deviation seen for this tank.

Table 7.3 Statistical properties of true strength for the four tanks

	Tank A	Tank B	Tank C	Tank D	Whole
Mean (MPa)	33.6	34.1	33.8	33.5	**33.7**
sd (MPa)	3.2	2.9	3.1	3.2	**3.1**
cv	0.096	0.086	0.093	0.095	0.093

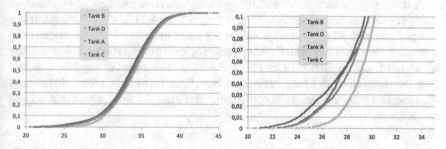

Fig. 7.4 Cumulative distribution function of the synthetic strength in the four tanks. The left tail of the functions is magnified in the figure on the right

7.3.5.2   Question 2. Provide Average and Standard Deviation of Strength

The expected answer is provided in Table 7.3. The overall average strength is 33.7 MPa, and the overall standard deviation is 3.1 MPa. The specific results of Tank A are very close to these average values (33.6 and 3.2 Mpa, respectively).

7.3.5.3   Question 3. Can You Identify Defective Areas in Tank A?

Two defective areas were generated in Tank A which is made of 36 batches. Each batch covers a surface of 25 m length × 0.6 m height. The location and properties of the two defective areas are:

– Zone A1, on the left upper corner (0 m < x ≤ 25 m, 4.20 m < y < 4.80 m):
  $f_{c\,mean} = 25.5$ MPa, sd = 2.2 MPa
– Zone A2, on the lower middle (25 m < x ≤ 50 m, 0 m < y < 0.60 m):
  $f_{c\,mean} = 26.8$ MPa, sd = 2.4 MPa

These two areas are visible in Fig. 7.5 which provides the synthetic test results of all the points of the grid for rebound on the internal side (left) and direct velocity (right). This figure shows that the contrast between batches can be easily "seen", but our eye is a very efficient sensor. The strength pattern resulting from the casting process and the batch-to-batch variability clearly appears. Although this pattern is clear when the full data set is considered, despite the measurement noise, it could be much less easy to identify when only few data are available, which is the case both in true life and in the benchmark. For instance, notice that some light blue values can be seen in all parts of the structure.

A third zone, which is not strictly speaking defective can be distinguished on the upper right corner (Zone A3, 75 m < x ≤ 100 m, 4.20 m < y < 4.80 m). It has less contrast with all other areas and does not correspond to a defective batch but to strength values that lie in the left part of the strength distribution curve. The properties of A3 are $f_{c\,mean} = 30.6$ MPa and sd = 2.4 MPa. This mean strength is 10% less than the mean strength of the whole structure. Because of the circular character of the tank, this zone appears to be contiguous with Zone A1 and might eventually be merged with it.

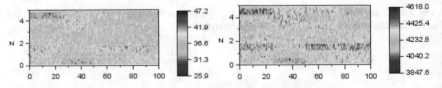

Fig. 7.5 Maps of NDT test results for Tank A: rebound on internal face (left), direct velocity (right)

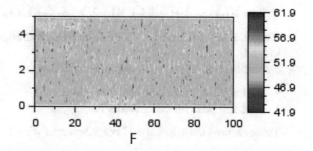

Fig. 7.6 Map of rebound test results on the carbonated side

It has been explained (Sects. 7.3.2 and 7.3.3) how the carbonation depth and its effects on rebound were simulated, pointing to an issue that led a compensation effect (see footnote 7):

- on the one hand, the lower the strength, the lower the rebound number,
- but on the other hand, the lower the strength, the higher the carbonation depth, and thus the higher the rebound number).

Figure 7.6 plots the rebound test results on the carbonated side, showing that the "defective areas" with low strength have vanished. The compensation effect is even such that these areas have the highest rebound values. Therefore, it would be very dangerous to derive strength estimates from rebound test results without having considered that the conversion model between rebound and strength had to be fitted.

7.4 Analysis of Investigation Methodologies

This section focuses on the first part of the assessment flowchart, more specifically on tasks corresponding to the definition of the testing methods, the choice of test locations and, as a result, the number of tests for each testing method. The main objective of this analysis is to understand what options can be chosen by experts, knowing there is no "right option". However, this analysis can be helpful to identify criteria that would correspond to a good practice or, reversely, to point out some inadequate options.

Three knowledge levels (KL1–KL3) were defined, corresponding to different amounts of resources (defined in time units, see Table 7.1). Furthermore, as Tank A deserves a specific attention (see Question 3), the contributors may logically choose an uneven resource distribution between the four tanks.

A total number of 24 contributions were received from 15 different experts (named A to O). The same code, KLi-J, is used for all contributions, where i is the knowledge level index and J indicates the contributor.

- 9 contributors have focused on a single level:

  - 2 on KL1 (KL1-D, KL1-G) and

- 7 on KL3 (KL3-H, KL3-I, KL3-J, KL3-K, KL3-L, KL3-N, KL3-O),

– 3 contributors (C, E and M) have adopted a progressive approach covering KL1 to KL3,
– 3 contributors have covered two different KLs:

- KL1 and KL2 for contributors A and B
- KL1 and KL3 for contributors F.

These choices led to 8, 5 and 11 contributions for KL1, KL2 and KL3, respectively

7.4.1 Comparison of Methodologies at the Three Knowledge Levels

Tables 7.4, 7.5 and 7.6 use a common presentation format to summarize how resources have been spent by all experts for the three KLs. For each contributor, the tables first indicate how resources were distributed between NDT, core testing

Table 7.4 Resource distribution for KL1, 8 contributors (available resource is 25 units)

	NDT	Core (nr)	Other	NDT			% on A
				R	UPV	PO	
D	11.4	13.4 (9)		4.3	7.1		47
G	3.6	16.8 (12)	4.6	3.6			25
A	14.6	10.2 (7)		9.3	5.3		66
B	15	5.6 (4)	4	15			63
C	25			25			50
E	24.6		0.4	1.6	21	2	34
F	11.3	13.6 (8)		4.3	7		26
M	7.6	16.8 (12)		2	5.6		61
Average	14.1	9.6 (6.5)	1.1	8.1	5.7	0.25	46

Table 7.5 Resource distribution for KL2, 5 contributors (available resource is 50 units)

	NDT	Core (nr)	Other	NDT			% on A
				R	UPV	PO	
A	35.1	14.4 (10)		9.3	25.8		50
B	16.8	29 (19)	4	16.8			71
C	38	9.9 (6)	2.1	27	1.5	9.4	48
E	38.3	11.3 (7)	0.4	6.6	27.2	4.5	45
M	20.7	28 (20)		5.7	15		50
Average	29.8	18.5 (12.4)	1.3	13.1	13.9	2.8	53

Table 7.6 Resource distribution for KL3, 11 contributors (available resource is 75 units)

	NDT	Core (nr)	Other	NDT			% on A
				R	UPV	PO	
C	54.3	18.3 (12)	2.2	32.4	10.6	11.4	49
E	61.9	12.7 (8)	0.4	17.6	38.5	5.8	40
F	34	40.8 (23)		13	21		33
H	52.7	17.6 (10)	2.5		47.2	5.5	64
I	54	20.8 (12)		25.6	28.4		39
J	36.8	36.8 (24)			36.8		37
K	34	39.2 (27)		34			36
L	39.6	34.4 (24)		16.8	22.8		28
M	33	42 (30)		9.6	23.4		42
N	48.2	26.8 (18)		48.2			36
O	50.6	21 (12)	3.4	13.3	25.7	11.6	65
Average	45.4	28.2 (18.2)	0.8	19.1	23.1	3.1	43

(i.e. drilling + strength measurement) and other tests (carbonation, NDT on cores). The number of cores taken by each contributor is also provided. Three columns detail how resources were distributed between the three types of NDT methods. The last column indicates what percentage of resources was devoted to Tank A, for which a more detailed assessment is expected. The last line of each table provides the average value for each parameter.

For the sake of clarity, Tables 7.4, 7.5 and 7.6 provide only global numbers. The reader is invited to refer to Appendix 7.1 that provides all the details in terms of type and number of tests.

7.4.2   Resource Distribution Between DT/NDT and Between Tanks

The first two questions address how resources are shared between DT and NDT on the one hand, and between tanks on the other. Tank A is only one of four, but more resources can be spent on it, as a more ambitious assessment is expected (see Q3, about the identification of defective areas). The average amount of resources devoted to NDT was seen to be relatively stable for the three KL since it varies between 56 and 61%, but there are large variations between contributors, as summarized in Table 7.7.

The extreme values (Table 7.7) correspond to investigators that prefer to take more cores or, reversely, to perform a larger number of NDT. The average number of cores increases linearly with the amount of resources. **At KL3**, 18.2 cores are taken on average for a total investigation domain which covers 2000 m², which corresponds to **one core for every 110 m²**.

Table 7.7 DT/NDT resource distribution ratio

	Percentage of resources used for NDT			Number of cores		
	Average	Minimum	Maximum	Average	Minimum	Maximum
KL1	56	14 (G)	100 (C, E)	6.5	0 (C, E)	12 (G)
KL2	60	34 (B)	77 (E)	12.4	7 (E)	19 (B)
KL3	61	44 (M)	83 (E)	18.2	8 (E)	30 (M)

These resources can be unevenly distributed between tanks. The average resource for Tank A is also relatively stable for the three KLs since it varies between 43 and 53%, but large variations are found between contributors. Some investigators tend to have a more even distribution, like G at KL1 or L at KL3, but the average tendency is to have a higher density of test results for Tank A. At KL3, the average number of cores for Tank A corresponds to one core for every 68 m². This surface corresponds to that of 4–5 batches, which clearly indicates that **coring cannot be used alone to identify tentative defective areas**.

7.4.3 Resource Distribution Between Different Types of NDT Methods

Among all the contributions, only two of them use UPV test results only for a KL3 assessment (Table 7.8). In all other cases, R test results are used, alone or in combination with UPV test results. The reasons are a the lower cost of R test results (see Table 7.1) and the fact that using the rebound is a more common practice among engineers. The lower cost also provides the ability to achieve a better spatial coverage of the investigation domain, which may be particularly useful regarding the identification of defective areas (Q3).

Pull-out was always combined with other NDT method (R, UPV or both). It must be noted that pull-out is very often considered as an alternative to cores. As can be seen from Tables 7.4, 7.5, and 7.6 for all KL, the investigators who use pull-out tests are those who have the fewest cores.

Table 7.8 Resource distribution between NDT methods

	R only	UPV only	R and UPV	PO (+other NDT)
KL1	3 (G, B, C)	–	5	1
KL2	1 (B)	–	4	2
KL3	2 (K, N)	2 (J, H)	7	4

7.4.4 What Would Be an "Average Investigation Program"?

Each investigator follows his own strategy. It may therefore seem meaningless to define an "average investigation program" which would only correspond to the mean number of tests for each technique and tank. Such a program may lack consistency. However, its purpose is to provide indicative information regarding how many tests of each type are carried out and how test locations are distributed in the investigation domain. Table 7.9 summarizes this "average investigation program", distinguishing the test locations in the lower and upper parts of each tank. The part of carbonation measurements and NDT on cores is marginal and is neglected for the sake of simplicity.

This program corresponds to an investigation with a duration of 74.85 h. Furthermore, it can be seen that 43% of the resources are devoted to Tank A, and the amount of resources devoted to NDT is 64%. Seven cores (+3 PO) are taken on Tank A, and four cores (+4PO) on the other tanks, which comes to 19 cores and 15 PO test results. The program is also balanced between rebound and velocity measurements. On Tank A, these data correspond to a total of 128 NDT (80 for rebound and 48 for UPV), whereas this number varied in practice from 32 (KL3-J) to 201 (KL3-C) NDT (Appendix 7.1 for details). Since each tank is made of 36 batches, the density of NDT results for Tank A is roughly 2 rebound test results and 1 velocity test result for each batch. This provides the possibility, by having both R and UPV test results at the same test location, to combine the two methods when identifying a conversion model. It can be noted that, in the three other tanks, the investigation is concentrated on the lower part of the structure, which is less resource consuming.

Table 7.9 Definition of a (hypothetic) "average investigation program" at KL3

	Number of tests				Resource spent (time units)			
	Tank A (low)	Tank A (high)	Others (low, × 3)	Others (high, × 3)	Tank A (low)	Tank A (high)	Others (low)	Others (high)
Cores	4	3	4	–	5.6	4.8	16.8	–
Pull out	2	1	2	2	1	0.65	3	3.9
Rebound	50	30	40	–	2.5	6	6	–
UPV direct	12	12	12	–	3	4.8	9	–
UPV indirect	12	12	12	–	1.2	3	3.6	–

Table 7.10 Classification of investigation programs (mostly means that DT or NDT takes more than 60% of the available investigation time, * refers to conditional coring)

		Mostly DT	Balanced	Mostly NDT
Assessment for KL1 or KL2		D*, G	A* (KL1), B* (KL2)	A* (KL2), B* (KL1)
Assessment for KL3	Direct		J*, K*, L*, F*	H*, I*, N, O*
	Progressive		M	C, E*

7.5   Description of Data Processing Methodologies

The first subsection (Sect. 7.5.1) briefly describes the main features of each contribution, providing some details about the measurement techniques, the number of cores and the definition of core locations, the type of conversion model and the conversion model identification process. This information is useful to understand why some options were, a posteriori, identified as less effective, but the reader may prefer to skip directly to the following subsections (Sects. 7.5.2 to 7.5.4) where all contributions are compared and subjected to a first analysis.[9]

7.5.1   Description of the Assessment Methodology for All Contributions

This section describes how contributors chose to collect data and how these data were processed. For each contribution, it provides information about:

- the main features of the investigation program (global indicators were summarized in Tables 7.4, 7.5 and 7.6),
- how the core locations were chosen, i.e. either predefined or only defined after the analysis of NDT results ("conditional coring"),
- the main features of the resulting datasets,
- the conversion models.

A total number of 24 contributions were received from 15 different experts. Table 7.10 summarizes how each contribution focuses (or not) on a given KL and pays more or less attention to NDTs (see DT/NDT resource distribution ratio, Table 7.7). It also identifies (*) the contributions that considered conditional coring. Only four contributors (C, G, M, N) have predefined the location of cores, independently of NDT results. Table 7.11 presents further details on this issue.

---

[9]Of course, the reader can come back at any time to Sect. 7.5.1 to check any information.

Table 7.11 Synthesis about the number of cores, NDTs and semi-destructive tests (SDT) for KL3

	F	J	K	L	H	I	N	O	M	C	E
Cores on A	8	11	9	6	10	8	5	6	12	3	2
Cores on others	5-5-5	5-3-5	6-6-6	6-6-6	–	2-1-1	5-4-4	2-2-2	6-6-6	3-3-3	2-2-2
Total cores	**23**	**24**	**27**	**24**	**10**	**12**	**18**	**12**	**30**	**12**	**9**
SDT on A	–	–	–	–	10	–	–	13	–	6	6
ND tests											
R on A	48	–	128	48	–	53	163	95	48	163	64
R on others	100	–	192	96	–	159	252	75	96	175	108
UPV on A	32	32	–	24	64	25	–	86	60	38	81
UPV on others	64	84	–	48	84	61	–	75	96	3	124
Total NDT	**244**	**116**	**310**	**216**	**148**	**298**	**415**	**331**	**300**	**379**	**377**

7.5.1.1 Investigation Programs at KL1 and KL2

Contributors D and G privileged cores, while the investigation is more balanced for B and A privileges NDTs.

Contributors D and G (for KL1)

These contributors worked with the very limited amount of resources of KL1. Contributor G got 12 cores (3 on each tank), while Contributor D got 9 cores, five of which being taken on Tank A. Their core strength data[10] were 33.4 MPa (mean) and 5.4 MPa (s.d.) for D, and 33.4 MPa (mean) and 3.8 MPa (s.d) for G.

Contributor G

Techniques: only rebound measurements, with an identical approach to all tanks: 12 measurements on the external side and 6 on the internal side, at identical locations. Carbonation measurements were carried out on the external side at all points where a rebound value was also taken on the inner side. All measurements and cores were taken at a maximum height of 2.50 m, in order to limit the duration of the investigation.

Conversion model: the model was built with rebound, with both linear and power models. The carbonation effect was considered using a prior influence model that was fitted to the present measurements. The four tanks were analyzed separately but after realizing that the data may belong to the same population, a new estimation procedure with a refined model was carried out, considering the four tanks as a whole.

Contributor D

Techniques: resources were distributed between rebound measurements, only on the interior face, and direct velocity measurements, with more measurements on the

---

[10]The numbers given in this section correspond to the mean value and standard deviation of all strength data obtained on cores. They are summarized for all contributors in Table 7.12.

lower part of the tanks than on the top part, and about twice more measurements on Tank A than on all other tanks.

Core location: it was defined after the NDT measurements, at points where both rebound and velocity were available. Three cores were taken at test locations with the lowest velocities, three cores at test locations with the highest velocities, and three additional cores were taken on Tank A at test locations chosen in order to correctly cover the central part of the distribution of velocities.

Conversion model: a bilinear regression model was identified on the set of (f_c, R, V) triplets:

$$f_{c\,est} = 0.0216\,V + 0.357\,R - 72.7 \quad \text{with a fitting RMSE error of 1.6 MPa.}$$

For points where only the rebound measurements were available, a linear regression model was also identified on the set on 9 (f_c, R) pairs which led to:

$$f_{c\,est} = 1.37\,R - 17.7 \quad \text{with a fitting RMSE error of 4.9 MPa.}$$

Strength assessment: the mean strength and standard deviation estimated from the nine cores were 33.4 and 5.4 MPa, respectively. However, this standard deviation is probably overestimated since the cores were preferentially taken at points were velocity provided extreme values. Thus, the nine direct strengths values on cores were combined with 14 strength values estimated with the bilinear conversion model and with 15 strength values estimated with the linear regression model. The final estimates provided an average strength of 33.6 MPa and a standard deviation of 3.6 MPa, the latter being a much lower value than the initial estimate based on cores. Contributor D also provided the mean strength and standard deviation estimated following the procedure established in Sect. 8.3.3 of EN13791:2007, which amounted were 29.6 and 4.8 MPa, respectively.

Contributor B (for KL1 and KL2)

Techniques and cores: for KL1, the investigation program was based exclusively on rebound and cores. Most rebound measurements were taken on the external side, with some carbonation measurements, and Tank A was privileged. Four cores were taken on Tank A at a height of 2.1 m. For KL2, the same strategy was used, with a very large number of cores (16 on Tank A and 3 more on Tank D). Rebound measurements were carried out on cores. The resulting average core strength and standard deviation were 33.2 and 3.2 MPa, respectively.

Conversion model: the carbonation effect was considered. The data processing was based on the methods defined in the Chinese standard for in-situ strength determination JGJ/T23-2011:

– the variability of test results was assessed and analyses were carried on to detect outliers,

- the rebound values on the carbonated face were multiplied by a correction factor whose value depends on the carbonation depth,
- a drift was calculated by comparing the average core strength and the average estimated strength at the same location (for 4 cores at KL1, and 16 cores on Tank A at KL2).

For KL2, the series of 20 rebound test results for each tank was compared to that of Tank A, to check for similarities. Concrete strength was estimated by finding the best fitted linear regression between cores and NDTs according the EN13791 standard. An additional calculation provided a safe estimation of strength at a specific location (Sect. 8.2.3 of EN 13791), which is about 7 MPa lower.

Contributor A (for KL1 and KL2)

Techniques: For KL1, Tank A was covered with rebound measurements on the internal face and direct velocity measurements along 8 regularly spaced vertical profiles. On the other tanks, only rebound was performed. For KL2, only direct velocities measurements were carried out, in order to get more measurement points on Tank A and to have some on all other tanks.

Core location: For KL1, seven cores were taken: five on Tank A at points of lowest or highest velocities, and two more at points where the rebound values were the lowest (both points on Tank D). For KL2, three more cores were taken at locations were velocity reached extreme values (2 at lowest value and 1 for the highest value). The resulting average core strength and standard deviation were 31.0 and 5.4 MPa, respectively.

Conversion model: a series of ten sets (f_c, R, V) was used to identify univariate and multivariate linear conversion models. The average estimated strength could be derived, as well as the standard deviation (including or not the model uncertainty).

Specific analysis: the rebound values distribution on all tanks was analyzed for KL1 and seen as being a non-Gaussian left-skewed distribution. This was confirmed by the separate analysis performed on Tank D. For KL2, the non-Gaussian character of the distribution was confirmed for both the R and V distributions. The identification of defective areas was based on both the estimated mean strength (or strength directly measured on cores) and the characteristic strength. Three defective areas were identified, whose extension was appreciated from the volume of loads/batches.

7.5.1.2   Investigation Programs for KL3 (Direct, Without KL1 or KL2 Prior Assessment)

Table 7.11 provides detailed information about the number of cores and tests for all contributors for KL3. The three columns on the right correspond to contributors (C, E, M) who made a progressive assessment (see Sect. 7.5.1.3). The other eight contributions can be distinguished between those involving a large number of cores (between 23 and 27) and fewer NDT results (F, J, K, L), and those involving less cores

(18 or less) and more NDT results (H, I, N, O). It may seem that some contributors carried out less tests (e.g. if one compares H and O), but there are several factor that do not appear in the table (like the fact that test locations at higher locations cost more, or that direct velocity tests cost more than semi-direct tests).

Contributors with a high number of cores and a low NDT/DT ratio

Contributor F

Techniques: a comprehensive investigation program was defined to cover all tanks, based on the variety of available NDT techniques. The logic was to have a variety of possible combinations between NDT and cores at different locations. The possible combinations are:

– cores with rebound on each face of the wall, direct and indirect velocity
– cores with one rebound and one velocity
– NDT only (one rebound and one velocity), without cores.

The measurements are taken alternatively on each face, according to a predetermined grid, and NDT measurements are also taken on all cores.

Core location: all cores are taken in the lower part of the tanks, based on the result of NDT measurements. A composite index is calculated from all NDT measurement values for each measurement point and the location of cores is defined in order to uniformly cover the distribution of this index (from points where NDT properties are the lowest to those where they are the highest). The resulting average core strength and standard deviation calculated from 23 cores are 34.1 and 3.9 MPa, respectively.

Conversion model: conversion models can be identified from many combinations between core strength and NDT results. The models can be identified either for all tanks simultaneously ("global") or for each tank independently ("partial"), they can consider either rebound (second order polynomial law), or velocity, or their combination (SonReb), and they can also combine measurements on the two faces. For each tank, the final output is a series of 23 different possible estimates, depending on the variant that is used. Among these estimates, the final contributor's preference was the SonReb approach calibrated on all cores.

Contributor J

Techniques: The experimental campaign is designed with two stages. The initial stage was common with that of contributor H (see in a subsequent page) and involved a screening of all tanks with direct velocity tests following a predefined regular pattern ("helical scheme") with 32 points on Tank A and 28 points on the other tanks. The analysis of the variances of the four tanks are very similar and the mean variance on one tank is about 98% of the total variance, which leads to the first conclusion that all tanks are similar.

Core location: The second stage consisted in destructive tests only, with 24 cores distributed across all tanks, among which 11 cores were taken on Tank A. The location of the cores was defined in order to uniformly cover the full range of velocities, as illustrated in Fig. 7.8. Velocity measurements were performed on each core before the strength measurement. The final data set is made of 24 (f_c, UPV_{is}, $UPV_{on\ core}$) triplets. The resulting average core strength and standard deviation are 35.0 and 3.0 MPa, respectively.

Conversion model: the comparison of velocity test results between on-site test results and tests on cores led to discard 4 values that were seen to have a difference larger than 6%. From the remaining 20 pairs, a linear conversion model was identified. The model error and model stability were then checked, by removing a certain number of data from the initial set (size = 20) and checking the RMSE or the remaining part. The prediction error was estimated to be about 4 MPa.

Contributor K

Techniques: the investigation methodology is based on an extensive rebound measurement program, on a regular grid (6.25 m × 0.6 m) which provides 128 rebound values on Tank A, and on a grid with half the density (12.5 m × 0.6 m) for the other tanks. All measurements were taken on the interior side of the wall. The denser grid on Tank A was selected in order to locate weak areas.

Core location: for DT measurements, 9 cores are selected for Tank A and 6 cores for each of the other tanks. The selection of core locations depends on the rebound number values: for each tank, the rebound number values are arranged from the minimum to the maximum and they are then subdivided into a number of groups equal to the number of cores required for the tank (for example 9 for tank A). The core location corresponds that of the NDT with a result equal to the median of each group. An exception to this rule was used for Tank A where the core location for the first group was selected as that corresponding to the lowest rebound value. The resulting average core strength and standard deviation are 32.7 and 3.4 MPa, respectively.

Conversion model: a specific linear conversion model is fitted for each tank according to the bi-objective method (see Sect. 1.5.5.2), i.e. in order to fit both the average strength and the standard deviation of cores. The model is defined by:

$$f_{c\ est} = a\,R + b$$

where $a = sd(f_{c,\ core})/sd(R)$ and $b = f_{c,\ core\ mean} - a\,R_{mean}$.

Strength assessment: the weaker areas on Tank A can be identified from simply plotting the estimated strengths from the R test result distribution.

Contributor L

Techniques: the investigation is based on the combination of direct velocity measurements and rebound measurements on both faces. The investigation program is the same for all tanks, with a slightly higher test number on Tank A for which NDT measurements are carried out for 24 points located along three horizontal lines and 8 vertical profiles (mesh grid is 12.5 m × 1.2 m). For tanks B, C and D, 16 NDT test locations were used for each tank along two horizontal lines with a mesh grid of 12.5 m × 2.4 m. Based on the statistical analysis of the NDT test results, concrete was considered to be substantially homogeneous both within each tank (considering the upper and the lower zones) and between the tanks.

Core location and conversion model: 6 cores were taken from each tank, with 2 cores in the upper zone and 4 cores in the lower one. The core location was selected based on the distribution of the NDT values. For each tank, a specific SonReb multivariate relationship ($f_c = a*R^b*V^c$) was calibrated by combining 6 datasets containing f_c, R_{ext} and V (note that only rebound values measured along the external surface, R_{ext}, were used).

Strength assessment: for each tank, the specific SonReb relationship was applied to the points where only NDT data were available. The final strength assessment considers both direct strength results (6 values for each tank) and estimated strength values (18 values on Tank A, 10 values on the other tanks). The resulting average strengths are 34, 33, 34 and 34 MPa for tanks A, B, C and D, respectively, and the corresponding standard deviation values are 3, 3, 2 and 1 MPa, respectively.

Contributors with a low number of cores and a high NDT/DT ratio

Contributor H

Techniques: the experimental campaign is designed with two stages. The initial stage was common with that of contributor J and involved a screening of all tanks with direct velocity tests following a predefined regular pattern ("helical scheme") with 32 points on Tank A and 28 points on the other tanks. In addition, 10 pull-out tests were also carried out across Tank A. The comparison of the distributions of velocity values showed that all tanks were similar. The second investigation stage involved 32 additional velocity measurements that were carried out on Tank A only, leading to 64 points (corresponding to a 6.25 m × 1.20 m grid).

Core location: ten locations of cores (on Tank A only) were identified, on the basis of NDT (direct velocity measurements) in order to get a statistical distribution, similar to that of the direct velocity measurements. This kind of procedure implies taking more cores at test locations corresponding to the most frequent velocities (referring to the whole data set of 64 NDT results). These locations also tried to evenly cover the whole tank surface, both along the perimeter and the elevation. The mean and the standard deviation values of the on-site velocity values for the 10 locations were 4301 and 105 m/s, respectively, against the corresponding values of 4306 and 111 m/s obtained for the whole data set of 64 NDT results. On each core, rebound and velocity were measured before the compression test. The final data set consists of ten (f_c, UPV_{is}, $UPV_{on\ core}$, $R_{on\ core}$) sets. The resulting average core strength and standard deviation are 34.0 and 3.6 MPa, respectively.

Conversion model: a univariate power law model was identified ($r^2 = 0.85$) that provided strength estimates from direct velocity measurement values.

Strength assessment: concrete strength in the tanks was estimated using a weighted average, with weights of 1 for direct values on cores (10 values) and 1/3 for strengths estimated from NDT (54 additional values for Tank A on the measurement locations where no cores were taken).

Contributor I

Techniques: the investigation is based on a combination of rebound and direct velocity measurements with cores. The number of non-destructive measurements is similar for all tanks. Rebound tests are carried out on the external face. The density of NDT measurements is about 1 rebound test for every 10 m^2 and 1 velocity measurement for every 20 m^2 (the surface corresponding to one batch is about 15 m^2). The NDT locations are evenly distributed across the surface, following an irregular pattern.

Core location: the location of cores is chosen according to the distribution of NDT values, in order to cover the full range of values. 8 cores out of 12 are taken from Tank A. The resulting average core strength and standard deviation are 32.7 and 4.3 MPa, respectively. NDT measurements were also carried out on some cores. They showed that the rebound values on cores were very different than those measured on site, while there was no significant variation in the estimates of the carbonation depth.

Conversion model: attempts were made to derive coefficients for the SONREB technique, but at the same time, correlation graphs were produced independently for the UPV and R data. The graphs shown a quite reasonable correlation for UPV, but no useable correlation for R. This fact was attributed to the influence of carbonation. Given that, directly, velocity is less likely affected by carbonation, a decision was taken to use the UPV measurements only to provide the required strength estimates. The identified conversion model is $f_c = 0.0239 \text{ UPV} - 70.4$ ($r^2 = 0.64$).

Strength assessment: to answer Question 1, the method described in the standard BS 6089 to compare two areas was used, and each tank was compared to Tank A. When applying the same method but comparing all tanks to Tank B, inconsistent conclusions were curiously obtained though. Still, in the end, it was concluded that all tanks belong to the same population. A tentative weaker area was identified on Tank A based on a single lower estimate at $x = 21.25 \text{ m}$, $y = 4.3 \text{ m}$.

Contributor N

Techniques: the investigation consisted in an extensive measurement program of rebound tests performed on the internal face, with 163 tests on Tank A and half (82) on each of the other tanks, defined using a Latin Hypercube Sampling approach to maximize the sample space coverage. The survey planning was defined after the information available from the structure. The idea was to capture the material variability thanks to a dense enough test location grid. On the basis of the batch volume, the surface corresponding to a batch could be estimated at $25 \text{ m} \times 0.6 \text{ m} = 15 \text{ m}^2$, and the 163 tests on Tank A corresponded to an average of 4 to 5 points for each of the 36 batches.

<u>Core location</u>: eighteen cores were taken following a regular predefined zig-zag pattern.

<u>Conversion model</u>: before any conversion the rebound value datasets were processed in order to build the estimate of rebound at any point of the tank surface. Two spatial interpolation models (bi-harmonic spline interpolation and Lowess interpolation) were used. The rebound value of the four tanks were compared using the Kruskal-Wallis and Levene tests to determine if the strength of the four tanks could be considered to be statistically identical. Results indicated that the four tanks could not be considered to have the same properties. For each tank, an empirical model ($CoV_{fc} = 1.042\ CoV_R + 0.123$) was used to derive the coefficient of variation of the strength from the coefficient of variation of the rebound. Two linear conversion models were identified between rebound values and strength estimates: the first model targeted the estimate of average strengths while the second one targeted the estimate of lowest strength, thus eliminating some outliers.

<u>Strength assessment</u>: The analysis of the rebound value statistical distribution (see above), especially the kurtosis, indicated the existence of tentative defective areas. Since the strength estimate could be provided on the whole surface of the tank, it was also possible to estimate the characteristic strength value (corresponding to a 5% percentile). The location of areas of lower strength was straightforward, from a visual examination of the reconstructed strength field.

Contributor O

<u>Techniques</u>: The investigation was based on cores and on a variety of NDTs: rebound on both sides, direct velocities, indirect velocities on both sides and pull-out, with more tests on Tank A (other tanks have no direct velocity tests and NDT locations in the lower parts only). Only 12 cores are taken (6 on Tank A, and 2 for each of the other tanks). A total of 22 pull-out tests are taken on the external side (13 on Tank A only). Particular attention is paid to the estimation of the test result repeatability since, for each NDT, one location is chosen on Tank A at which the test is repeated four more times at a close vicinity. Thus, a standard deviation of the test result is calculated from the five test results. A coefficient of variation for each technique can be deduced, and compared with the same parameter at the scale of the whole tank. Based on these results, direct velocity appears to be the most repeatable technique, whereas indirect velocity, pull-out and rebound follow in a decreasing order of repeatability. The consistency of the test results was checked at all locations where several NDT results are available. This analysis has shown that rebound measurements on the carbonated layer could not be used, since carbonation influences the rebound measurement in such a way that all possible correlation is lost. Therefore, these data were not considered in the analysis.

<u>Core location</u>: it is defined on the basis of NDT results. On Tank A, based on velocity and pull out values distributions, 2 cores are taken for the lowest values, 2 for average values and 2 for the highest values. On the other tanks, 2 cores are taken at points where the pull out forces are the highest and the lowest. It is also checked that the

resulting set of 12 cores uniformly covers the distribution of pull out force, direct and indirect velocity. The resulting average core strength and standard deviation are 34.0 and 6.2 MPa, respectively. For each core, velocity and rebound are also measured before testing.

Conversion model: A variety of conversion models can be defined, either from NDT on site, SDT, or NDT on cores. However, all correlations between the different types of tests cannot be identified since the test locations do not always coincide. The conversion models were compared regarding their quality of fit and six models (M1–M6) were identified. The best model correlates with pull out force, three models (respectively linear model with direct velocity, and double power law with rebound and either direct or indirect velocity) have an intermediate quality, and two models (linear models with rebound and indirect velocity) have a lower quality.

Strength assessment: depending on the available data at each measurement point, one or several of those models can be used to estimate the local strength. The last stage consisted in processing all estimates, by calculating a weighted average of the local strength between all models (with more weight for better models), and by calculating a standard deviation between various estimates at a given point, which provides some estimation of the model error. This model error is about 1.5 MPa. Finally, all local estimates are used in order to derive the average strength, standard deviation and to identify weakest areas.

7.5.1.3 Investigation Programs for KL3 (with KL1 or KL2 Prior Assessment)

As seen in Table 7.11, this strategy was followed by contributors M, C and E, with a very contrasted balance regarding the number of cores, as M took 30 cores against only 12 for C and 9 for E.

Contributor M

Techniques: the strategy was based on the combination of cores and a variety of NDT results: direct velocity, indirect velocity on both faces and rebound on both faces. For KL1, efforts were strongly focused on Tank A. For KL2 and KL3, the investigation addressed tanks B, C and D and the mesh of tests was refined for tank A. The priority was given to indirect velocity and rebound on the external side, with 32 measurements for KL2 and KL3.

Core location: it was predefined, e.g. with 4 vertical profiles of 3 cores on Tank A for KL3. The resulting average core strength and standard deviation are 33.5 and 3.6 MPa, respectively.

Conversion model: for KL1, a preliminary analysis showed that rebound measurement values on the external side are negatively correlated with strength while indirect velocity (the only NDT available at all core locations) have a good positive correlation once some outliers were discarded (the way outliers are identified and removed is mostly a matter of expert judgment). Thus, five models are systematically tested for the three KLs: quadratic and power regressions with indirect velocity on the external side, linear and power models with rebound on the external side (always showing a negative slope), a double power model with these two NDTs.

Strength assessment: the last model (double power model, SonReb type) was used in order to identify the local strength. The last step consisted in processing all individual strengths and to calculate a weighted average strength and its standard deviation from a composite dataset that mixed individual core strengths (with a weight of 2) and estimated strengths (with a weight of 1).

Contributor C

Techniques: the investigation process was divided into three stages. The definition of each stage was adapted from the analysis of the previous one:

- for KL1, only rebound on the external face was used, with a high density of measurements: on Tank A, rebound was measured with a 2 m horizontal spacing on two horizontal lines, amounting to 100 test results. The density was smaller on the three other tanks.
- for KL2, the program focused on pull-out tests (17 tests).
- for KL3, since rebound measurements on the external face appeared to bring little value because of the carbonation effect, the investigation was mostly based on measurements of rebound on the internal face and direct velocity on Tank A.

Core location: for KL1, no cores were taken. for KL2, three cores were taken on Tank A (and on Tank D) in a very close vicinity (at $+/-$ 25 cm distance), following direct velocity measurements at those same locations. Velocity was also measured on cores. For KL3, six more cores were taken on Tanks B and C. They were grouped in two sets of three, enabling a test repeatability assessment.

Conversion model: because of the carbonation effect, the large amount of resources spent for rebound measurement on the external face for KL1 and KL2 prevented the identification of a model between rebound and strength. A specific linear model was built for assessing strength from velocity and pull-out test results.

Contributor E

Techniques: the investigation process was divided into three stages:

- for KL1, the objective was mostly to compare the properties between the four tanks. As such, 32 surface velocity tests were taken on the internal face of each tank following a regular (25 m $\times$ 0.6 m) grid, and rebound tests were performed on four points, on both faces of each tank. Additional tests (direct velocity and pull-out) were taken only on Tank A at the same four points.
- for KL2, pull-out tests were carried on at the location of cores. The investigation was also refined on Tank A by performing additional velocity tests and an extensive rebound investigation on both faces.

– for KL3, the investigation was refined by getting a more extensive coverage of internal rebound and direct velocity test results. Particular attention was also paid to the tentative identification of defective areas on Tank A where two pull-out tests were carried out.

Core location: for KL1, no cores were taken. For KL2, 7 cores were taken from the four tanks (1 only on Tank D). Their location was defined from the available NDT and SD results of the previous stage, in order to cover the full range of properties (mean values, lowest values, highest values).

Conversion model: three NDT results were considered as being reliable indicators of the concrete strength: internal rebound, direct velocity and semi-direct (internal) velocity. The three corresponding conversion models were identified to estimate strength.

7.5.2   Synthesis About the Definition of Core Location

Four contributors followed a predefined pattern for the location of cores, which did not depend on the result of NDTs. All other contributors used conditional coring, i.e. they defined the location of cores after performing NDT. Most contributors tried to define the core location in order to uniformly cover the expected strength distribution, as can be hypothesized from the distribution of NDT results. This option is illustrated in Fig. 7.7 where the location of the 24 cores is defined by crosses evenly distributed along the UPV cumulative distribution of direct velocity measurements. The advantage of this option is that the core strength results are unbiased and can be used directly for strength assessment.

A possible problem can arise when these locations are not well-balanced regarding the spatial distribution (for instance, when grouped within a limited domain) or between tanks (in the previous chapter this problem could arise between beams and columns). The investigator can adapt the sampling, by taking the core where the value of the NDT is almost identical and where the additional spatial distribution criterion is also satisfied.

Fig. 7.7 Illustration of conditional coring based on the uniform coverage of the statistical distribution of NDT measurements after coring (contributor J)

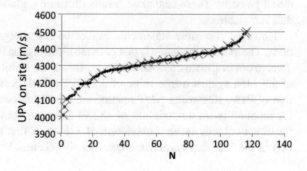

Another logic was followed by some contributors (A, D, O) who tried to magnify the contrast by privileging core sampling at locations where the ND test results provided extreme values (lowest or highest). The advantage of this strategy is that it can improve the stability of the conversion model). The drawback is that the core strength results cannot be used directly for strength assessment, since the strength distribution is biased due to the sampling process and the scatter is magnified.

Regarding the distribution of cores between tanks, it can be balanced between the four tanks (C, E, L) or involve a larger number of cores from Tank A. Contributor H is the only one who took all cores from Tank A.

7.5.3   Synthesis About the Data that Can Be Used for Correlation with Cores

Table 7.12 summarizes (for the most common measurements) what synthetic data are available for each contributor, and can be used for identifying conversion models. In this table, the statistical information regarding the NDT results is restricted to the ND test locations that also correspond to those of cores. The information on NDT results is given for rebound on the internal face (R_i), direct velocity (UPV_d), semi-direct velocity ($UPV_{surf,i}$) and pull-out (PO). The data for rebound on the external face was discarded because of problems with carbonation (see discussion in Sect. 7.6.1). The three last columns provide the values for core strength, rebound and velocity measured on the cores, respectively. The last row provides the true values (mean and standard deviation) for all parameters.

One can compare first the mean strength estimated from cores to the true value (33.6 MPa, see Table 7.3). The difference between the mean strength from cores and the true mean strength never exceeds 2.4 MPa (i.e. relative difference of 7%), the lowest being from contributor A (31.9 MPa with 10 cores for KL2) and the highest being from contributor C (35.2 MPa with 12 cores for KL3). The difference between the estimated and the true mean strengths tends to reduce as the number of cores increases and is less than 1 MPa for all contributors who have 20 cores or more. Regarding the mean values for the NDT results, the maximum difference between estimated and true values are 3 units (i.e. 8%) for rebound and 63 m/s (i.e. 1.5%) for direct velocity. These statistical results illustrate a global consistency between the different datasets.

However, the same statements cannot be made regarding the standard deviations. The general tendency is that of overestimating the true standard deviation (sd). Twelve out of fifteen contributors overestimate the sd on strength, with a maximum of 5.9 MPa against a true value of 3.2 MPa, which corresponds to an overestimation of 81%. Regarding NDT results, ten out of ten contributors overestimated sd for rebound (with a maximum of 182%) and nine out of eleven overestimated sd for direct velocity (with a maximum of 116%). A first reason for this overestimation is the fact that cores were consciously taken at locations of extreme values:

Table 7.12 Average value and standard deviation of test results, limited to test locations where correlation with cores can be identified, N_C = number of cores (*corresponds to a conditional coring which leads to taking more extreme values)

Level	Contr.	N_C	R_i	UPV_d (m/s)	$UPV_{surf,i}$ (m/s)	PO	$f_{c\,core}$ (MPa)	R_{core}	UPV_{core} (m/s)
KL1	G	12	38.5/3.1				33.4/3.9		
KL1	D*	9	37.2/3.5	4300/188			33.4/5.4		
KL2	A*	10	34.4/4.8	4238/185			31.9/5.9		
KL3	B	19	38.0/3.5				33.2/3.3	37.2/2.5	
KL3	C	12		4327/51		27.3/1.8	35.2/3.5		4366/97
KL3	E	8	35.8/4.4	4297/100	3860/255	27.4/3.7	34.7/3.1	37.4/2.3	4292/112
KL3	F	23	37.0/2.4	4274/95			34.1/2.8	37.3/2.7	4328/84
KL3	H*	10		4301/105		27.8/1.3	34.0/3.6	38.1/1.7	4324/76
KL3	I	12		4303/120			32.7/3.5	35.7/1.2	4264/62
KL3	J	24		4302/126			35.0/3.0		4298/141
KL3	K	27	37.2/2.8				32.7/3.4		
KL3	L	24	38.2/2.8	4310/82			33.9/3.3		
KL3	M	30	38.0/2.4	4304/101	3889/142		33.5/3.6		
KL3	N	18	37.2/2.7				33.3/4.0		
KL3	O*	12		4320/175	3848/146	27.4/4.4	34.0/6.4	37.1/4.1	4293/197
True values			37.4/1.7	4301/87	3871/78	27.5/2.4	33.6/3.2		

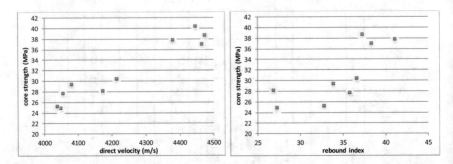

Fig. 7.8 Relation between NDT results and core strengths (direct velocity on the left—10 pairs, rebound on the right—9 pairs)

contributors A, D and O have effectively chosen this option and have, as a result, the largest estimates of sd whatever the property (strength, rebound, velocity, pull-out force). However, there is another reason, that applies to all NDT results, which is the measurement uncertainty, which adds some noise to test results and increases the standard deviation.

Table 7.12 also indicates that, if one considers only the two most common test results, conversion models can be established:

- between strength and rebound by 10 contributors,
- between strength and direct velocity by 11 contributors and
- between strength and both rebound and velocity by 6 contributors.

Figure 7.8 illustrates the existing correlations found by Contributor A for KL2, with 10 cores. The effect of the stratified sampling method, privileging extreme values for choosing core locations can be seen, in particular for direct velocity for which there are more velocity test results close to the maximum and minimum. This is less clear for rebound, because of the rebound measurement uncertainty (the figure shows only nine pairs, as rebound was not measured at the last core location). It is from such datasets that all conversion models can be identified.

7.5.4 Synthesis About the Identification of Conversion Models

The identification of the conversion model is the next step of the investigation process according to the flowchart of Fig. 7.9. This stage offers many degrees of freedom, both regarding the explaining variable (i.e. the type of test result: R, V, PO...), the mathematical expression of the model (linear, power law, exponential...), the choice of regression of a specific model or of calibration, the choice of a univariate model or of a bivariate one. As briefly described in Sect. 7.5.1, each contributor followed his own logic, and diversity is the main characteristic:

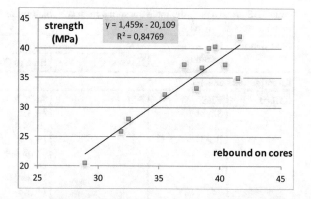

Fig. 7.9 Regression between rebound test results on cores and core strength

- all possible types of test-results were used. Table 7.12 indicates that rebound and direct velocity are the most common variables but semi-direct velocity and pull-out have been used by 3 and 4 contributors, respectively,
- linear models are the most common, but power and polynomial ones were not unusual,
- calibration was systematically used (this was identified as necessary after the first benchmark, see Chap. 10), but through different ways, including fitting a specific model or using a drift calibration (see Sect. 1.5.5.3),
- six contributors identified a bivariate conversion model, either with a bilinear model or with a double-power law model ("SonReb type").

Final assessments provided by the contributors will be discussed in the next section, but it is not relevant to directly compare the efficiency of the models identified by all contributors since these models result from different options developed on different datasets. In fact, it is more useful to analyze, on the basis of the data they have received, what could have been obtained if each of them had adopted the same options for the identification of the conversion model. This issue will be detailed in Sect. 7.7.

7.6 Analysis of Assessments Provided by the Contributors

Each contributor had to answer three questions (Q1, Q2, Q3). The right answers to these questions were given in Sect. 7.3.5. This section details all of the contributors' answers. The purpose is not to distinguish who was successful but why: what strategical and methodological choices prove to be more efficient for correctly assessing the concrete strength? Tables 7.13 and 7.14 provide the synthetic answers of all contributors for KL1/KL2 and KL3, respectively. The specific problem due to the influence of carbonation on rebound test results is discussed first (Sect. 7.6.1). Sections 7.6.2–7.6.4 analyze the answers to questions Q1, Q2, Q3, respectively. Lastly, the overall

Table 7.13 Answers to the three questions from contributors for KL1 and KL2 (bold characters when answers are right)

Contributor	Q1 (yes/no)	Q2		Q3 x/y/f_c
		If yes to Q1	If no to Q1	
Investigations at KL1				
KL1-D	**YES** Slight differences B > D > A > C	**33.6 +/− 3.6**	A: 33.8 + /− 4.1 B: 34.4 + /− 3.9 C: 32.2 + /− 3.5 D: 34.1 + /− 5.4	**0/4.5/26.7** 50/2.1/29.7 (unsure)
KL1-G	**YES**	**33.4 +/− 3.0**	A: 33.8 + /− 4.5 B: 33.9 + /− 4.9 C: 32.4 + /− 4.4 D: 34.9 + /− 5.1	50/0.5/26 (unsure)
Investigations at KL2				
KL2-A	**YES**/NO* (C > A > B) > D	**34.1 +/− 3.3** Distribution is not gaussian (left queued)	A: 34.7 + /− 2.8 B: 33.1 + /− 3.1 C: 34.8 + /− 3.3 D: 31.9 + /− 4.7	**0–18 and** **88.5–100/4.5/25.5–26.9** **26–44/0.3/27.6** **72–86/4.5/27.7**
KL2-B	**YES** D could be slightly lower	**33.9 +/− 1.8**		50–62.5/>2.7/22.1–25.1

* The answer depends on the analysis (univariate or multivariate model, type of regression)

performances of all contributors are compared (Table 7.17) and tentative conclusions are drawn.

7.6.1 Specific Problem Due to Carbonation Effect on Rebound Test Results

The way the synthetic simulation accounted for the development of concrete carbonation and of its consequences on the NDT results was described at Sect. 7.3.3. There are some coupling effects in true life, since a poor strength concrete carbonates

Table 7.14 Answers to the three questions from contributors for KL3 (bold characters when answers are right)

Contributor	Q1 (yes/no)	Q2 (fcmean and sd)		Q3 $x/y/f_c$
		If Yes to Q1	If No to Q1	
Investigations at KL3				
KL3-H	YES	**34.1+/− 3.5**		**0–18.25/4.5/28** **80–100/4.5/29** **37/0.3/30**
KL3-I	YES Marginal differences (A lower)		A: 31.5 +/− 1.8 B: 33.0 +/− 2.6 C: 32.8 +/− 2.6 D: 32.2 +/− 1.8	**21.25/4.3/26.4**
KL3-J	YES	**35.0 +/− 3.0**		**18.25/4.5/29.7** volume about 4.5 m^3
KL3-K	NO Slight differences		A: **33.1 +/− 2.9** B: 32.1 +/− 1.6 C: 34.8 +/− 4.7 D: 32.4 +/− 4.9	3–21.75/4.3/28.2 34.25 to 53/0.3/30
KL3-L	YES Slight differences		A: **34 +/− 3** B: 33 +/− 3 C: 34 +/− 2 D: 34 +/− 1	No weak areas identified
KL3-N	NO		A: 31.7 +/− 6.3 B: 34.6 +/− 6.6 C: 35.4 +/− 6.4 D: 31.9 +/− 6.1	**37.5/0–1.50/18–24** **95–100/>4/26** **0–10/>4/26**
KL3-O	NO*	34.7 +/− 3.4	A: **34.6 +/− 3.5** B: 34.1 +/− 2.6 C: 35.2 +/− 4.5 D: 34.9 +/− 3.0	**28–50/0.5/26–31**
KL3-C	NO (B > A) > (D > C)		A: 36.0 +/− **4.0** B: 38.0 +/− 3.1 C: 31.7 +/− 2.4 D: 32.4 +/− 3.4	32–46/1.5/<32
KL3-E	NO*		A: 35.4 +/− **2.6** B: 35.7 +/− 3.2 C: 36.3 +/− 2.7 D: 35.0 +/− 3.1	**11.75/4.5** 61.75/2.5 86.75/3.3 and 3.9
KL3-F	YES	33.9 +/− 4.2	A: 35.0 +/− 4.7 B: 33.4 +/− 4.1 C: 34.5 +/− 4.2 D: 32.0 +/− 2.8	No weak areas identified
KL3-M	YES	33.2 +/− 3.0		No weak areas identified

*The answer could have been YES, as the strength estimates are very similar for all tanks

Table 7.15 Synthesis about the number of rebound tests for KL3

	F	J	K	L	H	I	N	O	M	C	E
Total	148	0	320	144	0	212	415	170	144	338	172
Interior face	76	–	320	72	–	0	415	95	64	96	16
Exterior face	72	–	0	72	–	212	0	75	80	242	16

more and, as carbonation increases the surface density, the rebound index increases. The equations driving the simulation led to some type of compensation between a lower strength and a higher carbonation with the resulting effect of rebound values being unable to detect variations of the initial strength in a carbonated concrete (see Sect. 7.3.3, note 6). Some contributors did not anticipate this eventuality when they defined their investigation plan. Table 7.15 summarizes, for all contributors, the number of rebound tests that were carried out on the two sides of the tank wall. Various strategies are visible. Contributors J and H did not use rebound. Among the others, the balance of rebound tests between the two sides varies a lot, from 0 (contributors K and N) to 100% (contributor I).

The main consequence of the carbonation effect is that some of the test results are useless for the strength estimation, as it is impossible to identify a relevant conversion model between strength and rebound values. This is illustrated with the case of Contributor O who planned to add some NDTs in the lab, on cores. Figures 7.9 and 7.10 plot the regression between rebound test results and core strengths for laboratory rebound test results and on-site rebound test results on the external face, respectively.

It clearly appears that the right regression that can be evidenced when rebound is unbiased (Figs. 7.9) is masked because of the carbonation influence: it is impossible to identify a reliable conversion model from on-site rebound test results (Fig. 7.10). In such a case, the regression model from laboratory tests ($f_{c\ est} = 1.46\ R - 20.1$) can be used if new rebound test results are obtained on-site from the uncarbonated side of the wall, but all NDT results obtained from the carbonated side are useless.

A direct consequence is that a contributor who choose to rely mainly on this type of tests must find another way (see Sect. 7.5.1.2 for contributor I, with a conversion

Fig. 7.10 Regression between on-site rebound test results (external side of the tank) and core strength

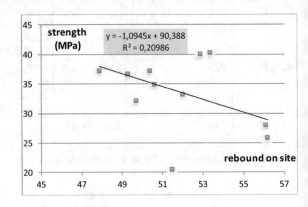

model based on V only). Other contributors who spent a significant part of their resources with rebound tests on the carbonated face were also impacted. The only way to deal with such a problem instead of simply rejecting test results is by using a progressive approach and checking test results at each step. This approach was followed by contributor C who, in a first stage, had only test results on the carbonated face. Having checked that these data could not be used, he therefore modified his investigation program accordingly for the subsequent stages.

However, the best way would have been to anticipate the possible effect of carbonation and to carry on rebound tests only on the uncarbonated face.

7.6.2   Answer to Question 1: The Four Tanks Have Similar Properties?

It was explained at Sect. 7.3.5.1 that the four tanks were simulated following a unique process, which resulted in very close material properties in all tanks (Table 7.3), with an average strength of 33.7 MPa and a standard deviation equal to 3.1 MPa. The difference between the mean strength of each tank was lower than 1 MPa and all tanks had a standard deviation of about 3 MPa. The right answer to Question 1 would then be YES.

As indicated in Tables 7.13 and 7.14, 7 contributors (D, G, B, H, J, F, M) answered YES and provided an unique value for all Tanks, 5 contributors (K, N, O, C, E) answered NO and provided specific estimates for each Tank, and 3 contributors (A, I, L) answered YES but however provided specific estimates. It can be added that contributors D, G and F also provided separate estimates, which leads to 11 contributors who provided a specific estimation. The estimated mean strengths in each tank are very close, with a maximum difference of 3 MPa for 9 out of 11 cases. The only two wrong answers to this question are those of contributors N and C with a 3.7 and 6 MPa difference, respectively. By trying to explain this relative prediction failure, it was found that:

- contributor C suffered from having carried an preliminary investigation with rebound tests on the external carbonated face (see Sect. 7.6.1) which revealed to be of poor value,
- contributor N carried an extensive rebound test program and statistical tests carried out on the results indicated that datasets corresponding to each tank could not be considered to have the same parent distribution. The strength of each tank was then obtained from the mean value of the core tests carried out in each tank and the differences found with respect to the true values can be assigned to sampling error (i.e. the statistical effect of chance).

As a general comment, one can consider that the most common (right) assessment is that all tanks have similar strength properties and can be considered as belonging to a single population. **A consequence is that data gathered on all tanks can be**

Fig. 7.11 Cumulative distribution of direct velocity test results for the four tanks (Contributor E, KL1)

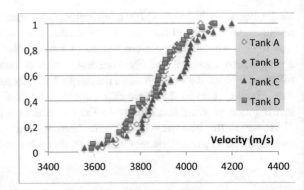

merged to build a single set, and that a unique conversion model can be used for all results.

Some contributors have refined their analysis regarding the similarity between tanks. A first possibility is a visual comparison between datasets. For instance, Contributor E carried out 32 direct velocity tests on each tank for KL1. When plotting the corresponding cumulative distribution (Fig. 7.11), some small differences appeared for Tank C leading to a shift to the right of its distribution function. As such, the mean value was slightly larger for Tank C (3910 m/s) than for the three other tanks (between 3850 and 3875 m/s). However, this possible difference was not confirmed when more data became available for KL2 and KL3.

Contributors A (for KL2) and I (for KL3) carried out more sophisticated analyses by comparing the statistical distributions on each tank through common statistical tests. Both were confronted with some difficulties since their conclusions depended on the data considered for the analysis (as an example, for Contributor I, the same test answered YES when using rebound data, but NO when using velocity data) or was inconsistent (for instance A = B, B = C, A ≠ C). These difficulties illustrate the practical limits of theoretical approaches when the measurement uncertainty is superimposed to a limited natural variability.

7.6.3   Answer to Question 2: Provide Average and Standard Deviation of Strength

The answers are here limited to the assessment of Tank A. The results illustrated on Fig. 7.12 are those of the whole population when the answer to Q1 was YES and those of Tank A when it was NO. Figure 7.12 plots the estimates of the average strength and that of the standard deviation for all final estimates. It also plots the partial estimates (M-1, M-2) at levels KL1 and KL2 for contributor M. The three KLs are indicated with different symbols. The reference value (33.6; 3.2) is indicated with a red square. Whereas the estimates may seem scattered along the x-axis, it must

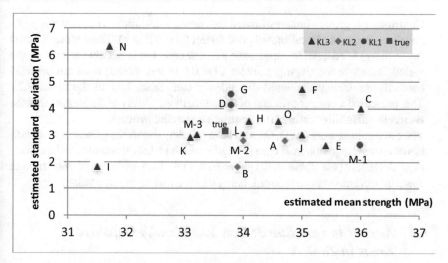

Fig. 7.12 Estimation of average strength and standard deviation on Tank A for all contributors

be pointed that differences are magnified by the scale: all estimates range in a +/−10% around the reference mean strength.

The main conclusions are the following:

– **the average strength is correctly estimated by all contributors**, since the error is never larger than 10%. However, this very positive result must be put into perspective, since the simple information provided by cores (see Table 7.12) is as good as this final estimate (with at least 9 cores, the direct estimate of the concrete strength is in fact robust),

– **the standard deviation is correctly estimated (with a +/− 30% difference) by most contributors.** This result is very interesting, since it illustrates what NDTs can bring in addition to cores,

– **the two targets (mean and standard deviation) are addressed with a very limited error by some contributors,** including M for KL2 and G and D for KL1 (even if these last contributors overestimated the standard deviation).

Some useful considerations can be added:

– Regarding larger errors on mean strength estimates, the reasons why contributors C and N were misleading have already been discussed. It also appears that contributor I underestimated the mean strength, which results from a **statistical effect of chance**: among the 10 cores used for correlation, only 2 cores had strengths larger than the true mean strength, which led the contributor to underestimate the strength (and the standard deviation too).

– Contributor N grossly overestimated the standard deviation. The reason is probably because he used an **uncalibrated prior model** (relationship between the cov of NDT results and the cov of concrete strength), which proved to be inadequate in this case.

- Contributor B grossly underestimated the standard deviation. The reason is probably because he used a calibrated prior model ($f_c = a\,R + b$). The drift calibration (by shifting) is adapted to capture the true strength, but is not able to capture the variability, since the slope parameter a ($= 0.6$ in this model) does not describe correctly the strength/rebound dependency (see Table 7.21 in Appendix 7.2). This prevented a correct estimate of the strength variability and confirms that the **concrete variability estimation requires a specific process.**
- The three contributors (A, D, O) who have privileged extreme values at the conditional coring stage obtained core strengths which grossly overestimated the standard deviation (see Table 7.12). However, their final estimate of the strength standard deviation was improved, being much closer to the true value.

7.6.4 Answer to Question 3: Can You Identify Defective Areas in Tank A

There are two or possibly three defective areas in Tank A, which have been detailed at Sect. 7.3.5.3. While Zone A1 and A3 have been generated independently but, due to the circular character of the structure, they can be merged into a unique zone. Table 7.16 synthesizes the results obtained by all contributors, regarding both the location of the three weakest zones, and their ability to quantify the local values (more details regarding their location and strength are provided in Tables 7.13 and 7.14).

Table 7.16 Efficiency of each contributor regarding identification and quantification of weakest areas

	Zone A1		Zone A2		Zone A3	
	Location	Strength	Location	Strength	Location	Strength
KL1-D	Yes	Yes				
KL1-G			Yes	Yes		
KL2-A	Yes	Yes	Yes	Yes	Yes	*
KL3-H	Yes	*	Yes	*	Yes	Yes
KL3-I	Yes	Yes				
KL3-J	Yes	*				
KL3-K	Yes	*	Yes	*		
KL3-N	Yes	Yes	Yes	No	Yes	No
KL3-O	Yes	*				
KL3-E	Yes	No				

*Means that the strength is only grossly estimated (more than 2 MPa difference with real value)

The location of the three defective areas was identified by KL2-A, KL3-H and KL3-N, while KL3-K has identified the two more defective areas and 6 other contributors have identified one defective area. Five contributors (B, L, C, F, M) failed to identify any defective area.

Figures 7.13, 7.14 and 7.15 offer a visual illustration of three interesting contributions at KL3 level regarding the identification of weaker areas on Tank A. They correspond to Contributors H, N and K, respectively. The first two contributors positively located the three defective areas whereas the third contributor identified zones A1 and A2 only. Velocity tests were carried on by contributor H and rebound tests by the two others. Figure 7.13 draws the strength estimate pattern at 64 NDT locations where the direct velocity was measured (see Table 7.11). The blue-gray values on the color scale correspond to the three defective areas whose location, extension and magnitude are visible. Figure 7.14 is a color map of the estimated strength, using

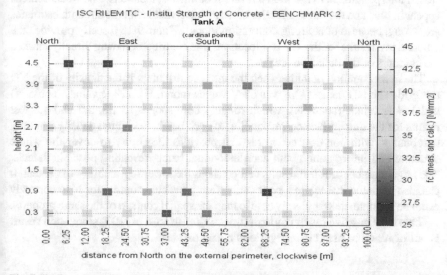

Fig. 7.13 Contributor H, velocity tests on a regular (6.25 m × 1.2 m grid, 64 ND test locations)

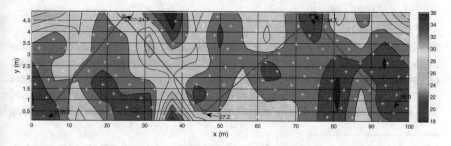

Fig. 7.14 Contributor N, spatial distribution of the converted strength values; the zig-zag pattern refers to the location of cores

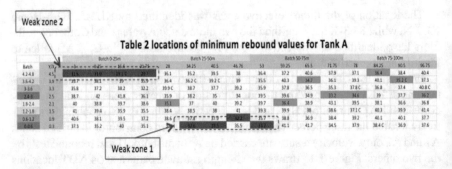

Fig. 7.15 Contributor K, using rebound screening on a regular (6.25 m × 0.6 m grid) (144 ND test locations)

spatial interpolation between the 163 NDT rebound test locations. The weakest areas appear in blue and the three defective areas are also visible but the strength estimates are wrong, because of a wrong conversion model. Figure 7.15 directly provides the rebound test results for the 128 test locations, the contributor having directly marked the contours of the defective areas.

The main common point between the three contributors is the density of the ND test grid. The ND dataset size was 64 V values (contr. H), 128 R values (contr. K) and 163 R values (contr. N), which roughly corresponds to 1.8, 3.5 and 4.5 test results for the unit surface corresponding to each batch. Such high density can only be obtained with non-destructive tests. It provides the ability to cover the whole structure and information about each individual batch, revealing possible defects. The larger number of R test results compensates the lower repeatability of the test. It can also be noted that while a poor calibration prevented contributor N to correctly estimate strength, he was clearly able to map the spatial contrast of the same property.

This very positive result confirms one major interest of NDT, since such result is out of reach in usual investigations based on cores.

7.6.5 Summary of Contributor Performances Regarding All Objectives

The benchmark was not designed as a competition. Its purpose was to compare the methodologies used by experts, in order to clarify what are the possible consequences of the strategic choices at the various stages of the concrete strength assessment process. Therefore, it may be helpful to compare the quality of assessment obtained by all contributors[11]. The details of each assessment are given in Table 7.12 regarding

---

[11]It must be pointed that, like for the first benchmark (Chap. 6), chance has played some role in the results, and that this comparison is not a ranking between contributors. Furthermore, some important features identified during this second benchmark will be, like for the first benchmark, further analyzed by randomly repeating the full process in a Monte-Carlo simulation (see Chap. 8).

Table 7.17 Synthesis of assessment efficiency and quality

KL		Number of cores	Mean strength	Standard deviation	Defective areas	Synthesis
KL1	D		***	**	*	6
	G		***	**	*	6
KL2	A	**	**	***	***	10
	B		***	*		4
KL3	H	***	***	***	***	12
	I	***	*	*	*	6
	J	*	**	***	*	7
	K		***	***	**	8
	L	*	***	***		7
	N	*	*		**	4
	O	***	**	***	*	9
	C	***	*	**		6
	E	***	*	**	*	7
	F	*	**	*		4
	M		***	***		6

the number of cores, in Tables 7.13 and 7.14 regarding the strength estimates and in Table 7.16 regarding the defective areas. Table 7.17 summarizes all the contributor performances on a unique ranking scale, by giving them from zero to three stars on each of the four main criteria:

- the limitation of the number of cores: i.e., for KL3, *** if $N_c \leq 12$, ** if $N_c \leq 16$, * if $N_c \leq 24$. These maximum numbers are proportionally reduced for KL2 (8, 12 and 16 cores, respectively) and KL1 (4, 6 and 8 cores, respectively);
- the ability to estimate mean strength of Tank A, with *** if error < 0.75 MPa, ** if error < 1.5 MPa, * if error < 3 MPa;
- the ability to estimate standard deviation of Tank A, with *** if error < 15% (or 0.5 MPa), ** if error < 30% (or 1.0 MPa), * if error < 50% (or 1.5 MPa);
- the ability of identifying defective areas, which includes their location and strength estimation, with *** if 3 zones are identified, ** for 2 zones, * for 1 zone.

A last column in Table 7.15 indicates the resulting number of stars.

This synthesis confirms that most contributors were efficient in estimating the concrete strength properties (mean value and standard deviation), even with a limited amount of resources[12]. Moreover, as it was illustrated at Sect. 6.4, NDT results are necessary to identify the spatial variations of concrete properties over the investigation domain.

---

[12]It must also be reminded that several contributors (I, C, F, L, O, M, see Table 7.15 in Sect. 7.6.1) suffered from the problem due to the influence of carbonation on rebound test results, which induced some wasting of resources.

7.7 Synthesis of What Can Be Derived for RILEM Guidelines

7.7.1 Example of a Successful Investigation with a Limited Amount of Resource (KL2)

Before drawing synthetic conclusions regarding the efficiency of the investigation strategies and the identification of good practice, analyzing the approach followed by Contributor A deserves further attention as he was successful in answering all questions of the benchmark with a limited amount of resources, as the concrete properties were estimated only at KL1 and KL2 levels, using a progressive approach (see Sect. 7.5.1.1 for details).

For KL1, the non-destructive investigation program combined 84 rebound tests (33 on Tank A, less on all others tanks) and 17 direct velocity tests on Tank A. Seven cores were taken (5 on Tank A), at locations where rebound or velocity had reached extreme values. By analyzing the distribution of rebound test results (Fig. 7.16), it was checked that they were not normally distributed (Shapiro-Wilk statistical test and a significance level of 5%).

For KL2, a larger coverage of the tanks was undertaken with velocity measurements (44 on Tank A, 14 on each of the other tanks) and three more cores were taken. The analysis of the test result distributions (considering all tanks together or separately) confirmed the left-skewed distribution of concrete properties, which could be viewed as an indicator of defective areas. Several conversion models were tested, using rebound results, velocity results, or their combination. The strength variability (including the error resulting from the uncertainty on the conversion model) was estimated to be between 3 and 5 MPa, being the highest value obtained for tank D.

The identification of defective areas was based on the analysis of lowest values of estimated strength and assumptions related to the prior knowledge about the structure. For instance, regarding Tank A, defective areas were considered to be those

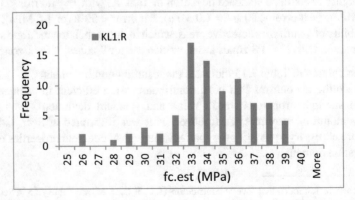

Fig. 7.16 Statistical distribution of strength estimated from rebound test results (all Tanks, KL1)

Table 7.18 Weakest areas identified for KL2

x (m)	y (m)	$f_{cm,is}$ (MPa)
0.0–18.0 and 88.5–100.0	4.5	25.5–26.9
72.0–86.0	4.5	27.7
26.0–44.0	0.3	27.6

exhibiting a strength below an estimated characteristic value (30,2 MPa, calculated with the velocity conversion model, applying the EN 13791 criteria). The expected extension of the areas is that corresponding to the volume that includes these locations. Assuming that the test locations are at the center of the volume, the extension of those areas are ± 12,5 m on each side and ± 0,30 m in the vertical direction from the original test location[13]. This led to the identification of the areas described in Table 7.18, which correspond to the right answer (see Sect. 7.3.5.3).

7.7.2 Lessons Regarding the RILEM Recommendations

The benchmark was carried on at a time where the experts of the RILEM TC-ISC 249 Committee were still discussing about the Recommendations. At that time the final consensus had not been reached but a series of good practices had been shared among the experts. Therefore, most of them were used in their contribution. This is the case, for instance, of a wide screening using NDT and of conditional coring, based on these NDT results. The final results of the benchmark can be analyzed having in mind the flowchart of Fig. 1.9 which describes the logical organization of the investigative program (Fig. 7.17).

A first important conclusion is that the identification of a relevant conversion model and the control of the quality of strength estimate must be the two major aims of the investigation program. Let us now summarize what works and is easy to put into practice.

– **carrying NDTs is the first step**. These tests enable a spatial coverage of the investigation domain at a low cost and provide the information necessary for the following stages, including the definition of core locations. The spatial grid of test locations can be regular or not, but it must account for the possibility of combining different types of test results),
– **devoting a minor part of resources to assess test result precision is useful** (see Appendix 7.3),
– **checking test results is mandatory** before going on. Building the statistical distribution of test results provides information about defective areas or where to take cores,

---

[13]In practice, during a real on-site investigation, refining the extension of these areas is easier because the limits between the different batches may be visible.

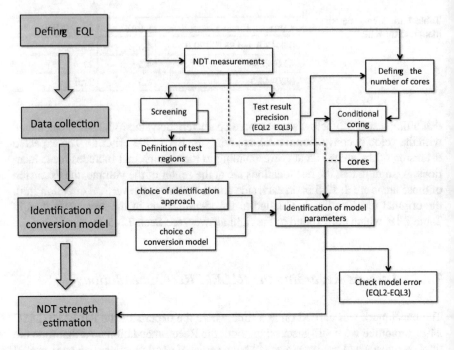

Fig. 7.17 Flowchart of the investigation and its mains steps (RILEM TC 249-ISC Recommendations)

- **as soon as cores are available, NDTs on cores can provide useful information**. For instance, it can inform about problems during drilling. In this benchmark, some contributors have understood, thanks to these laboratory tests, that some rebound test results on the carbonated side would lead to unreliable results, and they could adapt their strategy,
- **core strength is the best way to assess concrete strength!** Therefore, the major interest of NDTs is their ability to provide further information, like spatial variability, which cannot be detected with a limited number of cores. The final strength estimates can also merge some direct strength test results and a set of estimated strengths.

For the sake of simplicity, we have not developed the conversion model identification stage, which is detailed in other sections of this book. However, the reader will find in Appendix 7.2 useful information regarding this issue and the control of the predictive model error. It must be noted that the benchmark confirmed the need to be very cautious with the development of complex conversion models (i.e. that have more model parameters), as they usually do not result in a lower strength estimation uncertainty.

Acknowledgements The 2nd author would like to acknowledge the financial support by Base Funding—UIDB/04708/2020 of CONSTRUCT—Instituto de I&D em Estruturas e Construções, funded by national funds through FCT/MCTES (PIDDAC).

Appendix 7.1: Number of Tests of Each Type for KL1 and KL2 Investigations

These two tables provide the information about the number of cores and non-destructive tests for all contributors that considered the KL1 and the KL2 levels (see Table 7.11 in the main text for KL3 level) (Tables 7.19 and 7.20).

Table 7.19 Synthesis about the number of cores and semi-destructive tests for KL1

	D	G	C	E	M	A	B	F
Cores on A	5	3	–	–	6	5	4	2
Cores on others	0-1-3	3-3-3	–	–	2-2-2	0-0-2	–	2-2-2
Total cores	9	12	–	–	12	7	4	8
SDT on A	–	–	–	4	–	–	–	–
ND tests								
R on A	17	18	100	8	28	33	72	14
R on others	21	54	142	24	12	51	60	36
UPV on A	8	–	–	36	32	17	–	10
UPV on others	15	–	–	96	12	–	–	24
Total NDT	61	72	242	164	84	101	132	84

Table 7.20 Synthesis about the number of cores and semi-destructive tests for KL2

	A	B	C	E	M
Cores on A	5	16	3	2	8
Cores on others	1-1-3	0-0-3	0-0-3	2-2-1	4-4-4
Total cores	10	19	6	8	20
SDT on A	–	–	5	4	–
ND tests					
R on A	33	81	108	50	26
R on others	51	72	157	27	78
UPV on A	44	–	3	68	26
UPV on others	41	–	3	101	69
Total NDT	169	153	271	246	199

Appendix 7.2: Recommendations Regarding the Conversion Model Identification and Validation

During the benchmark, each contributor received its own set of test results, corresponding to its specific investigation program and identified its own conversion model(s). It is therefore very difficult to compare the models, as everything is different: data set, model type and identification strategy. To better understand what can be done (and to draw lessons for the RILEM TC 249-ISC Recommendations), some specific post processing work has been carried out separately, which is described here.

Comparison between conversion models

All datasets of test results were considered as they were obtained by the contributors (their main characteristics are summarized in Table 7.12). For each of these datasets, three conversion models were identified when it was possible, i.e. a linear model with rebound, a linear model with velocity and a combined power law model. A least squares regression was used for the model parameter identification. This led to 10, 11 and 6 datasets, respectively. Table 7.21 summarizes what was obtained for both rebound and velocity. For each model, it provides the model parameters a and b, the determination coefficient identified by the regression analysis and the number of degrees of freedom of the model. It must be noted that these models may differ from

Table 7.21 Characteristics of the linear regression models identified from the available data sets (dof = number of degrees of freedom of the conversion model)

	$f_{c\,est} = aR + b$					$f_{c\,est} = aUPV + b$				
	a	b	r^2	dof	RMSE$_{pred}$	a	b	r^2	dof	RMSE$_{pred}$
A	0.866	1.16	0.6	7	2.44	0.0305	−97.4	0.93	8	1.65
B	0.55	11.1	0.79	2	3.13					
C						0.0319	−103.3	0.74	4	1.69
D	1.372	−17.7	0.81	7	2.90	0.0275	−84.9	0.92	7	1.62
E	0.702	9.3	0.7	4	3.17	0.0306	−96.9	0.95	6	2.00
F	0.344	21.5	0.06	15	2.83	0.0201	−53.1	0.59	10	1.80
G	1.04	−6.6	0.66	10	2.81					
H						0.0313	−100.5	0.83	8	1.74
I						0.0279	−87.5	0.84	8	1.92
J						0.020	−50.5	0.37	20	1.70
K	0.515	13.6	0.18	25	2.63					
L	0.625	10	0.29	22	2.48	0.0332	−109.2	0.67	22	1.69
M	0.788	3.6	0.28	28	2.48	0.0305	−98.1	0.75	28	1.65
N	0.968	−2.7	0.42	16	2.48					
O						0.0383	−131.1	0.97	4	1.92

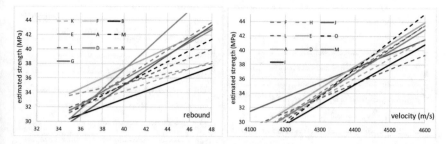

Fig. 7.18 Illustration of conversion models identified with the different sets on test results (rebound on left and velocity on right)

those identified by the contributors, as they may have used another model type or another calibration approach. The two remaining columns in Table 7.21 (RMSE$_{pred}$) provides, in MPa, the prediction error (see Sect. 1.5.6.2 for the definition of this parameter). This estimator is the best way to quantify the quality of the assessment since it measures the average uncertainty when using the conversion model as a predictor of an individual strength. In real investigations it can be estimated by using the leave-one-out procedure (see Sect. 11.2). However, with synthetic simulations, it is easy to calculate it by using the conversion model to estimate strength at all possible test locations on Tank A from the values of NDT results.

Most comments regarding the results in Table 7.21 are valid for both rebound and velocity test results. The first comment is about the variety of identified (a, b) pairs. Figures 7.18a, b illustrate linear conversion models for rebound (Fig. 7.18a) and velocity (Fig. 7.18b). They show large differences, both on the slope parameter a and on the intercept parameter b.

However, when the a and b model parameters are plotted together, a strange feature emerges, as these parameters are strongly correlated. This phenomenon, named "trade-off", was previously pointed as being a general property, and can be seen to be relevant for both on field data and on laboratory data[14] (Fig. 7.19).

Assessing the prediction error

The explanation lies in the mathematical inversion problem of identification of "best" (a, b) pair from a data set of N pairs (f$_c$, NDT) with random uncertainty on measurements (see Sects. 11.3 and 12.4 for details). Due to (a) the sampling uncertainty which leads to different data sets, (b) to the minimization of an error function (least squares method), each single identification process leads to a specific (a, b) pair, which belongs to a whole set of "equivalent error set". The conclusion is that the differences between the (a, b) pairs of the various conversion models is only the result of chance. However, these models have not the same prediction ability. This is why the issue of **identifying the best conversion model must be revisited in order to ensure the best predictive** ability of the model.

---

[14]Breysse and Fernández-Martínez [8].

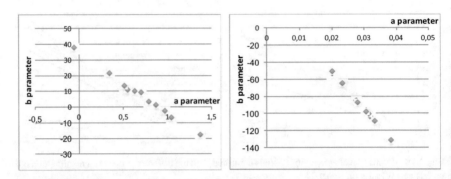

Fig. 7.19 Illustration of the trade-off between conversion model parameters (left: $f_c = a\,R + b$, right: $f_c = a\,V + b$)

The second interesting point is related to the criterion used to assess the quality of the model. The determination coefficient r^2 (Table 7.21) is the most commonly used parameter, but it is not a relevant indicator. In fact, $RMSE_{pred}$, which is the relevant indicator of the predictive ability, appears to be poorly correlated with r^2. Furthermore, r^2 is highly dependent on the number of cores, since it is easier to fit a model when the size of the dataset (in our case the number of cores) is reduced, but this is by no way a sign that the model can accurately estimate strength. Table 7.21 shows that the lowest $RMSE_{pred}$ values mostly correspond to larger datasets (i.e. higher dof). A larger dataset leads to a more stable and representative model, an aspect that the r^2 value provides nothing about.

It is also visible that conversion models identified with velocity test results are better (lower RMSE) than those identified with rebound test results. With velocity, the best models lead to RMSE values of about 1.6 MPa while with rebound, the best models yield RMSE values of about 2.4 MPa. This is directly linked with the random measurement error affecting NDT results. This error has a double influence: (a) on the dataset used for identifying the conversion model, (b) when the model is used for assessing strength for a new test result. **This clearly puts a limit for the estimation of any local strength value**, which can be assessed from the **test result precision** (Appendix 7.3). It can be added that values of the prediction error close to 3 MPa, as is obtained with some datasets, are of little interest, since it is approximately the magnitude of the standard deviation of strength for the whole tank.

More complex conversion models: nonlinear models, multivariate models

A usual question when identifying the conversion model is that of the mathematical shape to be chosen (linear, exponential, power…). Most contributors of the benchmark have chosen linear models. We have however tested alternative models (exponential model $f_c = a\,\exp\,(bV)$), and seen that they exhibit the same trade-off property between the a and b coefficients. While the determination coefficient may be a little bit better for nonlinear models, the RMSE values are roughly identical, for each data set, to those of linear models. It thus seems **relevant to consider only**

Table 7.22 Comparison of the predictive ability of univariate and bivariate models

	$f_c(R)$	$f_c(V)$	$f_c(R, V) = a\,R^b\,UPV^c$					
	$RMSE_{pred}$	$RMSE_{pred}$	b	c	ln (a)	r^2	ddl[15]	$RMSE_{pred}$
A	2.44	1.65	0.310	3.19	−24.3	0.96	6	1.62
D	2.90	1.62	0.454	2.77	−21.28	0.93	6	1.81
E	3.17	2.00	0.072	3.97	−29.91	0.99	3	2.56
F	2.83	1.80	−0.129	2.85	−19.87	0.63	9	1.90
L	2.48	1.69	0.204	3.72	−28.35	0.67	21	1.55
M	2.48	1.65	0.269	3.68	−28.22	0.80	27	1.95

linear models, as long as experimental data do not cover a large range and do not exhibit obvious nonlinear features.

A last issue commonly discussed in the literature is that of the possible combination of several non-destructive methods for improving the estimation of concrete strength. The efficiency of this combination remains an open issue in the literature. Table 7.22 synthesizes, for the 6 datasets which combined rebound and velocity measurements, the (a, b, c) model parameters that can be identified with a double power law (SonReb type) conversion model. The predictive $RMSE_{pred}$ of the combined model can be compared with those of the two univariate linear models.

Although the coefficient of determination r^2 of the combined models is slightly better than that of the single models (it is easier to fit a model when the number of model parameter increases), this is not the case for the RMSE values.

For five out of six datasets, the performance of the combined model is inferior to that of the univariate models. This comes from the fact that the lower repeatability of rebound measurements has a negative effect on the combined model efficiency. This confirms that:

- the coefficient of determination cannot be used as an indicator of the model predictive ability,
- the combination of several NDT test results in a single model may not be efficient.

Appendix 7.3: Repeatability of Test Results (or Test Result Precision, TRP)

The Test Result Precision (TRP) is a crucial parameter which indicates the magnitude of the measurement uncertainty associated to each test result. This parameter is related to both destructive and non-destructive tests, and both on-site and laboratory tests. As explained in Appendix 7.2, a higher TRP may be the reason why the final predictive RMSE is smaller. However, assessing the TRP is not common in engineering practice and only two contributors (O and E) paid some attention to its assessment. Their approach and the information that can be deduced is described in the following.

Test result repeatability for non-destructive tests (Contributor O, KL3).

The choice was to repeat 5 times the testing process at the same test location. In practice, for Tank A, the reference test location was x = 50 m, y = 1.5 m and measurements were carried out at 4 additional testing points within a close vicinity, 20 cm apart in all directions (i.e. vertical and horizontal). Table 7.23 provides the series of test results.

The standard deviation and, therefore, the coefficient of variation can be compared to that estimated from all tests carried out on the same tank. Table 7.24 synthesizes these results.

The value of information provided by a test result increases if the value of cv_{rep} is small compared with the overall cv, calculated on the whole tank. Particularly valuable information can therefore be drawn from these results:

- the semi-direct velocity tests on the internal face provided the larger amount of information, followed by the direct velocity tests,
- the semi-direct velocity tests on the external face and pull-out tests are not so good,
- the rebound number tests on the internal face have the lowest repeatability.

The main consequence of these results is their direct influence on the quality of correlations used in the conversion model identification stage. The effect of a lower test result precision (TRP) is a higher RMSE value obtained at the end of the process (see Table 7.22). Another effect is that a test result cannot be taken at face value, but should be considered in a wider context. It is only with a larger number of test results and the smoothing effect of larger samples that tests with a poor repeatability can be useful. It must be noted that, when compared to the total number of NDTs carried out on the same tank (Table 7.11), the additional effort for assessing repeatability remains marginal (4 more rebound tests *vs* 91, 12 more velocity tests *vs* 74).

Test result repeatability for non-destructive tests (Contributor E, KL3).

Contributor E took 12 cores (3 cores on each tank) following a similar process: the second and third cores were taken at +/− 25 cm from the first one. This process enables to quantify the local scale variability of core strength, which combines the effect of the local scale material variability, and that of the variability of the strength measurement process (including the drilling phase). The average variability for each of the four sets of three cores can be compared to the global variability (on the full set of twelve cores), which combines the variability of the strength measurement process to the effect of the material variability at a larger scale. The available results are summarized in Table 7.25.

The local cv varies from 4.7 to 7.3% (average value is 6%), while the global one amounts 9.8%, which comes to a global variance two to three times larger than the local one. Therefore, any local strength value must be considered with this +/− 6% uncertainty margin (i.e. roughly +/− 2 MPa), corresponding to repeatability of the

Table 7.23 Test results for repeatability assessment (the index indicates the internal or external side)

	Rebound$_i$	UPV$_{direct}$ (m/s)	UPV$_{semi-direct\ e}$ (m/s)	UPV$_{semi-direct\ i}$ (m/s)	X$_{c\ e}$ (mm)	PO$_e$ (kN)
Test 1	36.9	4361	3854	4013	3.5	27.6
Test 2	47.2	4479	3892	3992	3.5	27.7
Test 3	38.4	4455	3739	4025	4.0	27.6
Test 4	42.3	4473	3996	4030	3.2	27.9
Test 5	38.3	4371	3874	3963	3.3	24.4
st. dev.	4.2	57	92	27	0.3	1.5

Table 7.24 Test results precision assessment (cv is the coefficient of variation)

	Rebound$_i$	UPV$_{direct}$ (m/s)	UPV$_{semi-direct\ e}$ (m/s)	UPV$_{semi-direct\ i}$ (m/s)	$x_{c\ e}$ (mm)	PO$_e$ (kN)
cv$_{rep}$ (%)	11	1.3	2.4	0.7	8	5
cv (Tank A) (%)	7	2.6	2.8	3.4	3	7

Table 7.25   Core strength test results and variability assessment

Tank	A	A	A	B	B	B	C	C	C	D	D	D
Strength (MPa)	33.8	34.4	37.9	37.3	38	42.6	31.1	30.2	33.1	33.2	36.7	33.8
Mean	35.4			39.3			31.5			34.6		
cv_{rep} (%)	6.3			7.3			4.7			5.4		
Global cv (%)	9.8											

strength measurement process at a local scale. As for the non-destructive test results repeatability, this variability has also some impact on the accuracy of conversion models.

The large scale variability is estimated at about 10%, which is more than the local uncertainty. This means that the material variability has an important effect but these measurements cannot separate between the variability at the tank scale and the between-tank variability.

References

1. Soutsos et al.: In: D. Breysse (ed.) Non-destructive Aassessment of Concrete Structures: Reliability and Limits of Single and Combined Techniques. RILEM SOA TC-207, pp. 151–154 (2012)
2. Bob, C.: Durability of concrete structures and specification. In: Dhir, R.K., Dyer, T.D., Jones, M.R. (eds.) International Congress on Creating with Concrete. University of Dundee, Dundee, 6–10 September 1999, pp. 311–318
3. Parrott, L.J.: A Review of Carbonation in Reinforced Concrete. Cement and Concrete Association, Slough (1987)
4. Duval, R.: La durabilité des armatures et du bétons d'enrobage, in La durabilité des bétons, Collection de l'ATILH, pp. 173–226. Presse ENPC, Paris, France (1992)
5. Basheer, P.A.M., Russell, D.P., Rankin, G.I.B.: Design of concrete to resist carbonation: rate of carbonation of concrete. In: Lacasse M.A., Vanier D.J. (eds.) 8th International Conference on "Durability of Building Materials and Components, Vol. 1. NRS Research Press, Vancouver, Canada, 30 May-3 June 1999, pp. 423–435
6. Turgut, P., Kucuk, O.F.: Comparative relationships of direct, indirect and semi-direct ultrasonic pulse velocity measurements in concrete. Russ. J. Nondestruct. Test. **42**(11), 745–751 (2006)
7. Pham, S.T.: Étude des effets de la carbonatation sur les propriétés microstructurales et macroscopiques des mortiers de ciment Portland, Ph. D. Thesis, University of Rennes, France (2014)
8. Breysse, D., Fernández-Martínez, J.L.: Assessing concrete strength with rebound hammer: review of key issues and ideas for more reliable conclusions. Mater. Struct. **47**, 1589–1604 (2014)

Chapter 8
Illustration of the Proposed Methodology Based on Synthetic Data

Denys Breysse, Jean-Paul Balayssac, David Corbett, and Xavier Romão

Abstract This chapter illustrates first how the recommendations proposed by RILEM TC 249-ISC can be applied in practice. All the steps of the investigation process described in the recommendations are followed and the obtained results are discussed. The second objective is to demonstrate the practical interest of some of the options promoted or recommended in the recommendations. The case study, based on synthetic data, is a four-storey reinforced concrete frame structure. To be representative of a real case study, the synthetic data involves different types of variability, i.e. between mixes (each storey corresponding to a different mix), between components of a given storey and within each component, due to the casting process. Additional variability comes from concrete moisture that can vary slightly around its mean value corresponding to a saturation degree assumed to be 80%.

The purpose of this chapter is to address two issues:

1. to illustrate how the Recommendations can be followed in practice, by showing what must be done at each main step of the investigation process and what type of results are obtained,
2. to show, by using the same synthetic framework and illustrative example, what is the practical interest of some options that are promoted or recommended in the Recommendations.

---

[1] Alwash et al. [1, 2].

D. Breysse
University Bordeaux, I2M-UMR CNRS 5295, Talence, France

J.-P. Balayssac (✉)
LMDC, Université de Toulouse, INSA/UPS Génie Civil, Toulouse, France
e-mail: jean-paul.balayssac@insa-toulouse.fr

D. Corbett
Proceq SA, Zurich, Switzerland

X. Romão
CONSTRUCT-LESE, Faculty of Engineering, University of Porto, Porto, Portugal

© RILEM 2021
D. Breysse and J.-P. Balayssac (eds.), *Non-Destructive In Situ Strength Assessment of Concrete*, RILEM State-of-the-Art Reports 32,
https://doi.org/10.1007/978-3-030-64900-5_8

279

The data used in this chapter are synthetic data that have been generated in order to reproduce all of the main features that have some influence on the efficiency of the assessment strategy. The principles and advantages of the synthetic simulations are not described here and the reader is invited to refer to available literature[1] to get more detailed information on how data are generated and processed. Such simulations were used by RILEM TC 249-ISC during their work to address some theoretical points that could not be addressed with real data sets, because of the intrinsic random nature of any investigation program and of the resulting data set.

8.1 Description of the Case Study

8.1.1 The Synthetic Structure

The case study is based on synthetic data that are fully representative of what could be found on a real site. The interesting aspect of synthetic data is that their "true strength values" are known (although they are kept secret during the assessment process, they can be used in order to determine the error in the estimates, i.e. the deviation between reference (true) and estimated strengths). Additional information about the simulation of synthetic data can be found in the above cited references.

The case study under investigation is a four-storey reinforced concrete frame structure. Each storey has 20 ($= 4 \times 5$) columns and 16 ($= 4 \times 4$) beams (Fig. 8.1). The distance between columns is 5 m in each direction, which corresponds to a floor surface of 300 m^2 for each storey. The concrete properties are slightly different in beams and columns whose mean concrete strength is 28 MPa and 35 MPa respectively. Variability in the concrete strength is simulated as a result of 3 sources of variability: the variability between mixes (each storey corresponding to a different mix), variability between components of a given storey, and variability within each component, due to the casting process. Additional variability comes from concrete moisture which can vary slightly around its mean value corresponding to a saturation degree assumed to be 80%).

8.1.2 The Synthetic Investigation Program

The measurement uncertainty during non-destructive and destructive testing is considered and simulated by adding a random value, assumed to follow a normal distribution, to the "exact value" of a theoretical perfect measurement. To be in agreement with what can be found in common practice several options are considered for Test Result Precision (TRP) with rebound measurements and ultrasonic pulse velocity measurements:

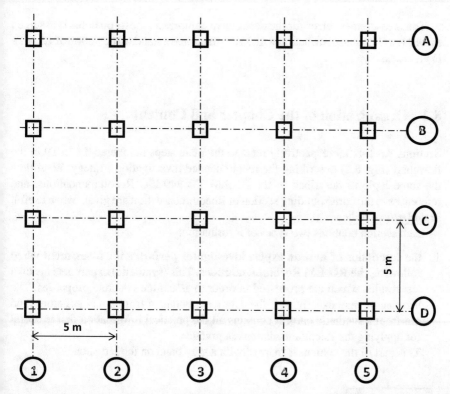

Fig. 8.1 Schematic distribution of columns on each story (where the letter denotes the line and the number denotes the row; beam A_i is between columns A_i and A_{i+1})

- for rebound measurements, the two possibilities are named R^- and R^+ and respectively correspond to sd_{rep} values of 2.5 units and 1.5 units respectively. These two cases correspond to medium precision TRP according to Table 5 of the RILEM TC 249-ISC Recommendations (i.e. $1 < sd_{rep} \leq 3$) even if the R^- case is close to the upper bound whereas the R^+ case is close to the lower bound;
- for UPV measurements, the two possibilities are named V^- and V^+ and respectively correspond to sd_{rep} values of 75 m/s and 35 m/s respectively. According to Table 5 of the RILEM TC 249-ISC Recommendations, the V^- case corresponds to medium precision TRP (i.e. $50\,m/s < sd_{rep} \leq 125\,m/s$) and the V^+ case corresponds to high precision TRP ($sd_{rep} \leq 50$ m/s).

Three possible test locations are defined for each component (bottom/middle/top in columns and left/centre/right in beams) and it is assumed that all measurements are taken at the centre of rebar mesh in order to have a negligible influence of rebar. It is possible to repeat the non-destructive test at any test location in order to estimate the TRP.

The measurement error for the destructive compression tests on cores is assumed to follow a Gaussian distribution with a zero mean and a standard deviation of 1.5 MPa in all cases.

8.2 Organization of the Chapter and Content

Sections 8.3.1–8.3.9 respectively refer to the main steps numbered T1 to T9 in the flowchart (Fig. 8.2) describing the recommended investigation strategy. We follow the same logic as described in the RILEM TC 249-ISC Recommendations, and references to the corresponding section of Recommen-dation are given, when useful, in *underlined italics*.

Each section contains two types of information:

1. the description of **how an expert investigator performs the assessment when following the RILEM Recommendations**. This (synthetic) expert gets his own test results, which are processed in order to assess the concrete properties. The text can be read exactly as if the data were obtained from a real structure (and not from a synthetic one). It contains all the practical information that is useful for applying the Recommendations in practice.

This part of the content is marked with a grey band on the left side.

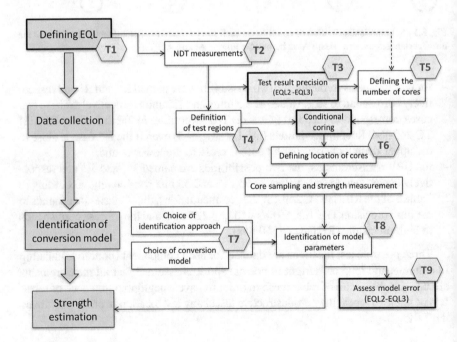

Fig. 8.2 Flowchart illustrating the overall methodology

2. **more detailed analyses** in which the test results, models and assessed values are compared, either between several (simulated) expert investigators using different means or strategies, or, thanks to synthetic simulations, by repeating the same investigation and assessment strategy many times (which is one of the main advantages of using synthetic simulations). This makes it possible to really compare the efficiency of different strategies or options, without disturbances resulting from the random nature of each individual measurement. This also illustrates how the different strategies are actually able to satisfy the prescribed assessment objectives.

   The interested reader will find useful information which quantifies the differences between the different options and that will help him to better understand the advantages and/or limits of his possible choices/options during the investigation. This part of the content is marked with a light grey band on the right side.

8.3 Developing the Investigation and Assessment

8.3.1 Task 1. Defining EQL

The first task described in the flowchart (Fig. 8.2) is the definition of the required Estimation Quality Level, that depends on the assessment objectives and on the resources available for the investigation.

Eight expert investigators (respectively numbered from E1 to E8) are simulated, to whom different objectives and resources are assigned. They correspond to more or less ambitious investigation programs. Table 8.1 synthetizes what EQL is assigned to each investigator and what type of NDT is used by each of them. One of the key differences is that some investigators choose predefined coring (PC) whereas others use conditional coring (CC).

Table 8.1 Main features defining the eight investigators, their objectives and technical options (R for rebound, UPV for ultrasonic pulse velocity)

Expert	E1	E2	**E3**	E4	E5	E6	E7	E8
EQL	EQL1	EQL2	**EQL2**	EQL2	EQL2	EQL2	EQL3	EQL3
ND technique	R	R	**UPV**	UPV	UPV	UPV	UPV	R and UPV
TRP	Medium (R^-)	Medium (R^-)	**Medium (V^-)**	Medium (V^-)	High (V^+)	High (V^+)	High (V^+)	High (V^+) and Mean (R^+)
PC/CC	PC	PC	**PC**	CC	PC	CC	CC	CC

Table 8.2 Target tolerance intervals for the three EQLs (*see Table 1 in §1. Scope*)

Estimated property		EQL1	EQL2	EQL3
Mean value of local strengths		±15%	±15%	±10%
Standard deviation of local strengths	Relative	Not addressed	±50%	±30%
	Absolute		±4 MPa	±2 MPa
Error on local strength	Relative		20%	15%
	Absolute		6 MPa	4.5 MPa

It can be noted that:

– Experts E2 to E6 have an EQL 2 objective with either medium TRP or high TRP, whereas E1 has an EQL1 objective and E7 and E8 have an EQL 3 objective. Table 8.2 provides a reminder about the assessment targets and the tolerance intervals corresponding to each EQL.
– Conditional coring is mandatory at EQL3 (experts E7 and E8) and recommended at EQL2. Some experts use PC whereas others use CC, which can therefore be compared.

We will follow the specific example of expert E3, who has an EQL2 requirement, using NDT tests with UPV measurements and has a predefined coring strategy.
Table 8.1 shows that the following comparative analyses will be possible:

– between E1 and E2 or between E6 and E7 to see the effect of changing the EQL while keeping the same type of NDT test and the same way of selecting core location,
– between E2 and E3 to see the effect of changing the type of NDT test while keeping the TRP level (medium),
– between E3 and E4 or between E5 and E6 to see the effect of the method used to select the core locations (PC or CC),
– between E3 and E5 or between E4 and E6 to see the effect of the TRP level (from medium to high) of NDT measurements,
– between E7 and E8 to see the effect of combining the two types of NDT tests.

Other comparisons will be possible but they correspond to options during the processing of NDT test results, i.e. regarding the type of conversion model (linear or other) or the possibility of using or not using a bi-objective approach (targeted at the concrete strength variability assessments). These comparisons will be considered later, as they have no impact on the first steps of the investigation.[2]

---

[2]This comparison and the analysis of the combination of ND methods is detailed at Sect. 12.9.

Table 8.3 Number of test locations for each EQL and the corresponding percentage of tested components (NDT measurements)

EQL	Experts	Total number of tests	Tests in columns		Test in beams	
			Number of tests	% of components	Number of tests	% of components
EQL1	1	28	16	20%	12	19
EQL2	2 to 6	56	32	40%	24	38
EQL3	7–8	84	48	60%	36	56

8.3.2 Task 2. Performing NDT Measurements

The number of NDT tests is not explicitly defined in the RILEM Recommendations (see Sect. 1.3.3 *Methodology for determining the locations and number of NDT test locations*).

As the total cost of investigation must be shared between taking and testing cores to obtain direct strength measurement and making non-destructive tests measurements, it is logical to increase the number of test locations with EQL. Furthermore, as the NDT test results are useful for (1) getting a full picture of the structure, (2) identifying the conversion model, (3) assessing strength where only NDT test results are available, it is also logical to have an even distribution of these test locations across the different floors.

Table 8.3 provides the number of test locations for the three possible EQLs and the corresponding percentage of components where NDT measurements are taken. All of these numbers can be divided by a factor of 4 to have the number of tests at each floor.

Expert E3 performs a total of 56 NDT tests, i.e. 14 tests at each floor with 8 tests distributed among the 20 columns, and 6 tests on the 16 beams.

These tests are evenly distributed at each floor following a predetermined pattern. For instance, on the fourth floor, the 8 tested columns are A3, A4, B1, B5, C1, C4, D2, D4, where the letter denotes the line and the number denotes the row (see Fig. 8.1). UPV test results are provided in Table 8.4.

The number of tests for the other experts is equal to (experts E2, E4, E5, E6), half of (expert 1) or 50% more than (experts 7 and 8) that of expert 3.

8.3.3 Task 3. Assessing the Test Results Precision (TRP)

Assessment of NDT test result uncertainty is a crucial stage (*see Sect. 1.5.2. NDT test result precision*) as the precision of input data has a major effect on the uncertainty of the strength estimate. Thus, a specific part of the investigation must be devoted to assess the magnitude of the test result precision (TRP), which is a key factor in the next stages of the compressive strength assessment, and determines the recommended number of cores at each EQL.

Table 8.4 Test results of Expert E3, for all beams and columns (UPV in m/s)

Floor	Columns			Beams			All	
	Test results	Mean	s.d	Test results	Mean	s.d	Mean	s.d
1	4395, 4548, 4257, 4338, 4474, 4529, 4347, 4446	4417	101	4132, 4156, 4408, 4200, 4321, 4241	4243	105	4342	133
2	4289, 4194, 4570, 4134, 4398, 4311, 4084, 4336	4290	155	4038, 4148, 4288, 4271, 3979, 4315	4173	141	4240	156
3	4375, 4285, 4097, 4395, 4411, 4478, 4372, 4111	4315	141	4148, 4087, 4034, 4218, 4087, 4100	4112	63	4228	152
4	4404, 4489, 4289, 4402, 4222, 4328, 4435, 4253	4353	94	3932, 4294, 4234, 4226, 4340, 4158	4197	144	4286	138
All		4344	129		4181	120	4274	148

It is based on a series of repeated measurements at some specific and predetermined test locations (2–5). **The determination of TRP is recommended at EQL1 and mandatory at EQL2 and EQL3** (*see 1.5 in Sect. 1.6.1.1. Recommended sequence of tasks*).

To assess TRP, expert E3 defines two test locations, respectively column C3 on the third floor and beam B4 on the second floor, where six NDT measurements have to be taken. These components are two of the components for which test results are presented in Table 8.4. The first UPV test result of the series for each component are marked in bold characters in Table 8.4, i.e. 4411 m/s and 4288 m/s, respectively.

The series of six NDT test results are, in m/s:

- in column C3: 4411, 4481, 4431, 4252, 4372, 4421
- in beam B4: 4288, 4319, 4278, 4229, 4147, 4301.

These results lead to standard deviations (sd) of 78 m/s and 63 m/s, respectively, which comes to a mean standard deviation of 70 m/s. According to *Table 5 in Sect. 5.2. NDT test result uncertainty of Guidelines*, this corresponds to a **medium precision TRP**.

All experts follow the same methodology, assessing the repeatability of the test results at 2 or 4 test locations, on columns and beams. Their results are summarized in the Table 8.5. It is noted that E8 combines two NDT techniques, whereby the UPV measurements have a high TRP whereas the R measurements have a medium TRP.

Table 8.5 Assessment of the TRP level (sd for standard deviation)

Expert	EQL	Technique	NTL$_{rep}$	sd	Assessed TRP
E1	EQL1	R	–	–	–
E2	EQL2	R	2	1.3	Medium
E3–E4		UPV	2	70 m/s	Medium
E5–E6		UPV	2	37 m/s	High
E7	EQL3	UPV	4	33 m/s	High
E8		R + UPV	4	1.3 and 33 m/s	Medium + High

8.3.4   Task 4. Identifying Test Regions

A test region is defined (*see Sect. 1.3.3.2.5*) as a given volume of concrete which can be considered to be homogeneous. It may be a part of a component, a whole component, a set of components or a larger zone, like a whole floor in a building or a whole structure.

In a multi-story building, tentative test regions can be a set of components (i.e. beams and columns) or different floors or groups of floors that are suspected to have similar characteristics. Several methods that can be used to identify test regions have been detailed in Chap. 4.

The NDT test results of Table 8.4 show that the datasets taken in the four different floors have more or less the same UPV values and that the only visible difference is between beams and columns, with slightly lower UPV values in beams. The mean UPV values are 4181 m/s in beams and 4344 m/s in columns, respectively. However, as can be seen in Fig. 8.3, the two distributions of UPV values overlap. UPV values ranging between 4100 and 4300 m/s are common for both types of components.

In the following, we will consider that a single Test Region covers all data.

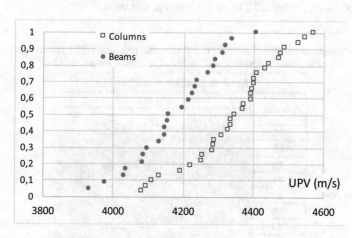

Fig. 8.3 Cumulative distribution of UPV test results in columns and beams (expert 3)

8.3.5 Task 5. Defining the Number of Cores

The number of cores mainly depends on the EQL and on the TRP level which are summarized in Table B3.1. The way of defining the number of cores is defined in the RILEM TC 249-ISC Recommendations (*see Sect. 1.6.2 Recommended number of cores*) and has been detailed in Chap. 2.

Expert E3 has an EQL2 requirement and wants to assess the concrete strength with medium precision UPV test results. This case is described in Sheet IIb of Chap. 2. The number of cores depends on how the requirements on RMSE and standard deviation are considered, i.e. in absolute or relative terms. For the sake of simplicity, we assume that these two requirements are given in absolute terms, which leads to 4 MPa for the standard deviation and 6 MPa for the RMSE. Therefore, the recommended number of cores must be taken from Fig. 2.35 of Sheet IIb. Table 8.6 reproduces the Recommendations in this specific case.

If there is no restriction on the range of mean concrete strength and the standard deviation, the number of cores can be as high as 33. However, this number can be lower if some restrictive criteria are considered. For instance, this number of cores is 12 if the mean strength is no more than 40 MPa and the coefficient of variation is no more than 20%.

Expert 3 therefore chooses to take 12 cores.

Note. Because of the restrictive conditions applied, it is **mandatory to check that they correspond to the concrete properties that will be estimated at the end of the assessment process** (see Sect. 8.3.9).

The same principles are applied for all experts, assuming that the mean strength is no more than 40 MPa and that the coefficient of variation is no more than 20%. They lead to the results summarized in Table 8.7.

The comparison between experts E3–E4 and experts E5–E6 illustrates how a better TRP (smaller within-test-repeatability) leads to a significant reduction of the number of cores, i.e. from 12 to only 5 cores in this case.

Table 8.6 Minimum number of cores for EQL2 (medium precision UPV measurements, absolute requirements for the standard deviation and the RMSE), derived from Chap. 2

f_c mean	Medium precision (TRP2)				
	cov (%)				
	10	15	20	25	30
10	6	6	6	6	7
15	6	6	6	6	6
20	6	6	6	6	6
25	6	6	6	6	6
30	6	6	6	7	9
35	6	6	8	11	13
40	6	8	12	15	18
45	7	11	16	20	25
50	9	15	21	27	33

Table 8.7 Minimum number of cores for all experts (reference sheets in Chap. 2)

Expert	EQL	Technique	Assessed TRP	Reference sheet	Number of cores
E1	EQL1	R	–	I, Fig. 2.29	6
E2	EQL2	R	Medium	IIa, Fig.2.31	7
E3–E4		UPV	Medium	IIb, Fig. 2.35	12
E5–E6		UPV	High	IIb, Fig. 2.35	5
E7	EQL3	UPV	High	IIIb, Fig. 2.43	8
E8		R + UPV	Medium + High	IIIb, Fig. 2.43	8

Regarding expert 8 who combines two non-destructive techniques that have a different TRP, the number of cores has been derived from that of expert 7 who uses only the UPV with high TRP test results. The number for medium TRP rebound test results (Sheet IIIa, Fig. 2.39) would be larger.

nNote. Because of the restrictive conditions applied on strength properties, it is **mandatory to check that they correspond to the concrete properties that will be estimated at the end of the assessment process** (see Sect. 8.3.9).

8.3.6 Task 6. Defining the Location of Cores

The way of defining the core location is defined in RILEM TC ISC 259 Recommendations (*see Sect. 1.4.1 Location of cores*) and two main options are available, a predefined pattern or a conditional pattern. Conditional coring is recommended, since it is expected to lead to a more reliable conversion model. Conditional coring does not induce any additional cost, and only requires that NDT test results have been made available before the time of taking cores in order to decide their locations (these NDTs may however be only a part of the full NDT program).

We follow the specific case of Expert E3, who has an EQL2 requirement, uses NDT tests with UPV measurements and has a predefined coring strategy.

He must take 12 cores. He chooses to take 6 cores in both columns and beams, with a sampling evenly distributed across floors, lines and rows. Table 8.8 synthesizes the core sampling predefined pattern.

Table 8.8 Core sampling pattern with predefined coring (PC) option

Columns		Beams	
floor	Component	floor	Component
1	A1, C3	1	A3
2	B4, D1	2	B4
3	C3	3	B1, C2
4	D2	4	A1, D3

This (arbitrary) sampling pattern leads to 3 cores for each floor. All structural alignments (see lines and rows of Fig. 8.1) are also considered.

We detail here the case involving conditional coring, as followed by expert E4, who uses the same NDT technique and test results as expert E3. The only difference is that expert E4 does not define the core sampling pattern before carrying out the NDT measurements. The core sampling pattern is defined on the basis of the NDT results (Table 8.4).

To provide the best possible coverage of the strength distribution in the structure, all UPV test results are ranked, from the highest value (4570 m/s on Column B2 of the second floor) to the lowest value (3932 m/s on Beam A1 of the fourth floor). Thus, the location of the 12 cores is defined from the ranking of all velocity values as detailed in Table 8.9. The resulting core sampling pattern is given in Table 8.10. It can be noted that the test location at rank 37 has been considered instead of that at rank 38, since the latter would have led to an uneven spatial distribution of cores, as it corresponded to a fourth beam on the fourth storey.

Results in Table 8.10 shows that this selection process leads, to have 6 cores in columns and 6 cores in beams, but that the distribution between floors, lines and rows is now uneven. The distribution of UPV test results according to the two options (i.e.

Table 8.9 Definition of core location based on the UPV ranking

Rank	UPV (m/s)	Component type	Floor	Component
1	4570	Column	2	B2
6	4474	Column	1	C2
11	4404	Column	4	A3
16	4375	Column	3	A2
21	4336	Column	2	D4
26	4294	Beam	4	A4
31	4271	Beam	2	C3
37	4222	Column	4	C1
41	4158	Beam	4	D3
46	4132	Beam	1	A1
51	4087	Beam	3	D3
56	3932	Beam	4	A1

Table 8.10 Core sampling pattern with the conditional coring (CC) option

Columns		Beams	
floor	Component	floor	Component
1	C2	1	A1
2	B2, D4	2	C3
3	A2	3	D3
4	A3, C1	4	A4, D3, A1

Table 8.11   Core sampling patterns for 5 cores (high precision test results)

Expert E5, predefined coring pattern			Expert E6, conditional coring		
Component type	Floor	Component	Component type	Floor	Component
Columns	1	A1	Columns	3	C5
	2	B2		3	D1
	3	C3			
Beams	2	B4	Beams	2	C3
	4	D2		2	D2
				3	B1

PC and CC) can be compared. The mean values are respectively 4261 m/s for PC and 4271 m/s for CC, both very close to the mean value of the whole data set, which is 4274 m/s (Table 8.4). However, the standard deviation is somewhat larger with CC (178 m/s against 149 m/s for both the PC sampling and the whole data set) which is a direct consequence of the CC sampling pattern, which considered extreme values of UPV test results.

As explained in Table 8.7, with high precision test results, experts E5 and E6 only need to take 5 cores. Expert 6 makes his choice based on the ranks of the UPV distribution, choosing the test locations which have given respectively the 1st, 5th, 28th, 42th and 56th test results. Table 8.11 compares the core sampling pattern between these two experts, the former using predefined coring and the latter using conditional coring.

Table 8.11 illustrates (as Table 8.10 did) that the conditional coring may lead to an uneven spatial distribution of cores within floors, as it is determined by the NDT test result distribution. However, the statistical properties of the NDT test results obtained on the 5 TL remain very close to those obtained with 11 cores, with a mean value equal respectively to 4276 m/s and 4285 m/s for PC and CC respectively and a standard deviation of 180 m/s and 228 m/s for PC and CC respectively (the latter value is higher, as CC prioritizes extreme values which have a direct effect on the standard deviation of the small size sample).

8.3.7   Task 7. Choosing a Conversion Model

Once cores have been taken, destructive testing must be carried out, which provides compressive strength test results. These results are used to establish a conversion model between NDT test results (here R value or UPV values) and strength.

The investigator must now choose the mathematical expression of the conversion model (see Sect. 5.4.3 *Mathematical shape of the model*).

Table 8.12 synthesizes the non-destructive and destructive test results obtained by expert E3 and Fig. 8.4 plots the corresponding values.

On viewing the data scattering in Fig. 8.4, a linear model is logically chosen

Table 8.12 Test results with 12 cores (expert E3)

Component type	Floor	Component	UPV test result (m/s)	Core strength (MPa)
Columns	1	A1	4395	37.7
	1	C3	4529	41.9
	2	B4	4134	39.3
	2	D1	4084	31.1
	3	C3	4411	45.0
	4	D2	4435	36.1
Beams	1	A3	4156	29.8
	2	B4	4288	32.1
	3	B1	4087	23.2
	3	C2	4218	30.5
	4	A1	4197	28.4
	4	D3	4158	33.9

Fig. 8.4 Core strength and UPV test results from columns and beams (expert E3, PC)

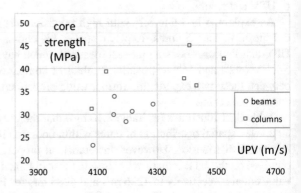

$$f_c = a + b\,UPV$$

whose parameters a and b must be identified.

Figure 8.5 plots the same type of data as Fig. 8.4 for expert E4 who used conditional coring with the same NDT test result precision (i.e. medium) and same number of cores as expert E3.

These results look like those of Fig. 8.4 with the same range of variation of the test results and the same type of scatter within the cloud of test results. A linear model can be also proposed:

$$f_c = a + b\,UPV$$

whose parameters a and b must be identified.

Figure 8.6 plots the test results for UPV and strength for both experts E5 and E6

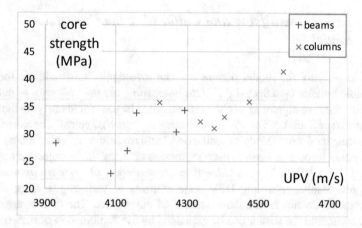

Fig. 8.5 Core strength and UPV test results from columns and beams (expert E4, CC)

Fig. 8.6 Core strength and UPV test results from Expert E5 (PC) and Expert E6 (CC)

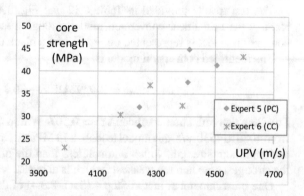

who both use high precision NDT test results. As they have only 5 cores, results for beams and columns are merged.

These results have significant differences with respect to those of Fig. 8.4 and Fig. 8.5. The reduced sample size (only 5 cores) increases the random effect of the statistical fluctuations. It comes that the UPV test results of expert E5 have a more limited range (between 4240 and 4500 m/s) whereas the conditional coring prevents this effect and leads to a UPV test result range which is similar to the one obtained with 12 cores. One can also see that the correlation between the two series of test results (i.e. E5–E6) seems to be better than with medium precision test results (i.e. E3–E4).

Finally, the logical choice for experts E5 and E6 is also that of a linear model

$$f_c = a + b\,UPV$$

whose parameters a and b must be identified.

8.3.8   Task 8. Identifying and Calibrating the Conversion Model

Before identifying the model parameters, the investigator must address the issue of possible outliers (see Sect. 1.5.3 *Data processing, the possible case of outliers*). Therefore, the investigator must choose the method he uses for identifying the model parameters (see Sect. 1.5.5 *Model identification approach*). Typically, he can calibrate a preexisting model or identify a specific one. In this example, a specific model is calibrated, either through a least-squares regression or using the bi-objective approach (see Sect. 1.5.5.2 *Specific model identification approach, bi-objective approach*).

The test results, regarding UPV measurements as well as regarding strength measurements do not reveal the existence of any outlier. The dataset appears to be consistent, and the scatter can be explained by the combination of material variability and measurement uncertainty. Therefore, all test results are kept for the model identification.

The test results provided in Table 8.12 and Fig. 8.4 are processed in order to identify the best linear conversion model through a least-square regression. This conversion model is represented on Fig. 8.7 by a black line.

The identified conversion model is:

$$f_{c,est} = 0.0292\,UPV - 90.3$$

with a determination coefficient $r^2 = 0.507$. Within the measured range of UPV from 4080 to 4530 m/s, this model leads to a 13 MPa difference between the maximum and minimum strength, which is much less than the difference measured between the strongest core and the weakest (which is about 22 MPa).

One must remember that at Step 4 (Sect. 8.3.4) of the investigation process, we wondered whether the investigation could consider a single test region or two test regions (Fig. 8.3). If the choice had been to consider two test regions, one for the beams and one for the columns, two conversion models would have been obtained:

Fig. 8.7 Test results and identified conversion model, expert E3

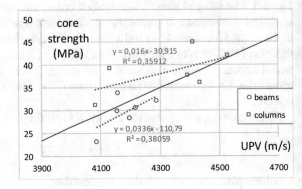

- a conversion model for the columns, identified from 6 pairs of (UPV, strength) test results: $f_{c, est, columns} = 0.0245$ UPV $- 73.3$, with $r^2 = 0.211$.
- a conversion model for the beams, identified from 6 pairs of (UPV, strength) test results: $f_{c, est, columns} = 0.0336$ UPV $- 110.8$, with $r^2 = 0.381$.

These two models are, however, determined from a more limited data set. They show a strength contrast between the two types of components. Moreover, the strength estimated for the same UPV test result in beams and columns is significantly different, i.e. 30.3 MPa and 36.3 MPa from the beam and column models, respectively, for the same UPV value of 4200 m/s.

Figures 8.5 and 8.6 provided the test results obtained by other experts (E4, E5, E6). Each data set of test results leads to a different conversion model. The results of the conversion model identification and calibration process are summarized in Fig. 8.8 and in.

Table 8.13. They contain an alternative option for expert E6 who identified two conversion models: the first one using least-squares regression, and the second one (expert E6′) using the bi-objective approach. It must be remembered that this approach is recommended to better capture the concrete strength variability and that this option is only applied during the data processing stage (Task 7 of the investigation flowchart of Fig. 8.2), and does not affect the previous stages of the process.

Fig. 8.8 Linear conversion models identified by all experts (from E3 to E6)

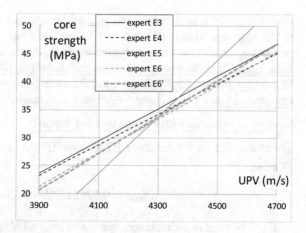

Table 8.13 Conversion models identified by all experts

Expert	TRP	PC/CC	Conversion model	r^2
E3	Medium	PC	$f_{c, est} = 0.0292$ UPV $- 90.3$	0.507
E4	Medium	CC	$f_{c, est} = 0.0275$ UPV $- 84.0$	0.612
E5	High	PC	$f_{c, est} = 0.0509$ UPV $- 185.1$	0.737
E6	High	CC	$f_{c, est} = 0.0303$ UPV $- 96.9$	0.857
E6′ (bi-obj.)	High	CC	$f_{c, est} = 0.0328$ UPV $- 107.3$	0.857

Figure 8.8 shows that most models are close together, with a difference of about 2–4 MPa on the predicted strength for the full range of variation of UPV. Only the model of expert 5 exhibits significant differences with respect to the others. In fact, with high precision measurements, only 5 cores were taken and, by the effect of chance, predefined coring resulted in cores taken only from the central part of the UPV range (Fig. 8.6). The consequence of this situation is that the model does not correctly capture the strength/UPV sensitivity and that using the model outside the range for which it has been calibrated for (i.e. here for UPV < 4200 m/s or UPV > 4500 m/s) leads to wrong estimates. This shows the significant advantages of CC, particularly when the number of cores is small.

The bi-objective approach leads to a model whose slope (value of parameter b of the conversion model) is larger, with a reliability which seems comparable to that of the other models. However, it is impossible to classify the quality/reliability of the conversion models based only on the value of determination coefficient r^2 and further analyses are mandatory (see Sect. 8.3.9).

8.3.9 Task 9. Estimating Concrete Strength and the Uncertainty on Strength Estimates

Once the model is identified, it can be used for strength assessment at each location in the structure where NDT test results are available. Additionally, checking the prediction error of the strength estimation is highly recommended in all cases (see Sect. 1.5.6 *Using the conversion model and quantifying its error*) and mandatory for EQL3 (see Table 1.5 in the Guidelines). The leave-one-out-cross-validation process (see Sect. 1.5.6,4. *Quantifying prediction error*) is used.

Strength estimation, with one and two test regions

The conversion model for expert E3 and one TR has been identified in the previous stage as:

$$f_{c,est} = 0.0292\,UPV - 90.3$$

It can be used to estimate strength where only UPV test results are available, i.e. at 44 test locations (Table 8.4). In fact, UPV was measured at 56 test locations but the core strength was directly measured at 12 test locations (Table 8.12). These 44 points are spread between columns (26 test locations) and beams (18 test locations). Therefore applying the model to these 44 UPV values provides 44 strength estimates. Mean values and standard deviations of the strength estimates can also be derived. The results are summarized in Table 8.14.

Applying the same conversion model to all components, one can derive the strength properties either for the building as a whole or for each type of component separately. Whereas the estimated mean strength of the building is 34.6 MPa, the mean strength in columns is 4.3 MPa higher than in beams. The overall standard

Table 8.14 Strength estimates (mean strength, strength standard deviation), excluding information from strength test results from cores

	Components	Model	$f_{c, est, mean}$ (MPa)	$sd(f_{c, est})$ (MPa)
One test region	All	0.0292 UPV − 90.3	34.6	4.0
	Columns		36.3	3.4
	Beams		32.0	3.4
Two test regions	All	0.0160 UPV − 30.9 in columns	35.1	5.1
	Columns	0.0336 UPV−110.8 in beams	38.6	1.9
	Beams		30.2	3.9

deviation is estimated at about 4 MPa, which comes to a coefficient of variation of 12%.

Note. Restrictive conditions were assumed at Stage T5 (Sect. 8.3.5) for defining the number of cores (strength below 40 MPa and coefficient of variation of strength not exceeding 20%). The estimated properties validate these prior assumptions.

If two test regions had been considered and two conversion models identified, as seen on Fig. 8.7, the statistical results are somehow different:

- the mean strength is only slightly changed, and is equal to 35.1 MPa,
- the difference between beams and columns is multiplied by a factor 2, with a 8.4 MPa difference between mean strength of columns and beams,
- the concrete strength standard deviation is twice as large in beams (3.9 MPa) than in columns (1.9 MPa).

Strength assessment uncertainty with one test region and one conversion model

One must be aware than nothing can be said, at this stage, about the model prediction error. That must be addressed specifically. The leave-one-out-cross-validation method is used to estimate this prediction error. It consists in iteratively removing each of the 12 pairs of data (UPV test result, core strength) and identifying a new conversion model from a dataset containing the remaining 11 pairs of data. This leads to a series of 12 alternative conversion models. For each of these models, it is easy to quantify the difference between the estimated strength for the sample that has been removed from the dataset (from its UPV test result) and the measured core strength on the same sample. This difference is the model error. The basic data used in this process are those of Table 8.12. The results of the iterative leave-one-out-cross-validation process are given in Table 8.15 where each conversion model is given with its determination coefficient.

It can be seen that:

- The expression of the conversion model significantly changes when removing one data pair, as the result of statistical fluctuations. One must be aware that, when ONE model is derived with 12 data pairs, the same pattern would have been observed if the process had been repeated, with a new dataset of the same size.

Table 8.15 Results of the leave-one-out cross validation process, expert E3

Excluded data	Measured core strength (MPa)	Conversion model	r^2	Estimated strength (MPa)	Error (MPa)
C1–A1	37.7	0.0295 UPV − 91.5	0.490	38.2	0.5
C1–C3	41.9	0.0294 UPV − 91.1	0.413	42.1	−0.2
C2–B4	39.3	0.0343 UPV − 113.0	0.702	29.0	−10.3
C2–D1	31.1	0.0311 UPV − 98.6	0.509	28.5	−2.7
C3–C3	45.0	0.0244 UPV − 70.5	0.459	37.2	−7.8
C4–D2	36.1	0.0320 UPV − 101.9	0.531	40.1	3.9
B1–A3	29.8	0.0286 UPV − 87.7	0.487	31.3	1.5
B2–B4	32.1	0.0296 UPV − 91.7	0.524	35.2	3.2
B3–B1	23.2	0.0242 UPV − 68.2	0.437	30.6	7.4
B3–C2	30.5	0.0288 UPV − 88.3	0.506	33.2	2.6
B4–A1	28.4	0.0282 UPV − 85.5	0.506	32.8	4.4
B4–D3	33.9	0.0304 UPV − 95.9	0.527	30.8	−3.1

This is commonly neglected in the strength assessment process, and it is why the prediction model error is usually unknown.

– The determination coefficient varies between 0.413 and 0.702 and its highest value is reached when the prediction error is the largest! This clearly confirms that **r^2value cannot be considered as an adequate parameter to estimate the conversion model quality**.

– The model prediction error varies from −10.3 to +7.4 MPa and its mean value is zero. However, individual values of the prediction error (i.e. estimation of the local strength) can be large, namely corresponding to about +25% of the true strength. It can be added that the conversion models tend to overestimate the strength in beams and to underestimate the strength in columns (as it is visible also on Fig. 8.7).

– **A relevant estimation of the model error is the prediction error RMSE$_{pred}$,** whose value is easily computed from the 12 errors, and amounts here to 4.9 MPa. **This corresponds to about 14% of the mean strength and satisfies the requirements of EQL2** (see Table 8.2).

Strength assessment uncertainty with two test regions and two conversion models

Let us analyze now the same performances if after having carried out the measurements, the investigator had chosen to consider two test regions and therefore identified two conversion models, as described in Sect. 8.3.8. In this case, the datasets involves 6 data pairs for columns and 6 data pairs for beams. The leave-one-out cross validation process must be carried out separately for each type of component, and leads to respectively 6 models for columns and 6 models for beams. Their statistical results are summarized in Table 8.16.

It can be seen from these analysis that:

– The expression of the conversion model significantly changes with the removal of one data pair, which is the result of statistical fluctuations. These fluctuations are exacerbated when compared to those of Table 8.15 because of the smaller size of the dataset.

Table 8.16 Results of the leave-one-out cross validation process with two test regions

Excluded data	Measured core strength (MPa)	Conversion model	r^2	Estimated strength (MPa)	Error (MPa)
Specific models for columns					
C1-A1	37.7	0.0170 UPV − 34.5	0.392	40.0	2.3
C1-C3	41.9	0.0156 UPV − 29.0	0.272	41.5	−0.4
C2-B4	39.3	0.0242 UPV − 67.3	0.582	32.6	−6.6
C2-D1	31.1	0.0043 UPV − 21.1	0.033	38.7	7.6
C3-C3	45.0	0.0128 UPV − 17.9	0.385	38.4	−6.5
C4-D2	36.1	0.0194 UPV − 44.7	0.539	41.4	5.3
Specific models for beams					
B1-A3	29.8	0.0353 UPV − 118.4	0.404	28.4	−1.4
B2-B4	32.1	0.0471 UPV − 166.8	0.302	35.0	3.0
B3-B1	23.2	0.002 UPV + 22.7	0.002	30.7	7.5
B3-C2	30.5	0.0341 UPV − 121.8	0.373	30.9	0.3
B4-A1	28.4	0.0348 UPV − 115.6	0.417	30.5	2.1
B4-D3	33.9	0.0407 UPV − 41.9	0.799	27.5	−6.4

- The determination coefficient varies a lot and is close to zero in two cases, when C2-D1 test results are removed from the column dataset, and when B3-B1 test results are removed from the beam dataset. By looking at Fig. 8.4, one can check that these data pairs correspond to the points located at the left side of the diagram. When they are removed, the slope of the regression model built on the remaining test results is close to zero, as is the determination coefficient (in some cases, the slope could even be negative). This illustrates the case, which can occur with a small number of cores, when the statistical fluctuations due to measurement uncertainty masks the (expected) physically based positive correlation between the NDT test results and the strength.
- The model prediction error varies from -6.6 to $+7.5$ MPa and its absolute value exceeds 5 MPa in 6 cases out of 12. **A relevant estimation of the model error is the prediction error RMSE$_{pred}$**, whose mean value is 4.7 MPa, i.e. quite the same value than that resulting from the analysis with a unique conversion model for all data.

Because of the large statistical fluctuations due to the limited size of the dataset, it appears that there is no interest in practice to try to identify a specific model in each possible test region.

Table 8.17 synthesizes the performances corresponding to an investigation of the same case-study performed by four experts (E3 to E6), with an extra variant for the latter (involving the bi-objective method). Figure 8.9 provides the same results regarding the two main dimensions of assessment, i.e. concrete mean strength and standard deviation. Let us focus first on the information that is available to any investigator, as we have explained in this chapter. The first three columns emphasize what the main characteristics of the investigation are. it should also be noted that experts E3 and E4 have 12 cores, whereas experts E5 and E6 have only 6 cores, thanks to high precision NDT test results (as they have checked at stage T3). The fifth and sixth columns provide the estimated mean strength ($f_{c, mean}$) and the estimated strength standard deviation sd(f_c) that are calculated from the compressive strength

Table 8.17 Comparison of the performances of the different experts

Expert	TRP	PC/CC	from cores			from ND test results and conversion model		
			N_c	$f_{c, mean}$ (MPa)	sd(f_c) (MPa)	$f_{c, est,mean}$ (MPa)	sd(f_c) (MPa)	RMSE$_{pred}$ (MPa)
3	Medium	PC	12	34.1	6.2	35.0	4.0	5.0
4	Medium	CC	12	32.2	4.9	33.7	3.9	3.9
5	High	PC	5	36.8	6.9	31.8	7.5	5.1
6	High	CC	5	33.3	7.5	32.8	4.2	3.2
6' bi-obj.	High	CC	5			32.8	4.6	2.9
True value						33.2	5.1	

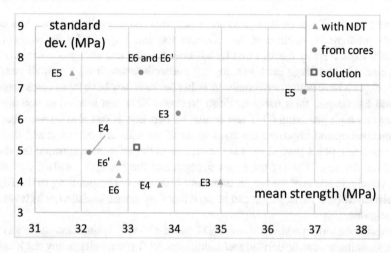

Fig. 8.9 Assessment for each expert, either with cores only or by using cores and NDT

of cores. Columns 7–9 provide all the results that are available thanks to NDT test results and conversion model:

1. Column 7: the estimated mean strength.
2. Column 8: the estimated strength standard deviation.
3. Column 9: the estimated predictive RMSE, which represents the mean error on the local estimation of concrete strength when an NDT test result is available at the same test location. This value is estimated by the iterative leave-one-out cross validation method (see Sect. 5.6.4. *Quantifying prediction error*).

Processing core strength results provides, by simply averaging the core test results and calculating their standard deviation, a first estimate of the material within the building,. Using NDT with cores and a conversion model aims at improving the concrete assessment and its reliability. Whereas in real practice the investigator can never know how far their assessment is from the "true" strength properties, one advantage of synthetic simulations is that the "truth" is available to anyone who performs the simulation. In this specific example, the synthetic concrete properties are simulated according to rules summarized in Sect. 8.1, with mean strengths of 28 MPa and 35 MPa in beams and columns, respectively. The result of this simulation was an overall mean strength equal to 33.2 MPa and a standard deviation equal to 5.1 MPa, corresponding to the blue square in Fig. 8.9.

Figure 8.9 shows that the strength based on cores and that the variability are overestimated by most experts.[3] The result is better for experts E3 and E4 who have 12 cores than for experts E5 and E6 who have only 5 cores, which is a logical consequence of sampling statistics. Globally, the estimates provided by the combination

---

[3]No firm conclusion must be drawn from this statement as these properties could have been underestimated for the same statistical reasons. The only relevant statement regards the magnitude of error (the reader is invited to refer to Chap. 12 (Sect. 12.1) and Table 12.1 for more details).

of cores and NDT (green triangles) are better than those resulting from cores only, and the predictions are closer to the reference solution. Expert E5 is the worst of all, which is explained by the fact that he has only 5 cores and uses predefined coring, which may induce large random sampling uncertainties with such a small number. Other experts have better performances, either because they have more cores (experts E3 and E4, despite their using medium precision NDT test results) or conditional coring and high precision NDT test results (expert E6). It can also be seen that the bi-objective option improves the assessment of the standard deviation and that the corresponding point (expert E6$'$) is the closest to the reference solution: the errors are about 1% and 10% for the mean strength and the strength standard deviation, respectively. In terms of the **target tolerance intervals corresponding to the EQL2 requirement** (see Table 8.2), it can be seen that they **are all satisfied, whatever the investigator**.

Another important added-value of NDT-based concrete strength assessment is that strength estimates can be derived from additional NDT test results, at any test location where no core has been taken. Of course, in real practice, investigators can estimate strength by simply using the conversion model they have calibrated, but the only way to check how their estimate is close to the true value is to estimate the predictive RMSE as was explained above (with the leave-one-out cross validation, RMSE$_{pred}$ values in column 9 of Table 8.17). All these estimates **fall within the target tolerance interval** of Table 8.2 **for the error on local strength for the EQL2 requirement**, which are respectively 6 MPa and 4.5 MPa for EQL2 and EQL3 respectively.

Note.

The reader who would better study some specific features of the investigation program is invited to read Chap. 12 which provides more detailed explanations about the interest of options that are promoted in the RILM TC ISC-249 Recommandation.

Acknowledgements The 4th author would like to acknowledge the financial support by Base Funding - UIDB/04708/2020 of CONSTRUCT - Instituto de I&D em Estruturas e Construções, funded by national funds through FCT/MCTES (PIDDAC).

References

1. Alwash, M., Breysse, D., Sbartaï, Z.M.: Using Monte-Carlo simulations to evaluate the efficiency of different strategies for nondestructive assessment of concrete strength. Mater Struct **50**, 90 (2017). https://doi.org/10.1617/s11527-016-0962-xhttps://doi.org/10.1617/s11527-016-0962-x
2. Alwash, M., Breysse, D., Sbartaï, Z.M.: Non destructive strength evaluation of concrete: analysis of some key factors using synthetic simulations. Constr Build Mater **99**, 235–245 (2015)

Chapter 9
Illustration of the Proposed Methodology Based on a Real Case-Study

Angelo Masi, Denys Breysse, Hicham Yousef Qasrawi, David Corbett,
Arlindo Gonçalves, Michael Grantham, Xavier Romão,
and André Valente Monteiro

Abstract This chapter aims to apply the recommendations of RILEM TC 249-ISC on a real structure, a four-level reinforced concrete framed structure. The flow chart describing the methodology proposed in the recommendations is applied and each stage is detailed. The proposed procedure is compared to the usual practices. The application of the new procedure proposed in the recommendations does not induce significant difficulties compared to usual ones. It is demonstrated that this new procedure can provide mean strength and local strength by saving a significant number of cores. The concrete variability is assessed with a higher reliability compared to usual practices. The unavailability of the test result precision (TRP) in the case study is discussed and its influence on the prediction error is quantified. The definition of test regions by using NDT methods is also proposed. The relevance of test region identification is analysed in relation with the identification of the conversion models. In particular, the possibility to identify either a unique conversion model for all the test regions or a specific model for each test region is discussed.

A. Masi (✉)
School of Engineering, University of Basilicata, Potenza, Italy
e-mail: angelo.masi@unibas.it

D. Breysse
University Bordeaux, I2M-UMR CNRS 5295, Talence, France

H. Y. Qasrawi
Hashemite University, Zarqa, Jordan

D. Corbett
Proceq SA, Zurich, Switzerland

A. Gonçalves · A. Valente Monteiro
Laboratório Nacional de Engenharia Civil (LNEC), Lisboa, Portugal

M. Grantham
Consultant - Sandberg LLP, London, UK

X. Romão
CONSTRUCT-LESE, Faculty of Engineering, University of Porto, Porto, Portugal

© RILEM 2021

303

D. Breysse and J.-P. Balayssac (eds.), *Non-Destructive In Situ Strength Assessment of Concrete*, RILEM State-of-the-Art Reports 32,
https://doi.org/10.1007/978-3-030-64900-5_9

The purpose of this chapter is to illustrate and to explain how the RILEM TC 249-ISC Recommendations can be used in practice with real in-situ data. We have chosen to consider a real case that has already been published [1]. The real investigation did not follow the RILEM TC 249-ISC Recommendations (that did not exist in 2016) but was at the best level of engineering practice at the time. This investigation developed several important ideas:

1. the number and location of cores was defined after a prior series of non-destructive tests (NDT),
2. the possibility of identifying "homogeneous areas" (that are named "Test Regions", TR, in the RILEM TC 249-ISC Recommendations),
3. the calibration of a multivariate conversion model, combining ultrasonic pulse velocity (UPV) and rebound test results in a SonReb model [2, 3].

The case study will be briefly presented with its main results (the interested reader is invited to refer to the original publication for more extensive information about the case study). These results will be discussed, and submitted to a critical analysis, in light of the RILEM TC 249-ISC Recommendations. The third section will develop what could be achieved on this same structure by following the RILEM TC 249-ISC Recommendations. The experimental database obtained from the real investigation will be used to show: (1) how the test results can be processed following the RILEM approach, (2) what information can be finally derived about the concrete properties and the precision of the estimation. Finally, some additional analysis of all results will be carried out in order to compare the efficiency of the different approaches (expert practice *vs* RILEM TC 249-ISC Recommendations).

We must underline that the dataset that is used is exclusively the one obtained from the investigation. This has two limiting consequences for the comparison: (1) some data that are important regarding the RILEM TC 249-ISC Recommendations may be missing, (2) the number and location of test results is fixed, whereas the RILEM TC 249-ISC Recommendations might have led to some other choices. For the sake of clarity, the text corresponding to these stages will be identified by a vertical line on the right side.

9.1   Description of the Case Study—Original Methodology

The structure (Fig. 9.1) is a four-level reinforced concrete framed structure, designed to support vertical loads only in accordance with the Italian design code in effect at the time of construction (1972–74). The structure has a rectangular in-plan shape with total dimensions 23.10×10.30 m and a surface of about 240 m^2 for each floor with 18 columns and 21 beams on each floor. The original design drawings show that the structure has lateral load resisting frames only along the longitudinal direction X. Specifically, along the X direction, three resisting frames with bay lengths varying in the range 2.8–5.0 m (the shorter one corresponding to the staircase width) are present. The beam cross-section dimensions are constant throughout the structure

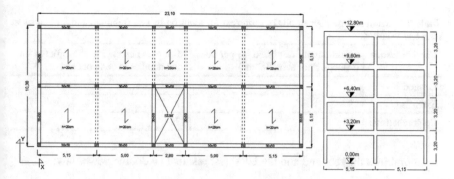

Fig. 9.1 In-plan (left) and in-elevation (right) of the building under study

and equal to 30×50 cm². Along the transversal direction Y, the structure has two bays with only the exterior frames having a rigid continuous beam (30×50 cm). In the interior frames, columns are connected through the one-way RC slab with a thickness equal to 22 cm. The staircase structure is placed in a central position and is made up of two inclined cranked beams at each story, arranged in two adjacent frames along the Y direction. The cross-section dimensions of staircase beams are equal to 30×50 cm². The smallest dimension of the column cross-sections is 30 cm while the other dimension depends on the story number and the in-plan position of the column, thus varying from 30 to 45 cm (internal columns at the first story). With respect to the constituent materials, design specifications are related to the strength classification at the time of its construction, and provide $R_{bk}250$ concrete (which corresponds to a characteristic strength in cylinders of about 20 MPa) and FeB38k steel (which corresponds to a characteristic yielding strength equal to 380 MPa).

To estimate the in-situ concrete strength at any point throughout the structure, the logical steps were to first identify homogeneous areas, and then to identify in each homogeneous area a relevant conversion model. The investigation and assessment program therefore consists of five main steps:

(a) scheduling and planning stage: analysis of available technical documentation, preliminary site survey, identification of the main components, cost analysis and selection of test locations for NDTs,
(b) non-destructive investigation: at each test location, both rebound test results and UPV test results were obtained,
(c) statistical analysis of the NDT results to identify relevant test regions (TR). This step involves identifying components (beams and columns) with similar values of NDT results to include them in the same TR. The validation criterion for each TR is to have a coefficient of variation (CoV) lower than 15% for both rebound and UPV test results,
(d) definition of the number of cores in each TR, coring, obtaining destructive test results on cores,
(e) identification of conversion models for each TR and the estimation of local in-situ concrete properties. A multivariate conversion model is used which

Table 9.1 Minimum number of NDTs proposed for identifying TRs (rebound and UPV tests at each test location)

Level of testing	Minimum number of NDTs per floor (average floor area 300 m²)	Minimum percentage of NDTs (on the total number of components), in %
Limited	6	8
Extended	9	11
Comprehensive	14	17

combines the rebound and the UPV test results in a double power law:

$$f_c = a R^b UPV^c \tag{9.1}$$

where the coefficients a, b and c are derived for each TR.[1]

In order to assess the actual in-situ concrete strength value, a test campaign was performed by considering non-destructive (both rebound and direct UPV tests) and destructive (core) tests. All the in-situ tests were carried out according to European standards [4–6]. The number of NDTs was tentatively established for identifying homogeneous concrete areas (i.e. TRs) as a function of the NDT results variability and a predetermined maximum error. The authors proposed a minimum number of NDTs for three levels of testing (Table 9.1).

Firstly, assuming an extended level of testing, at least nine NDTs for each floor were selected on elements having different roles and positions within the structure. As an example, on the first floor, eight columns were investigated: two interior columns, three peripheral columns, two corner columns and one belonging to the staircase structure (globally, leading to eight tests on the columns). As a consequence of the greater difficulties in performing tests on beam members, only three beams were investigated in each story (globally, leading to three tests on beams). Some additional measurements were carried out in order to increase the significance of the collected samples, e.g. by including some extra columns according to their location. The overall non-destructive investigation program resulted in 47 test locations (TL), made of 14 TL in beams and 33 TL in columns, which leads to about 12 TL for each floor with corresponding sampling rates of 17 and 46% in beams and columns.

One crucial stage is the identification of TRs. The basic assumption is that a tentative TR contains all beams (or columns) on the same floor, which leads to 8 tentative TR, denominated B1–B4 and C1–C4 for beams and columns respectively, where 1–4 represents the floor number. Thus, a series of statistical tests (Student's test, ANOVA analysis) are carried out in order to find a solution which merges some of these tentative TRs into composite ones while retaining the homogeneity criterion. Further information can be found in the original publication [1]. Chapter 3 of this book also describes in detail what can be done for identifying relevant TRs. The final result that was obtained is summarized in Table 9.2 which also provides the main statistical properties regarding NDT results. Figures 9.2 and 9.3 plot the cumulative

---

[1] The a, b and c coefficients can be identified by using a spreadsheet like that developed by Proceq.

Table 9.2 Composition of each TR and statistical properties of NDT results (N_{TL} = number of test locations, UPV in m/s)

TR		N_{TL}	Rebound R			Ultrasonic pulse velocity UPV		
			Mean	s.d.	CoV (%)	Mean	s.d.	CoV (%)
TR 1	C1	8	40	4.1	10.2	4107	417	10.2
TR 2	B1, C3, B4, C4	24	25	2.1	8.4	3316	421	12.7
TR 3	C2, B2, B3	15	21	2.3	10.8	2623	302	11.5

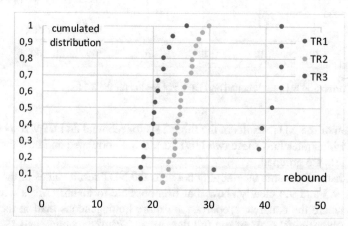

Fig. 9.2 Cumulative distribution of rebound test results in the 3 TRs

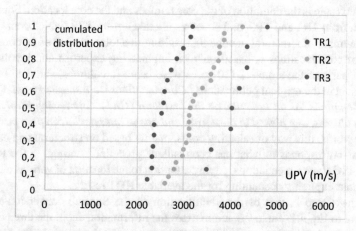

Fig. 9.3 Cumulative distribution of UPV test results in the 3 TRs

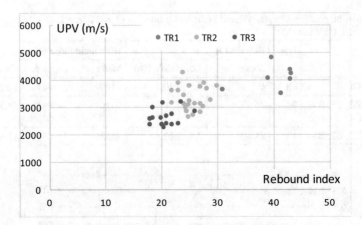

Fig. 9.4 Correlation between rebound and UPV test results for the 3 TRs

distribution of the NDT results in the three TRs for rebound and UPV respectively. Figure 9.4 illustrates how these two NDT results are correlated and how the values for the 3 TRs are grouped.

Notwithstanding some overlapping between the NDT result values of the 3 TRs, Figs. 9.2, 9.3 and 9.4 clearly show that it is possible to identify areas within the structure where the concrete properties are more homogeneous than in the whole structure (the original CoV of the full data set is 27% for rebound and 19.6% for UPV).

The next step is the definition of the core number and the core location. Four cores are taken from TR1 and TR3, and six cores are taken from TR2 (which contains the highest number of structural members), resulting in a total number of 14 cores. The choice of core locations is based only on engineering expertise and involves the different parts of the structure (10 cores were taken in columns, and 4 in beams). As was said in the original publication: *"Core locations were selected from different in plan locations. As an example, on the first floor (HA1), core testing involved two central columns, one placed in the corner and one in intermediate in plan position"*. In other words, this way of selecting cores, mainly based on structural and building expertise, corresponds to what has been called "predefined coring" in the RILEM TC 249-ISC Recommendations, paying attention to the location of various elements but without explicitly accounting for the available NDT results.

The next step is that of the identification of the SonReb conversion model, according to Eq. 9.1. Table 9.3 summarizes the information about the three conversion models identified on each TR. These results show high values for the coefficient of determination r^2 and a very low RMSE$_{fit}$ value: the mean difference between measured core strength and estimated strength at the same point by using the conversion model is lower than 1 MPa.

Table 9.3 Conversion models identified in each TR

TR	a	b	c	r^2	$RMSE_{fit}$ (MPa)
TR 1	1.251	0.817	0.031	0.989	0.38
TR 2	3.533	0.553	0.012	0.800	0.64
TR 3	45.29	1.101	−0.522	0.996	0.08

These models are finally used to estimate the concrete strength at all the TLs where only NDT results are available. In this case study, this estimation can be done at 33 TL ($NTL = 47$, $N_{core} = 14$).

9.2 Critical Analysis of the Expert Methodology and of Its Results—What Could Be Improved

Note. The following statements are not written to criticize what was done in the paper [1] by experts in the field of structural and earthquake engineering. In fact, some of them have contributed to the current analysis. The original study denotes a high level of expertise and practice in the field of non-destructive assessment of concrete strength. It contains several very interesting ideas, including the identification of homogeneous areas (i.e. test regions). However, RILEM TC 249-ISC has analyzed the most innovative ideas which emerged from research and practice and checked if they can be applied in order to improve the practice, without neglecting their drawbacks. Therefore, the work and results of §9.1 are analyzed with the purpose of suggesting possible improvements in the proposed methodology that have been identified thanks to the TC work and also identifying issues that would have deserved further attention.

9.2.1 Interesting Ideas Developed in the Original Study

The main innovative idea of [1] was to validate a **semi-automatic process in order to identify homogeneous areas (or test region, TR)**. Thus, on each TR, a specific conversion model is identified, which is further used to estimate the concrete strength at some additional test locations.

The authors have also pointed out the issue of **the number of cores** to be taken. This number of cores increases with the "level of testing" defined by the investigator (limited, extended or comprehensive) in accordance with the European seismic code for existing buildings (EC8-3),[2] and may be given as a strict number or as a percentage

---

[2]EN 1998-3:2005, Eurocode 8: Design of structures for earthquake resistance—Part 3: Assessment and retrofitting of buildings, European Committee for Standardization (CEN), Brussels, Belgium, 2005.

of the number of components. The expected result is a specific conversion model that is capable of capturing the concrete features in the TR for which it has been identified.

Lastly, in [1], the authors have also endorsed a **multivariate conversion model**, which has the mathematical shape of a double power law (SonReb). The basic idea is that, by combining NDT results with two techniques (rebound and UPV) which have a different sensitivity to concrete properties, it is possible to improve the accuracy of the assessed strength.

9.2.2 What Would Have Deserved Further Attention

The first and most important point that can be improved in the case study is that, at the end of the process, the **predictive capacity of the identified conversion models** is not really known. As we have explained and demonstrated (see Chaps. 10 and 12), the coefficient of determination r^2 and the fitting error $RMSE_{fit}$ (Table 9.3) provide poor indications about the predictive capacity of the model. The real predictive error $RMSE_{pred}$ can be estimated only with a specific procedure, and is the relevant indicator of the uncertainties that persist at the end of the process.

Another important issue is that the idea of **identifying several TRs**, each one requiring its own conversion model for calibration, is a very complex problem. Its relevance is affected by a series of driving factors, among which are the measurement uncertainties and the range of variation of concrete strength in the TR (Chap. 4). We have shown, on the basis of several examples, that the further refinement which would theoretically result from an increasing number of TRs may be counterbalanced by the more limited range of concrete strength on which the conversion model is identified and applied. To take an example, in TR 2, the standard deviation of rebound values is only 2.1 units and that of core strength is only 1.7 MPa. If one looks at Fig. 9.4 where UPV and rebound values are plotted together, a positive correlation can be observed between UPV and rebound values. Although a strong positive correlation is visible between results from the whole structure, a much less marked correlation is observed if the results from each TR are considered separately. The slope of the linear regression is even close to zero for TR2 and all the corresponding coefficients of determination are very small (0.16 for TR1, 0.05 for TR3 and less than 0.001 for TR2).

Figure 9.5 plots the concrete strength of cores against NDT results (rebound in Fig. 9.5 left and UPV in Fig. 9.5 right). For rebound, the three regression lines have a slope factor between 0.5 and 0.7 and the coefficient of determination r^2 is above 0.75. The results are very different for UPV. The slope of the linear regression varies significantly, being close to zero for TR3, and the coefficient of determination is very low for both TR2 and TR3. This may indicate that the precision of the UPV test

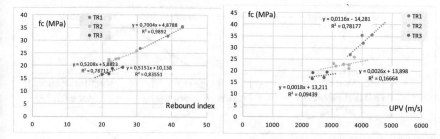

Fig. 9.5 Correlation between concrete strength and ND test results for the 3 TR: for rebound (left), for UPV (right)

results is not as good as that of the rebound test results.[3] This confirms that, in this case-study, the tendencies and correlations that can be seen for the whole structure can be weaker if each TR is considered separately.[4]

The two main causes for these poor results are probably:

– the more limited range of strength (but also of NDT results) in each TR when compared to the full range in the structure (note that this is the proper aim of identifying TRs, but it has direct consequences on the accuracy of strength estimates),
– the effect of measurement uncertainties. As we have explained in ([7], see also Chap. 2), the values of each NDT can be considered as the output of a random process, which would have delivered different values if it had been reproduced. This crucial fact is usually neglected in engineering practice, but has been firmly established by the RILEM TC 249-ISC work.

Figure 9.6 presents how all these factors contribute to the final predictive capacity of the conversion model. The choice of a pattern of TRs after the analysis of NDT results has mutual interactions (influences/is influenced by) on the concrete variability and the number of cores (that is constrained by the investigation cost). These two factors, for a given conversion model shape, govern the identification of the conversion model. We have added with dotted contour and arrows the Precision of NDT results which, as can be seen, is the most influencing among all factors.

However, the influence of the precision of NDT results was not formalized until recently and non-destructive investigators are usually limited to general considerations. The authors of [1] discussed the "reliability of core tests and methods" and they

---

[3]Of course, the TRP value is case-specific, and no conclusion can be drawn for another study. For instance, the UPV test results are very sensitive to the presence of steel reinforcement and it is possible that, in the case study, some tests were carried out too close to rebars, thus influencing the wave propagation mode and the test result.

[4]This statement is case-specific. It can depend on the specific structural type under examination (e.g. bridges, buildings, tanks) and the related peculiar construction process. The fact that it is better to consider that a unique conversion model is considered for several TRs or to calibrate a specific model in each TR does not obey general laws and deserves a careful analysis in each situation. This issue is developed in Sect. 4.3.

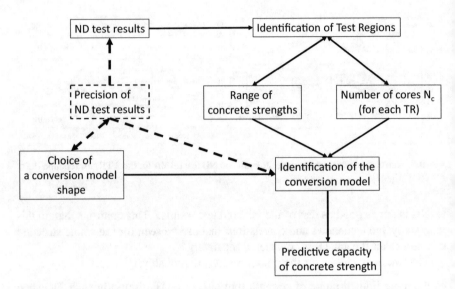

Fig. 9.6 Factors governing the predictive capacity of the concrete strength

assumed that, for assessing strength, the reliability of core tests, rebound number, UPV and combined methods was respectively good, low, moderate and good. This justified the choice of the SonReb approach, but they did not analyze in detail how this "reliability" impacts on the final predictive error. For instance, if the combination of NDTs is chosen, a higher number of model parameters must be identified in the conversion model (typically three parameters for a SonReb model, against only two in a univariate model). Due to the random nature of measurement uncertainties (and that of NDT results), it is now known there is some trade-off between the different model parameters (see Sect. 12.4). This effect is exacerbated when the number of parameters increases, and in particular when the number of cores is small (which is the case when several TRs are treated separately instead of considering a unique dataset containing all cores). This explains the dotted arrow on the left side of Fig. 9.6, which means that the choice of a conversion model shape cannot be done without knowing the precision of the NDT results.

The last point to be noted is that the authors of [1] defined the location of cores on the basis of structural considerations: the objective of getting a representative picture of the whole structure was combined with a focus on the most critical structural elements. To do so, they took cores from both beams and columns, on each floor and in different parts of the structure. However they did not explicitly adopted the conditional coring approach [8] which could have been very effective due to the small number of cores.

9.3 Concrete Properties Assessment of the Same Structure Following the RILEM TC 249-ISC Recommendations

Figure 9.7 reproduces the flowchart of the mains steps of this Recommendations, that we will follow here. However, as we work with data from a real structure that were not collected according to the RILEM TC 249-ISC Recommendations, some information will be missing, and some assumptions will be needed, as it will be explained herein.

9.3.1 T1. Defining EQL

The first task described in the flowchart (Fig. 9.7) is the definition of the required Estimation Quality Level (EQL), that depends on the assessment objectives and on the available resources for the investigation. In the RILEM TC 249-ISC Recommendations, choosing a more ambitious EQL corresponds to a better accuracy of the final estimates of concrete properties, i.e. a smaller uncertainty on the estimates of mean strength, strength standard deviation and local strength. This has consequences on the resources that must be devoted to the investigation (among which the number of cores and the precision of NDT results) and, therefore, on the cost of the investigation.

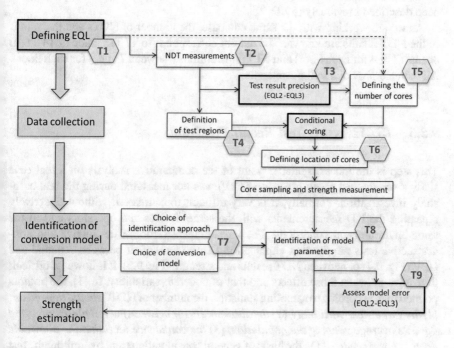

Fig. 9.7 Flowchart illustrating the overall methodology

The EQL is a new concept that did not exist before the publication of the RILEM TC 249-ISC Recommendations and therefore the investigator in a previous case-study could not have chosen one. However, the "level of testing" concept referred to by the investigators in [1], in accordance with the European seismic code for existing buildings (EC8-3), may be roughly seen as something close to the EQL concept, where the limited, extended and comprehensive levels would respectively correspond to EQL1, EQL2 and EL3, as a larger number of tests leads to a smaller predictive error.

An "extended level of testing" was chosen in [1] and we will assume that it corresponds to the EQL 2.

9.3.2   T2. Carrying Out NDT Measurements

The number of NDTs in [1] was chosen in order to correspond to the "extended level". This number is important for three reasons: (1) it provides a large coverage of the structure and reveals its possible contrasts, (2) in the original study [1] as in the RILEM TC 249-ISC Recommendations, it governs the identification of TRs, (3) in the RILEM TC 249-ISC Recommendations, it also governs the identification of the core locations if conditional coring is used. The way the investigators have defined the test locations (TL) for ND measurements, on the basis of structural expertise, has been described previously (§9.1).

As we are working with the same database, the number of NDTs and the values of the NDT results are known. The dataset corresponds to 47 TL, made of 14 TL in beams (3 or 4 for each floor) and 33 TL in columns (between 7 and 9 for each floor). All data are provided in the Appendix.

9.3.3   T3. Assessing Test Result Precision

This step is the most disputable point of the comparative analysis on a real case study since the test result precision (TRP) was not measured during the real case-study investigation. Formally, it is very difficult to estimate it without effectively repeating the ND measurements with the same devices and operators and in the same environmental conditions. The measurement errors of destructive and non-destructive tests are the first explanation for the lack of fit between rebound and UPV (Fig. 9.4) or between NDT results and strength (Fig. 9.5). It is however difficult to resolve between those effects and that of concrete variability. In [1], the authors pointed that, according to sampling statistics, the number of NDT "*should depend on: (1) the within-test-variability, (2) the allowable error between the sample average and the true average value of the population, (3) the confidence level that the allowable error is not exceeded.*" On the basis of general considerations and by limiting the test result variability (C_v) inside each TR to 15%, they proposed a minimum number of tests that enabled the proper determination of the concrete properties in each TR. But, as they stated, C_v *"derives from both within-test-variability and normal variability of in-situ concrete strength"*. Of course, some values of C_v can be found in the literature (e.g. [9, 10]), but they may not have general validity.

According to the RILEM TC 249-ISC Recommendations, the assessment of TRP is recommended at EQL1 and mandatory at EQL2 and EQL3. However, it remains unknown in this case-study. To proceed, **we must make an assumption, and we will consider that both techniques considered in this case study have a medium precision.**

9.3.4 T4. Identifying Test Regions

This task was the main objective of [1] who developed the statistical analysis of the NDT results and identified three TR (see Sect. 9.1). We have already explained in this book (see Sect. 4.2), both from a theoretical point of view and through examples derived from real on-site investigations, how TRs can be identified. Therefore, this process is not reproduced here and we will consider exactly the same TRs as identified by [1]. The interested reader can easily refer to the original paper to get full details. Table 9.2 summarizes how structural members are grouped into these 3 TRs and what are the statistical distributions of NDT results in each TR.

Table 9.4 Minimum number of cores for EQL2 requirement (absolute/absolute), with medium precision rebound test results, for different combinations of concrete mean strength and coefficient of variations (see Chap. 2)

fc mean	Medium precision (TRP2) cov (%)				
	10	15	20	25	30
10	6	6	6	6	7
15	6	6	6	6	6
20	6	6	6	6	6
25	6	6	6	6	6
30	6	6	6	6	6
35	6	6	6	7	8
40	6	6	7	9	10
45	6	7	9	10	12
50	7	9	11	12	14

9.3.5 T5. Defining the Number of Cores

The number of cores depends mainly on the EQL and on the TRP level, but also on the concrete strength properties (mean and variability), as explained in detail in Chap. 2. When several TRs are identified, two options are possible: either identifying a conversion model in each TR or identifying a unique conversion model that would work for all TRs (see Sect. 1.5.1):

"The choice between these options is based on:

– the consideration of prior knowledge about the structure and the concrete,
– information provided by NDT results and their statistical distribution,
– the number of cores that will be used for identifying the conversion model(s), since the uncertainty of strength estimates decreases when the number of cores increases while each model requires a minimum number of cores."

The minimum number of cores can be taken from the Tables available at Chap. 2. Table 9.4 corresponds to an EQL2 requirement (absolute/absolute), with medium precision rebound test results.

If no information about the concrete properties was available, the number of cores would be 14. However, as stated in §9.1, the analysis of documentation on the structure has revealed that the specified strength class of concrete was $R_{bk}250$,[5] which corresponds to a characteristic cylinder strength value equal to 20 MPa. We can thus assume that the mean strength should not exceed 30 MPa. It is difficult to make any assumption about the coefficient of variation of concrete[6] and we **therefore decide to take 7 cores,** which is the highest value in the restricted range.

9.3.6 T6. Defining the Location of Cores

Table 9.5 provides the characteristics corresponding to the 14 cores in the original case-study, with their location and the NDT results. As these are the only points where strength values are available we are constrained in our comparative analysis to choose the 7 TL for coring from this dataset (this restriction would not exist in a real investigation following the RILEM TC 249-ISC Recommendations and all the 47 ND TL could have been considered).

In the RILEM TC 249-ISC Recommendations, there are two possibilities for choosing the core locations (see Sect. 1.4.1) according to a predefined pattern or a conditional pattern. Conditional coring is recommended, since it is expected to lead to a more reliable conversion model. Conditional coring does not induce any additional cost and only requires that NDT results have been made available prior to coring. **We choose a conditional coring pattern**. Table 9.6 contains the same data

---

[5] Which was the denomination at the time of building, and roughly corresponds to a C20/25 concrete.
[6] The knowledge of the TRP could help estimating the CoV of the concrete strength as it was demonstrated in [11].

Table 9.5 Location and NDT results for the 14 cores taken in the original study

Floor	Component	Rebound	UPV (m/s)
I	C	31	3611
I	C	39	4046
I	C	43	4010
I	C	43	4350
I	B	23	3576
II	C	20	2345
II	C	26	2850
II	B	22	2742
III	C	22	3140
III	C	22	3587
III	B	23	2378
IV	C	25	3072
IV	C	30	3756
IV	B	24	3405

Table 9.6 Location and NDT results for the 7 cores taken with the RILEM approach

Core number	Floor	Component	Rebound	UPV (m/s)
1	II	C	20	2345
	II	B	22	2742
2	III	C	22	3140
	III	C	22	3587
3	III	B	23	2378
	I	B	23	3576
4	IV	B	24	3405
	IV	C	25	3072
5	II	C	26	2850
	IV	C	30	3756
	I	C	31	3611
6	I	C	39	4046
	I	C	43	4010
7	I	C	43	4350

as Table 9.5, simply ranked according to the rebound test results, from the lowest to the largest value. The first column identifies at which TL the 7 cores are taken, in order to have an even coverage of the distribution of rebound test results.[7] Figure 9.8

---

[7]Rebound test results have been privileged against UPV test results for reasons explained further, as they appear to have a higher TRP.

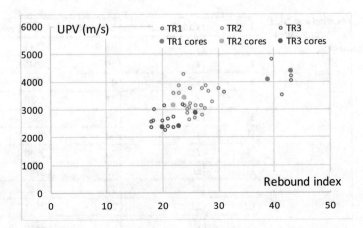

Fig. 9.8 Correlation between rebound and UPV test results for the 3 TR. Plain circles correspond to locations where cores have been taken

illustrates how cores are sampled among all original test locations and among the 3 TRs. The TRs that were identified by [1], see Table 9.2, were not used at this stage to define the core locations. However, the fact that conditional coring aims at covering the existing contrast in concrete properties gives interesting results: cores 6 and 7 belong to TR1, cores 2 and 4 belong to TR2 and cores 1, 3 and 5 belong to TR3. This confirms that both the weakest areas and the strongest are correctly sampled.

9.3.7 T7. Choice of a Conversion Model

After having performed the destructive tests, one gets 7 values of the compressive strength on cores, which can be plotted against either rebound test results (Fig. 9.9a) or UPV test results (Fig. 9.9b). The next step is that of choosing a mathematical shape for the conversion model. In practice, the investigator can choose to test several

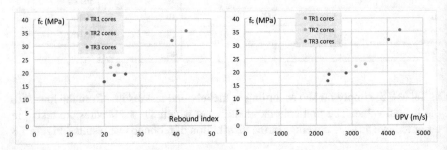

Fig. 9.9 a, b Correlation between core strength of the 7 cores and the ND test results: with rebound (left), with UPV (right)

models, as this does not induce any experimental cost, all test results being already given. We choose to test three different conversion models:

– a multivariate conversion model, as was chosen by [1]:

$$f_c = a\,R^b\,UPV^c \tag{9.2}$$

– a univariate conversion model, with rebound:

$$f_c = a + b\,R \tag{9.3}$$

– a univariate conversion model, with UPV:

$$f_c = a + b\,UPV \tag{9.4}$$

These choices are justified by the fact that: (1) univariate models are the simplest ones and Fig. 9.9a and Fig. 9.9b do not show nonlinear tendencies, (2) the SonReb model may improve the assessment, and has been chosen by [1]. In all cases, the same model is used for the three TRs (identifying a specific model for each TR would require $3 \times 7 = 21$ cores).

9.3.8 T8. Model Identification and Calibration

Among the several possibilities described in the RILEM TC 249-ISC Recommendations we have chosen to identify the model parameters of Eqs. 9.2–9.4 with a simple regression. For the case of the SonReb model, the linear regression is established between the logarithm of strength and the logarithms of NDT results and the equation is finally written in the form of a double power law. The results of this identification are synthesized in Table 9.7 which contains the estimated value of each parameter, its confidence interval and two additional indicators, the coefficient of determination r^2 and the fitting error (RMSE$_{fit}$). The fitting error (mean fitting error on local strength) is calculated from the difference between the estimated strength (using the conversion model) and the core strength test result at the same test location.

Results in Table 9.7 show that the coefficient of determination r^2 as well as the fitting error RMSE$_{fit}$ seem to indicate that the SonReb model is the best. This is a

Table 9.7 Identification of model parameters (coefficients a, b and c are given with their standard deviation, *for ln(a))

Model	a	b	c	r^2	RMSE$_{fit.}$ (MPa)
$f_c = a\,R^b\,UPV^c$	−3.75 (1.21)*	0.418 (0.168)	0.683 (0.206)	0.969	0.933
$f_c = a + b\,R$	2.40 (2.83)	0.755 (0.096)	–	0.924	1.805
$f_c = a + b\,UPV$	−4.76 (3.49)	0.0088 (0.0011)	–	0.932	1.701

logical consequence of the fact that the same dataset is fitted with a model which has 3 free parameters, against 2 parameters for the univariate conversion models. Any statement must be avoided at this stage, as the precision of the strength estimation can be assessed only with $RMSE_{pred}$.

9.3.9 T9. Strength Estimation and Estimation of Strength Assessment Uncertainty (Model Prediction Error)

9.3.9.1 Estimation of Local Concrete Strength

The identified conversion models are used first for estimating the local strengths, then for quantifying the prediction error and finally to assess the concrete properties over the whole structure. As NDT results are available at 40 test locations where no cores have been taken, the three models can be applied to estimate the local strength at all of these NDT locations. As an example, if we consider a NDT location in a column of the third floor where the NDT results are 25 for rebound and 2846 m/s for UPV, applying the models of Table 9.7, leads to strength estimates of 20.5 MPa for the SonReb model, 20.9 MPa for the linear model with rebound and 20.4 MPa for the linear model with UPV. The same process can be repeated for all test locations, and provide a series of 40 estimated strengths. Figure 9.10 plots the cumulative distribution of the estimated strength at all (40) NDT locations according to the three conversion models of Table 9.7.

The three curves have the same shape, with lower values for the $f_c(R)$ model and higher values for the $f_c(UPV)$ model, while the SonReb model lies in between. All

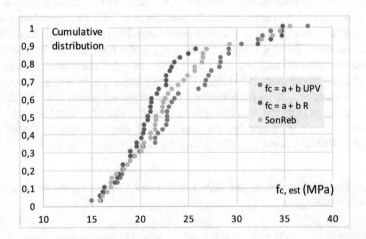

Fig. 9.10 Cumulative distribution of the estimated strength at all NDT locations according to the three conversion models

Table 9.8 Mean strength (± strength standard deviation) and coefficient of variation (CoV) for the three conversion models

Domain	$f_c = a\,R^b$ UPV^c	$f_c = a + b\,R$	$f_c = a + b$ UPV
whole building	22.9 (±5.1) MPa—22%	22.2 (±5.2) MPa—23%	23.8 (±5.4) MPa—23%
TR1	32.2 (±3.5) MPa—11%	32.7 (±3.5) MPa—11%	31.3 (±4.2) MPa—14%
TR2	23.1 (±2.2) MPa—10%	21.7 (±1.6) MPa—7%	24.6 (±3.9) MPa—16%
TR3	18.0 (±1.7) MPa—10%	17.7 (±1.4) MPa—8%	18.6 (±2.8) MPa—15%

curves also show a bimodal distribution, with the highest strengths corresponding to what was identified as TR1 (i.e. columns of the first floor, see Table 9.2).

9.3.9.2 Estimation of Concrete Properties in the Building

Statistical parameters can be derived for each of the three models, for the whole building or for a subset that is limited to a given TR (please remember that TRs have been identified as such but that the same conversion model is adopted for all TRs). Table 9.8 compares these strength distributions for the three models at two scales, that of the whole building and that of each TR.

These results reveal a strong consistency, both for the statistical distribution for the whole structure and for that of each TR. At the structural scale, the three conversion models predict a mean strength ranging between 22 and 24 MPa with a standard deviation of about 5 MPa and a CoV slightly above 20%. For all models, the contrast between the 3 TRs is marked. The $f_c(R)$ tends to give the highest estimates for TR1 but the smallest ones for TR3, whereas the $f_c(UPV)$ model makes the opposite. The SonReb model, in which the effects of rebound and UPV test results somehow compensate, lies in between. All coefficients of variation identified in the 3 TRs are much lower than those for the whole structure. **This justifies the interest of identifying such TRs, particularly if the investigator wants to estimate the strength properties of each TR** (mean strength, characteristic strength[8]). The main difference between the three conversion models concerns the CoV attached to each TR, whose estimates are much larger with the $f_c(UPV)$ model.

---

[8]The determination of the characteristic strength is not covered by the RILEM TC 249-ISC Recommendations, but it may be very important for structural calculations. Following the RILEM TC 249-ISC Recommendations provides estimates of mean strength and coefficient of variations, which can be processed in order to estimate the characteristic strength.

Table 9.9 Fitting and prediction error for the three conversion models

Model	r^2	RMSE$_{fit.}$ (MPa)	RMSE$_{pred}$ (MPa)
$f_c = a\,R^b\,UPV^c$	0.969	0.933	1.704
$f_c = a + b\,R$	0.924	1.805	1.743
$f_c = a + b\,UPV$	0.932	1.701	3.628

9.3.9.3 Estimation of Strength Assessment Uncertainty

At this stage, the investigator has established a picture of the concrete properties, but does not know how far the estimated strengths are from reality. The quantification of model uncertainties (RMSE$_{pred}$) is mandatory at EQL3 and recommended for EQL1 and EQL2. The estimation of RMSE$_{pred}$ enables us to determine the mean error on the estimation of local strength is, which is not possible with other estimators such as r^2 and RMSE$_{fit}$ quantified in Table 9.7 (see Sect. 1.5.6.1). The RILEM TC 249-ISC Recommendations provide several ways to estimate RMSE$_{pred}$, either through direct checking, on a number of cores used for validation, or through cross-checking (see Sect. 1.5.6.2).

In this example based on a real case-study, we took only 7 cores, in agreement with theRILEM TC 249-ISC Recommendations, whereas 14 cores had been taken in the original investigation. Therefore, 7 cores provided strength test results that have not been used for calibrating the conversion model, and they can be used for a direct checking of the model uncertainty.[9] The values of fitting and prediction errors for the three conversion models are compared in Table 9.9. These results confirm that the first two indicators (r^2 and RMSE$_{fit}$) do not provide a relevant estimate of the prediction error. The most surprising result is the much larger value of the prediction error (roughly double) for the f_c(UPV) model. This probably indicates that the measurement error, in that study, was larger for UPV test results than for rebound test results. Despite the fact that the TRP was unfortunately not assessed, we can hypothesize that UPV test results probably have a poor TRP. A consequence is that the investigator would be more confident in results provided either by the f_c(R) model or by the SonReb model. Table 9.8 has shown that these two conversion models lead to very close estimates regarding mean strength and strength variability at both structural and TR scales. Table 9.9 shows that they also lead to a very close prediction error (about 1.7 MPa on the local strength).

---

[9]In most cases, this is not possible in practice, as the investigator wants to limit the number of cores and prefers to use all destructive test results at the calibration stage. It is why the RILEM TC 249-ISC Recommendations recommends the cross-checking process that can be carried out in such a situation.

9.4   Comparison Between the Original Approach and the RILEM Approach

The results derived by applying the RILEM methodology to existing data can be compared with those obtained in the original investigation [1]. It must be remembered that there are two major differences between the expert's original approach and the RILEM approach:

- the first identifies a specific conversion model for each TR, whereas the RILEM approach identifies the same TR but uses a unique model for all TRs,
- the RILEM approach uses conditional coring to choose the location of cores.

Table 9.10 summarizes the estimated mean strength and standard deviation of strength at all scales, which are given according to the different ways. These estimates are those of Table 9.8 for the RILEM option. The "mixed" estimates are according to [1] "for each homogeneous area (i.e. TR) the average between the strength values from core tests and the strength values estimated by applying the SonReb relationship to the points where only NDTs were performed".

We can also note that some assumptions were made about the concrete mean strength at stage T5 (in order to limit the number of cores): the mean strength was assumed to be below 30 MPa, and this assumption is validated by the results. At first sight, all results are very close and all assessment procedures seem to work well, but these results deserve more careful attention:

- it must be pointed out that **only half of the cores are used in the RILEM approach**, but both approaches provide very similar results,
- the **RILEM approach, with NDT results only, estimates the strength standard deviation very closely to that provided by the 14 cores**. This is not the case for the original approach (second line in Table 9.10), which seems to indicate that the original SonReb approach did not capture correctly the concrete strength

Table 9.10 Mean strength (± strength standard deviation) in MPa as estimated in the original approach and with the RILEM Recommendations approach (*mixed as indicated in [1])

Approach	Data	Number of data	Whole building	TR 1	TR 2	TR 3
Original [1] 14 cores	cores	14	24.0 (±6.2)	32.2 (±4.0)	22.7 (±1.7)	17.7 (±1.5)
	SonReb	33	21.1 (±5.9)	34.1 (±1.1)	23.5 (±0.9)	16.3 (±1.0)
	mixed*	14 + 33	-	33.2 (±2.9)	23.1 (±1.2)	17.0 (±1.3)
RILEM 7 cores	SonReb	40	22.9 (±5.1)	32.2 (±3.5)	23.1 (±2.2)	18.0 (±1.7)
	$f_c(R)$	40	22.1 (±5.2)	32.7 (±3.5)	21.7 (±1.6)	17.7 (±1.4)

variability at the scale of each TR (this weakness is not visible at the building scale, due to the larger contrast between the weakest and strongest areas).

To explain the reasons why the original SonReb approach applied to each TR fails to capture the proper variability in each TR, one must analyze the prediction error of the models that have been identified and that were given in Table 9.9. Fourteen cores having been taken from the whole building, with 4, 6 and 4 cores in TR 1, TR 2 and TR 3 respectively. This means that the SonReb model which has 3 free model parameters must be calibrated from datasets of size 4 or 6, and that the trade-off effects between model parameters may be exacerbated, with all its consequences on the model prediction error (see Sect. 12.4). This can easily be verified by estimating the $RMSE_{pred}$ of each model (as suggested by the RILEM Recommendations), following the leave-one-out procedure.

If, as an example, the leave-one-out procedure (see Sect. 1.5.5) is applied to the original test results of TR 2, which are given for the 6 cores in Table 9.11 (and were plotted on Fig. 9.5), it is possible to iteratively identify 6 conversion models, each having its own set of (a, b, c) model parameters, which are summarized in Table 9.12 with the resulting estimated strength. The parameters of the original model were given in Table 9.3: a = 3.533, b = 0.553, c = 0.012.

The results of Table 9.12 confirm that the values of the model parameters can vary a lot. In three out of the six cases, the exponent of UPV is even negative. Suh a value would indicate that a larger UPV value corresponds to a lower strength. This seems to go against physics and can be found only in some cases when some additional

Table 9.11 Destructive and non-destructive test results in TR 2

Core	Rebound	UPV (m/s)	Core strength (MPa)
1	23	3576	20.8
2	22	3140	21.8
3	22	3587	22.3
4	24	3405	22.6
5	25	3072	22.7
6	30	3756	25.9

Table 9.12 Model parameters for the SonReb model of TR 2, original model with 6 cores and models identified for cross-checking

	a	b	c	$f_{c, est}$ (MPa)
leave core 1	0.978	0.431	0.219	22.7
leave core 2	1.71	0.572	0.093	21.1
leave core 3	10.13	0.705	−0.179	20.7
leave core 4	3.19	0.547	0.026	22.5
leave core 5	8.48	0.600	−0.114	23.5
leave core 6	41.35	0.196	−0.153	22.7

unknown parameter (like the humidity or the degree of carbonation) changes between the test locations of the same set. In addition, whereas the strength of the sixth core has been measured at 25.9 MPa and departs from all others (see Fig. 9.5 left), it can be estimated correctly by the model which predicts only 22.7 MPa, with a 14% relative error. The calculated prediction error $RMSE_{pred}$ is 1.66 MPa. This value is equivalent to 98% of the standard deviation of direct strengths as estimated from the six core strengths (i.e. 1.69 MPa). The logic conclusion must be that **the prediction error is so large for the SonReb model of TR 2 that this model is not reliable for the local strength estimation in that TR.**

9.5   Conclusions

The purpose of this chapter was to show how the RILEM Recommendations can be used in a real context in order to improve the methodology of non-destructive strength assessment and its results. For this reason we have chosen to consider experimental data that had already been published in a recent paper where innovative ideas regarding the identification of homogeneous test regions were developed. Conclusions can be drawn at two levels, first regarding what and how the RILEM approach improves the assessment, secondly regarding some specific issues.

By applying the methodology as summarized in the flowchart of Fig. 9.7, we have demonstrated that the RILEM methodology:

- is **easy to apply in practice, and does not induce strong changes to common engineering practice**;
- can give results (mean strength estimates, local strength estimates) that are equivalent or more accurate than those provided by previous approaches;
- **can save a significant number of cores** (in our example, half of the cores), thanks to the conditional coring option, without any additional cost. This is expected to have a strong impact for a wider application of ND techniques;
- **can provide more reliable estimates of concrete variability** than former methodologies. In order to be concise, we have not chosen the bi-objective approach for identifying the conversion models, even if it has been demonstrated that it further improves the estimation of concrete variability (see Sects. 1.5.5.2 and 12.8).

The main weakness of our comparison is that we were required to make an assumption regarding the precision of NDT results (TRP) in the original approach, as it had not been measured. It must be repeated that TRP is a crucial parameter, which has a strong impact both on the required number of cores and on the final model error. **Following the RILEM Recommendations requires that TRP is assessed at the very first step of on-site non-destructive investigation.** TRP also appears as a strong limiting factor regarding the precision of the final estimation. If the investigator aims at reducing the model error, improving the ND measurement (i.e. reducing the TRP) can be more effective than increasing the number of cores.

The two last issues are more specific and concern the conversion models and the identification of test regions:

- In common practice, investigators may prefer to choose a more sophisticated conversion model with a higher number of free model parameters, which would (logically) offer a better fit and a lower error. We have pointed out the risks of such an option, that increases the problems of model uncertainty due to the trade-off between model parameters. An apparently better model, with a higher coefficient of determination and a lower fitting error can in fact have a larger prediction error. In all cases, **the prediction error $RMSE_{pred}$ must be estimated before drawing conclusions about the relevance of the conversion model**. The advantage is that this work is independent of all on-site measurements and can be carried out afterwards, at low cost.

- Once local strength can be estimated at many test locations (by applying the conversion model to the NDT results) it is possible to map the spatial distribution of strength in the structure. Test regions (TRs) may have been identified by processing the NDT results, and it is then possible to **estimate the strength properties in each TR**. This can be particularly useful for instance if further structural analysis is planned, in which each TR will have its own material properties. However, one must be careful regarding the choice between either identifying a unique conversion model which covers all TRs or identifying a specific conversion model for each TR. The second option was preferred in the original expert approach of this chapter, but we have shown its possible drawbacks. The choice of more specific models is sometimes justified or necessary (particularly when comparable values of NDT results in two TRs correspond to different concrete strengths) but may be counter effective as it: (1) leads to an increase in the total number of cores and (2) leads to reduce the range of variation of concrete properties in each TR, which increases the model error. These issues are discussed in more detail in Sect. 4.3.

In any case, the investigator must pay attention to all consequences of his choices, by carefully checking his assumptions and estimating the precision of his results.

Acknowledgements The 7th author would like to acknowledge the financial support by Base Funding—UIDB/04708/2020 of CONSTRUCT—Instituto de I&D em Estruturas e Construções, funded by national funds through FCT/MCTES (PIDDAC).

Appendix

This table summarizes the test results that were obtained in the original study [1] and were used in this chapter. They indicate the core location (from floor I to floor IV), and the component (beam or column, B/C), the NDT results, and the compressive strength of the 14 cores.

Set	Floor	Beam or column	Rebound test result	UPV test result (m/s)	f_{core} (MPa)
C1	I	C	41	3500	31.6
	I	C	39	4046	
	I	C	43	4202	
	I	C	31	3611	26.8
	I	C	40	4788	
	I	C	43	4350	35.3
	I	C	43	4350	
	I	C	43	4010	35.1
B1	I	B	23	3576	20.8
	I	B	29	3257	
	I	B	27	2805	
C2	II	C	26	2850	19.2
	II	C	18	2599	
	II	C	20	2578	
	II	C	18	2985	
	II	C	20	2345	16.3
	II	C	18	2367	
	II	C	24	3181	
B2	II	B	18	2545	
	II	B	21	2659	
	II	B	22	2742	16.7
C3	III	C	23	3857	
	III	C	25	2846	
	III	C	22	3140	21.8
	III	C	24	2986	
	III	C	24	3121	
	III	C	22	3587	22.3
	III	C	28	3644	
	III	C	28	3852	
B3	III	B	20	3148	
	III	B	21	2244	
	III	B	23	2378	18.8
	III	B	21	2389	
	III	B	22	2341	
C4	IV	C	25	2622	
	IV	C	26	2697	
	IV	C	24	4250	
	IV	C	25	3072	22.7

(continued)

(continued)

Set	Floor	Beam or column	Rebound test result	UPV test result (m/s)	f_{core} (MPa)
	IV	C	30	3756	25.8
	IV	C	28	3017	
	IV	C	27	3126	
	IV	C	24	3124	
	IV	C	26	3123	
	IV	C	25	3215	
B4	IV	B	25	3758	
	IV	B	24	3405	22.6
	IV	B	27	3745	

References

1. Masi, A., Chiauzzi, L., Manfredi, V.: Criteria for identifying concrete homogeneous areas for the estimation of in-situ strength in RC buildings. Constr Build Mater **121**, 576–587 (2016)
2. Grantham, M.: Using combined NDT methods and the Sonreb method in concrete dispute resolution. In: 14th International Conference on Structural Faults and Repair, Edinburgh, United Kingdom, 3–5 July 2012
3. Papworth, F., Corbett, D., Barnes, R., Wyche, J., Dyson, J.: In-situ concrete strength assessment based on ultrasonic (UPV), rebound, cores and the SonReb method. In: Concrete 2015, 27th Biennial National Conference of the Concrete Institute of Australia, Melbourne, Australia, 30 Aug–2 Sept 2015
4. EN 12504-1:2012, Testing concrete in structures—Part 1: cored specimens—Taking, examining and testing in compression. In: European Committee for Standardization (CEN), Brussels, Belgium (2012)
5. EN 12504-2:2012, Testing concrete in structures—Part 2: non-destructive testing—Determination of rebound number. In: European Committee for Standardization (CEN), Brussels, Belgium (2012)
6. EN 12504-4:2004, Testing concrete—Part 4: determination of ultrasonic pulse velocity. In: European Committee for Standardization (CEN), Brussels, Belgium (2004)
7. Breysse, D., Balayssac, J.P.: Reliable non-destructive strength assessment in existing structures: myth or reality? RILEM Tech Lett **3**, 129–134 (2018). https://doi.org/10.21809/rilemtechlett.2018.73https://doi.org/10.21809/rilemtechlett.2018.73
8. Pfister, V., Tundo, A., Luprano, V.A.M.: Evaluation of concrete strength by means of ultrasonic waves: A method for the selection of coring position. Constr Build Mater **61**, 278–284 (2014). https://doi.org/10.1016/j.conbuildmat.2014.03.017https://doi.org/10.1016/j.conbuildmat.2014.03.017
9. Federal Emergency Management Agency (FEMA): FEMA 356—Prestandard and commentary for the seismic rehabilitation of buildings. Washington D.C., Nov 2000
10. ACI 214.4R-10: Guide for obtaining cores and interpreting compressive strength results (2010)
11. Pereira, N., Romão, X.: Assessing concrete strength variability in existing structures based on the results of NDTs. Constr Build Mater **173**, 786–800 (2018)

Part III
Appendix

Chapter 10
Statistics

Vincent Garnier, Jean-Paul Balayssac, and Zoubir Mehdi Sbartaï

Abstract This chapter gives definitions and basics on statistics, as mean standard deviation, trueness, accuracy, uncertainty … The relation of test result precision with the number of test readings is emphasized. Theoretical considerations about the minimal distance between two test readings to ensure their dependency are presented.

10.1 List of Definitions

10.1.1 Repeatability, Reproducibility and Variability

This appendix clarifies the definitions of both terms and concepts used in the guidelines, and also identifies the parameters that have an influence on measurements and measures.

Measurement is the act of measuring and the **measure** is its result. **In these guidelines the concept of "measure" corresponds to the test result**.

In general terms, the parameters influencing the measure can be described by the following Ishikawa diagram (Fig. 10.1).

These parameters are process, device, operator, material and environment. It is useful to know how much they influence a test reading or a test result. The three concepts of repeatability, reproducibility and variability cover the influence of these parameters as illustrated in Fig. 10.1.

V. Garnier
Laboratory of Mechanic and Acoustic, L.M.A - AMU/CNRS/ECM - UMR7031, Aix Marseille University, Marseille, France

J.-P. Balayssac (✉)
LMDC, Université de Toulouse, INSA/UPS Génie Civil, Toulouse, France
e-mail: jean-paul.balayssac@insa-toulouse.fr

Z. M. Sbartaï
University Bordeaux, I2M-UMR CNRS 5295, Talence, France

© RILEM 2021
D. Breysse and J.-P. Balayssac (eds.), *Non-Destructive In Situ Strength Assessment of Concrete*, RILEM State-of-the-Art Reports 32,
https://doi.org/10.1007/978-3-030-64900-5_10

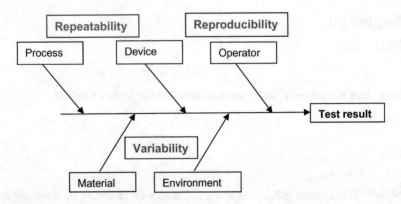

Fig. 10.1 Ishikawa diagram of the parameters influencing a measure

10.1.1.1 Test of Repeatability

The test of repeatability consists in repeating the measurement at the same location, with the same device, with the same process and by the same operator.

10.1.1.2 Test of Reproducibility

The test of reproducibility consists in repeating the measurement by changing one of the following parameters, location, device, measurement condition or operator. The reproducibility linked to the change of operator is considered in this case.

10.1.1.3 Variability

The variability includes the influence of the material and environment, being the process, device and operator fixed.

At a given TL the material is assumed to be fixed, so the variability only includes the influence of environment.

In a TR, material changes induce additional variability.

10.1.2 Precision, Accuracy and Uncertainty

These concepts are used in the RILEM TC 249-ISC Recommendations (see §1.3.3.2.6).

10.1.2.1 Mean Value μ

It is the mean of all the test readings or test results.

10.1.2.2 Standard Deviation SD

Standard deviation which is used to quantify the amount of statistical dispersion of a set of data values (x_i) is calculated by:

$$sd = \sqrt{\frac{\sum_{i=1}^{n}(x_i - \mu)}{n - 1}}$$

in which x_i is the ith test reading, n the number of test readings and μ is the mean value (test result) of the n test readings.

$$\mu = \frac{\sum_{i=1}^{n}(x_i)}{n}$$

Another quantity related to the statistical dispersion is the coefficient of variation COV.

$$COV(\%) = \frac{sd}{\mu} * 100$$

The mean value obtained from test readings or test results is generally not exactly the true value of the evaluated property, or reference value. Figure 10.2 illustrates

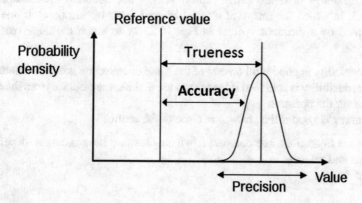

Fig. 10.2 From ISO 5725–1, distribution of the measures (GUM Joint Committee for Guides in Metrology, Evaluation of measurement data—Guide to the expression of uncertainty in measurement, JCGM 100:2008, 2008)

a distribution of measures around its mean value and permits to define precision, trueness and accuracy.

10.1.2.3 Precision

Precision is linked to the distribution of the measures around their mean. Precision is the result of repeatability and reproducibility. It is influenced by random errors.

10.1.2.4 Trueness

Trueness is the closeness of agreement between the mean of measures and the reference value of the evaluated property. It is influenced by systematic errors.

10.1.2.5 Accuracy

Accuracy is the closeness of agreement between one (single) measure and the reference value of the evaluated property. Accuracy results from a combination of precision and trueness, therefore from both random and systematic errors.

10.1.2.6 Uncertainty

Uncertainty is the combination of precision (repeatability and reproducibility) and trueness.

The meanings of repeatability, reproducibility and accuracy are illustrated on Fig. 10.3, in which the image of a target is used. Four operators, each one being represented by a different symbol in Fig. 10.3, try to touch the target with four arrows:

- Repeatability is good if all arrows of the same operator are grouped together,
- Reproducibility is achieved if the arrows are at the same distance from the centre, whatever the operator,
- Accuracy is good if the arrows are close to the centre.

Four configurations are compared, which illustrate how accuracy depends on precision and trueness.

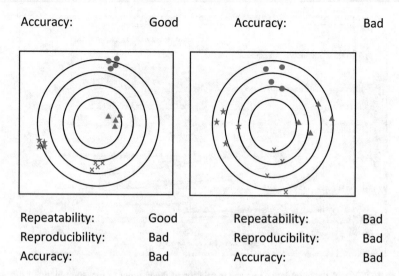

Accuracy: Good   Accuracy: Bad

Repeatability: Good   Repeatability: Bad
Reproducibility: Bad   Reproducibility: Bad
Accuracy: Bad   Accuracy: Bad

Fig. 10.3 Illustration of repeatability, reproducibility and accuracy

10.2 Test Result Precision (TRP) as a Function of the Number of Measurements

This section allows estimating the minimum number of measurements to obtain a given precision of the test result inside a test location. The reader must be aware that this "minimum number of measurements" is not the "minimum number of cores" that is discussed in §4.2 of Chap. 1. This "minimum number of measurements" addresses the issue of the "test result precision" (TRP, see §5.2 of Chap. 1) and relates to the influence of the number of experimental results on the estimation of a given quantity.

The precision impacts the repeatability and reproducibility of the measurements at a test location (§0). Nondestructive tests can be repeated at the same test location and the precision of the measurements in this condition can be easily estimated.

Increasing the number of measurements reduces the random error on their mean average. As a consequence, if there is no systematic measurement error, the distance between the mean average and the reference value is reduced. For NDT methods (UPV and RH) additional measurements do not induce a very high cost. So, the improvement of the assessment precision can be easily achieved by increasing the number of measurements.

By assuming a normal distribution of the measurements, the Student law (Eq. 10.1) permits to define the probability that the true value of the measure will be inside a confidence interval, if the standard deviation is known.

$$\Pr\left(\overline{x} - t\alpha_{/2} \cdot \frac{sd}{\sqrt{2}} < M < \overline{x} + t\alpha_{/2} \cdot \frac{sd}{\sqrt{2}}\right) = 1 - \alpha \qquad (10.1)$$

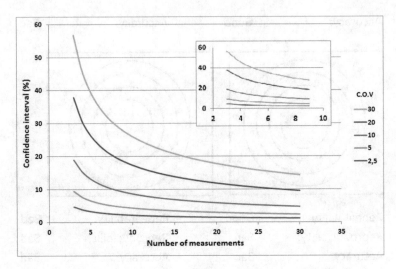

Fig. 10.4 Width of the confidence interval in relation to the number of measurements for different coefficients of variability (COV) (confidence level is equal to 95%)

with

- $t_{\alpha/2}$: percentile
- $\bar{x}$: mean
- sd: standard deviation
- n: number of measurements
- M: true value of the measure
- $(1-\alpha)$: confidence level

The percentile depends on the confidence level and on the number of measurements. For example, for 10 measurements (n = 10), by considering a confidence level $(1-\alpha)$ of 90% the fractile is equal to 1.383 (1.833 for 95%). Standard deviation is an input which also depends on the number of measurements. It can be estimated by using feedback of experience.

By analyzing Eq. 10.1, one can see that for a fixed confidence level, the width of the confidence interval increases with the increase of the standard deviation and with the decrease of the number of tests. If the standard deviation is high, a large number of measurements will be necessary to reduce the confidence interval. If performing a large number of tests is not possible, the quality of the assessment can be enhanced for instance by improving the measurement device to reduce the standard deviation[1] (see also Fig. 10.4).

For instance, for a width of the confidence interval of 10%, if the coefficient of variation (COV = sd/M*100) is equal to 20%, at least 27 measurements are necessary

---

[1] J.-P. Balayssac, S. Laurens, G. Arliguie, D. Breysse, V. Garnier, X. Dérobert, B. Piwakowski, Description of the general outlines of the French project SENSO—Quality assessment and limits of different NDT methods. Constr Build Mater 35, 131–138 (2012).

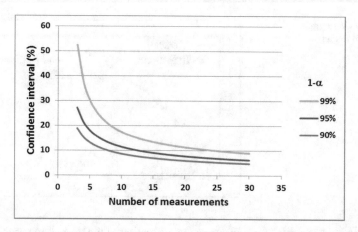

Fig. 10.5 Width of the confidence interval in relation to the number of measurements for different confidence levels (1-α) (COV = 10%).

to state that there is a probability of 95% that the true value is inside the confidence interval. If the coefficient of variation decreases to 10%, only 12 measurements are necessary.

Figure 10.5 illustrates the influence of the value of the confidence level (99, 95, and 90%) on the width of the confidence interval (in this case the coefficient of variation was fixed at 10%). The higher the confidence level, the larger is the confidence interval. Hence, if the confidence level is 90%, as the percentile decreases, the confidence interval is narrower and so the minimum number of measurements decreases for a given width of the confidence interval.

10.3   Minimum Distance Between Test Locations

Concrete is an heterogeneous material at all scales. Therefore, even with a perfect measurement device and process, two test results resulting from measurements at two different test locations TL separated by a short distance (typically few centimeters) will provide different test results. When the distance between these two test locations increases, the difference between the test results also tends to increase. If the concrete is "globally homogeneous", the long distance difference tends towards a maximum which corresponds to the concrete variability. This variability is usually assumed to be intrinsically related to the material fabric.

Reversely, this means that two test results taken from two test locations at a short distance are somehow correlated, as the investigated volumes considered by the two tests overlap. This section describes how can be identified the minimum distance between two test locations from which the values of test result can be considered as independent. The hypothesis of independence is necessary in statistics to calculate the mean value and the variability (standard deviation). Computing the variogram of

Fig. 10.6 Variogram
(spherical model)

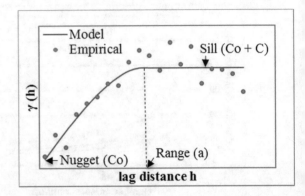

NDT can be then useful for improving the evaluation of the mean and variability of the test results.

There is usually some spatial variability in a concrete structure because of the concrete inhomogeneity and the variability of the exposure conditions.[2] However, most of the time, it can be said that, for a given number of assessments distributed on a surface, those that are close together have a certain similarity when compared with more distant ones. This spatial dependence can be represented by a statistical function known as the variogram which is defined as half the variance of the difference between two data items from two different locations separated by a distance h (Eq. 10.2).

$$\gamma(h) = \frac{1}{2} Var[Z(x) - Z(x + h)] \tag{10.2}$$

For a series of observed test results with a limited series of pairs separated by a distance h (N(h)), an empirical variogram can be determined as (Eq. 10.3):

$$\gamma_e(h) = \frac{1}{2N(h)} \sum_{i=1}^{N(h)} [Z(x_i) - Z2(x_i + h)]^2. \tag{10.3}$$

For each empirical variogram, a model can be fitted by the least squares method to obtain a mathematical function that can be used later to compute the expected value of the test result at any additional point.

Figure 10.6 shows an example of the spherical model fitted to an empirical variogram. Here, three main parameters can be inferred: nugget (Co) that describes the variance of a test result made several times at the same test location (linked to repeatability and reproducibility), sill (S) that represents the global half variance, and range (a) which represents the maximum distance where data are correlated. This distance is considered as the minimal distance between two TL to be sure that test results are

---

[2]New optimization algorithm for optimal spatial sampling during non-destructive testing of concrete structures, C. Gomez-Cardenas C., Z.M. Sbartaï, J.-P. Balayssac, V. Garnier, D. Breysse, Eng Struct 88, 92–99 (2015).

independent. A study from Nguyen et al. [3] on slabs stored outside determined a range of about 40 cm for both Rebound Hammer and indirect UPV. In a first approach, these results can be used as the minimum distance between two test locations.

---

[3] N.T. Nguyen, Z.M. Sbartaï, J.F. Lataste, D. Breysse, F. Bos, Assessing the spatial variability of concrete structures using NDT techniques – Laboratory tests and case study. Constr Build Mater 49, 240–250 (2013).

Chapter 11
Model Identification and Calibration

Zoubir Mehdi Sbartaï, Maitham Alwash, Denys Breysse, Arlindo Gonçalves, Michael Grantham, Xavier Romão, and Jean-Paul Balayssac

Abstract This chapter provides additional information about the "identification of conversion model" step and the "strength estimation" step as defined in the flowchart summarizing the RILEM recommendation. The advantages and limits of the various options are illustrated by analyzing the results of synthetic simulations. Based on the developed synthetic database, a comparison of the performance of different possible univariate conversion models identified and bi-objective method using linear regression is presented and discussed. It is shown here that the issue of the "best conversion model" is secondary, and that the choice of the conversion model has only negligible effects on the final uncertainty of the final strength. In fact, prediction error is the only way to address correctly. It is also shown why the bi-objective approach must be privileged as it provides, without any additional cost, a better estimation of concrete variability, without reducing the performance regarding the assessment of mean strength and local strength.

Z. M. Sbartaï (✉) · D. Breysse
University Bordeaux, I2M-UMR CNRS 5295, Talence, France
e-mail: zoubir-mehdi.sbartai@u-bordeaux.fr

M. Alwash
Department of Civil Engineering, University of Babylon, Babylon, Iraq

A. Gonçalves
Laboratório Nacional de Engenharia Civil (LNEC), Lisboa, Portugal

M. Grantham
Consultant - Sandberg LLP, London, UK

X. Romão
CONSTRUCT-LESE, Faculty of Engineering, University of Porto, Porto, Portugal

J.-P. Balayssac
LMDC, Université de Toulouse, INSA/UPS Génie Civil, Toulouse, France

© RILEM 2021

341

D. Breysse and J.-P. Balayssac (eds.), *Non-Destructive In Situ Strength Assessment of Concrete*, RILEM State-of-the-Art Reports 32,
https://doi.org/10.1007/978-3-030-64900-5_11

11.1 Introduction

This chapter provides additional information about the "identification of conversion model" stage and the "strength estimation" stage as defined in the flowchart summarizing the recommended assessment process (see Fig. 1.9). At these stages of the assessment process, all destructive and non-destructive test results have been made available. The three last Tasks defined in the flowchart (namely Tasks T7, T8 and T9) correspond to the post-processing of the available data and can be carried out without any significant additional cost. Therefore, the assessor can compare several options regarding the choice of the conversion model and that of the identification approach (T7 and T8) and has to finally estimate the model error (T9).

The merits and limits of the various options are illustrated by analyzing the results of synthetic simulations. The synthetic example on which these simulations are based is exactly the same than in Chap. 8. The reader must refer to this chapter to have all the details regarding the investigated structure and the various investigation strategies.

The three following Sects. 11.2, 11.3 and 11.4 are respectively paying attention to:

- The assessment of the model error (Task T9), in order to point at some common (and wrong) ideas and inadequate engineering practice,
- The uncertainty which is attached to the identification of the conversion model parameters (Task T8), and its effects on the model error (Task T9),
- The choice of the conversion model, which can be univariate or multivariate and that of the identification approach (Task T7). The advantage of the bi-objective identification approach is discussed.

11.2 Assessment of Model Error

This Appendix provides additional information about assessment of model uncertainty which corresponds to Task T9 in the flowchart summarizing the recommended assessment process (see Fig. 1.9). It shows that what is often done in engineering practice (and in academic publications) may be completely misleading. It justifies the choices that are taken in the RILEM TC 249-ISC Recommendations and justified in these Guidelines and shows how some major factors govern the model uncertainty.

11.2.1 Context

Once the conversion model is identified (Stage T8), the model error must be assessed (Stage T9). This can then be compared with the target tolerance intervals corresponding to the chosen EQL. The three quantities that can be estimated are respectively the mean value of local strengths (or the concrete mean strength), the standard deviation of local strength (or concrete strength standard deviation) and the local error on local strength.

Whatever the option used for identifying the model parameters (i.e. using a specific model or a calibrated model), it is recommended to check the accuracy of strength estimates. This is mandatory for EQL3. The recommended measure of model error is the root mean square error (RMSE) which provides the statistical lack of fit, expressed in MPa, i.e. the "mean distance" between the reference value of local strength and the local strength estimated by the model.

11.2.2 Definitions. r^2 Versus RMSE: How to Quantify the Model Uncertainty?

Let us consider a set of n values marked $f_{c\,i}$, $i = 1, n$, corresponding to the n measured values of strengths that have a mean value (arithmetic mean $f_{c\,mean}$). From NDT measurements and strengths measured on a calibration set, a model $f_{c,est}$ (NDT) has been identified and a set of n values $f_{c,est\,I}$ has been estimated having the same mean value (arithmetic mean) of the measured (observed) data.

The variability of the data set can be measured using three sum of squares formulas:

- The total sum of squares (equal to n times the variance of the data set):

$$SS_{tot} = \sum_i \left(f_{c,\,i} - f_{c\,mean}\right)^2 SS_{tot} = \sum_i \left(f_{c,\,i} - f_{c\,mean}\right)^2 \tag{11.1}$$

- The explained sum of squares

$$SS_{exp} = \sum_i \left(f_{c,\,est,\,i} - f_{c\,mean}\right)^2 \tag{11.2}$$

- The residual sum of squares:

$$SS_{res} = \sum_i \left(f_{c,\,i} - f_{c,\,est,\,i}\right)^2 \tag{11.3}$$

In Eqs. 11.1–11.3, $f_{c,\,i}$ is the individual core strength (the result of the compressive strength measurement), $f_{c\,mean}$ is the experimental mean strength (the mean of the individual values obtained on the n cores) and $f_{c,\,est,\,i}$ is the estimated local strength that results from the estimation of strength by applying a conversion model to the NDT result.

The most general definition of the <u>coefficient of determination r^2</u> is

$$r^2 = 1 - SS_{res}/SS_{tot} \tag{11.4}$$

This coefficient is directly related to the residual (unexplained by the model) variance. If the model was identified through a linear regression, one also has

$$SS_{tot} = SS_{exp} + SS_{res} \qquad (11.5)$$

which leads to

$$r^2 = SS_{exp}/SS_{tot} \qquad (11.6)$$

and r^2 can be seen to be the part of the total variance which is explained by the model.

The Root Mean Square Error (RMSE) can be calculated as:

$$RMSE = \sqrt{\frac{\sum_{i=1}^{n}(f_{c,est,i} - f_{c,i})^2}{n}} \qquad (11.7)$$

which measures the differences between values estimated by the model and the values that were actually observed, which are the residuals. The RMSE is an adequate measure for comparing the forecasting errors of different models.

One must understand the difference between the fitting error ($RMSE_{fit}$) and the prediction error ($RMSE_{pred}$):

- $RMSE_{fit}$ relates to the dataset used for model identification. It is calculated on the identification dataset, by comparing $f_{c,i}$ with $f_{c,est,i}$ for all n points[1] at which a data pair (Tr_i, $f_{c,i}$) was used to identify the model.
- $RMSE_{pred}$ relates to a new dataset. It is calculated by comparing measured values and strength estimates at points that have not been used to identify the model.

The prediction error is always larger than the fitting error, which is due to the random nature of the estimated parameters of the conversion model. Any estimation of the model error based only on the fitting error, as is common in practice, is meaningless. Two main possibilities exist for quantifying $RMSE_{pred}$: (a) direct assessment on a control dataset, (b) cross-validation analysis (see § 4.3).

11.2.3 Example. Comparison Between Different Investigation Strategies

To compare the merits and limits of the different ways of quantifying the model error, we will use the example given in Chap. 8. The reader must refer to this chapter to have all the details regarding the investigated structure and the various investigation

---

[1] This is the general expression. As the conversion model for strength estimations is concerned, n corresponds to the number of cores N_c used for calibrating the model.

Table 11.1 Experts, their objectives and means

Expert	E3	E4	E6	E7	E3A
EQL	EQL2	EQL2	EQL2	EQL3	EQL2
TRP	Medium	Medium	High	High	Medium
Number of cores	12	12	5	8	5 (*)
PC / CC	PC	CC	CC	CC	PC

*Number based on a wrong assumption

strategies. Table 11.1 summarizes what experts have more specifically analyzed here, with their objectives and means.

Both experts E3 and E4 share an EQL2 and a medium TRP (UPV test results), with 12 cores, their only difference being that the former uses predefined coring while the latter uses conditional coring. Experts E6 and E7 have a high TRP. However, since expert E6, with EQL2, needs only 5 cores, expert E7, with EQL3, needs 8 cores. Expert E3a must be considered apart, although he has the same objectives and means as expert E3, he wrongly assumes to have a high TRP, and takes only 5 cores, which is not enough.[2]

Thanks to synthetic simulations, the example can be used for statistical comparisons based on a large number of simulations (here 200 simulations were considered). The assessment of the performance reached by each expert can be analyzed through different measures, which are respectively:

- The determination coefficient r^2 defined by Eq. 11.6,
- The RMSE value defined by Eq. 11.7. This can be the fitting error $RMSE_{fit}$ or the prediction error $RMSE_{pred}$.

In common practice, r^2 and $RMSE_{fit}$ are used and considered by engineers as relevant estimates of the model error. It will be shown that this is WRONG and can induce misleading statements. The RILEM Recommendations suggests quantifying the $RMSE_{pred}$ value either with direct assessment (but this requires a control dataset, which increases the cost of investigation) or with a cross-validation analysis (see Chap. 2—Task 9). Thanks to synthetic simulations, another way of quantifying the prediction error is to compare (using Eq. 11.7) the local strength predicted by the conversion model with the true strength at the same test location (which is never available in practice but is available in synthetic simulations). Table 11.2 synthesizes the results obtained by the five experts for three types of error estimates: the determination coefficient r^2, the fitting error $RMSE_{fit}$ and the predictive error $RMSE_{pred}$ identified from the true strength values. For each of them the 50%-percentile of the distribution (among the 200 simulations) and three percentiles (10, 5, 1%) are

---

[2]Expert E3a is an additional case that is considered in order to point at the crucial issue of correctly checking the real TRP value of NDT results. This Expert has in fact the same TRP than Expert E3, i.e. medium precision test results (TRP2), but he wrongly assumes that he has TRP1, and thus takes a lower number of cores.

Table 11.2 Percentiles of r^2, $RMSE_{fit}$ and $RMSE_{pred}$ obtained by each expert

	Percentile	E3	E4	E6	E7	E3A*
r^2	50%	0.70	0.77	0.93	0.89	0.78
	10%	0.42	0.59	0.81	0.76	0.24
	5%	0.29	0.46	0.77	0.73	0.17
	1%	0.10	0.41	0.67	0.56	0.02
$RMSE_{fit}$ (MPa)	50%	2.81	2.91	1.66	2.23	2.15
	90%	3.67	3.84	2.84	3.02	3.70
	95%	4.01	4.06	3.18	3.30	4.29
	99%	4.64	4.51	3.65	3.92	5.27
$RMSE_{pred}$ (MPa)	50%	3.05	2.99	2.40	2.26	3.38
	90%	3.65	3.51	2.97	2.67	4.42
	95%	3.92	3.67	3.08	2.82	5.41
	99%	4.26	3.85	3.43	3.15	6.82

*Number of cores based on a wrong assumption

obtained. These percentiles correspond to worst cases and therefore have lower values for r^2 and higher values for RMSE.

Before analyzing these results, one must point what is logically expected for these five experts, from what is known about how they proceeded:

– Experts E3 and E4 used medium TRP, but they had a large number of cores. The CC option is expected to be beneficial, but its added-value may be marginal when the number of cores is large.
– Expert E6 had the same EQL 2 objective, with two differences that had adverse consequences: a better precision of test results (high against medium TRP) and a lower number of cores. Therefore, it is difficult to forecast if this will finally lead to a larger or a smaller error than that of the two first experts.
– Expert E7 had high TRP, but also more ambitious objectives so they took 8 cores. This is expected to reduce the model error. This expert would logically obtain the more accurate estimations.
– Expert E3A took a number of cores that is too low based on the real precision of the test results. The resulting estimates would logically be the worst ones.

Let us now look at what could be deduced from the values of the determination coefficient as given in the first four lines of Table 11.2. The medium values (50% percentile) highlight that experts E6 and E7 have the best model, which could be expected, but expert E3A surprisingly appears to perform better than experts E3 and E4. This is a direct consequence of the very small number of pairs of data for expert E3A. It is also important to analyze the full distribution and not only the 50% percentile: given that, in real practice, one gets only one set of results, it is as if the investigator is randomly picking one value from the full distribution, and this value may lay among the higher or the lower values of the overall distribution. When looking at the lower percentiles of the r^2 statistical distribution, it is clearly visible that experts E6 and E7 are less sensible to statistical fluctuations than experts E3 and E4 (this is the consequence of the lower fluctuations of the test results) and, obviously, than expert E3A who has not enough cores. Even in the worst cases, the r^2 values of experts E6 and E7 remain above 0.55, which is not the case for the others. To summarize, any conclusion about the expert efficiency on the basis of r^2 values deserves a careful analysis.

The information provided by the $RMSE_{fit}$ is consistent with what was provided by the coefficient of determination. Here experts E6 and E7 clearly appear to have better results (lower fitting errors) than experts E3 and E4, which is a direct consequence of the higher precision of their test results but also of a lower number of cores. However, irrespective of the percentile that is considered, experts E6 and E7 have very close results and expert E6 is slightly better, which goes against the expectation of a better performance for Expert E7 (given that he has more cores). Lastly, expert E3A would seem to have $RMSE_{fit}$ values that are comparable to those of experts E6 and E7 (at least for the 50% percentile). We must explain why assessing the expert performance by looking at this indicator ($RMSE_{fit}$) is irrelevant.

Let us now look at the last results, those concerning $RMSE_{pred}$. They clearly discriminate between the five experts. These values[3] (in MPa) can be compared with the standard deviation of concrete strength in the whole structure, whose value is 5.30 MPa. Therefore, any value close to or above this threshold means that the investigator's prediction is meaningless (or that the "prediction" is even worse than a simple random estimate). The values in Table 11.2 show that expert E7 is the most efficient and has the lowest prediction errors ($RMSE_{pred}$ is smaller than 50% of the concrete strength standard deviation in about 90% of all cases, enabling him to identify spatial variations in concrete strength in different areas of the building). These results also differentiate between experts E6 and E7 as the expert E6 estimates (who uses less cores) have a higher risk of reaching large values (above 3 MPa). Experts E3 and E4 are not performing very well, because of the medium quality of test results (TRP2), which may lead to conversion models that are not reliable enough. Lastly, the results also show that the effective performance of expert E3A is very bad, whereas this was not directly visible when looking only at r^2 and $RMSE_{fit}$ value. This confirms that $RMSE_{pred}$ is the only relevant indicator regarding the predictive performance of the assessment.

Short summary

RILEM TC 249-ISC Recommendations and Guidelines require the model error to be checked.

---

[3]It should be remembered that these results have been obtained thanks to simulations and that, in real practice they can only be estimated by the "leave-one-out procedure" (see §1.5.6.4 for theory and §8.3.9 for practical examples).

This Appendix explains **why the commonly used measures** (the coefficient of determination r^2 and the fitting error $RMSE_{fit}$) **are inadequate and can lead to misleading statements**.

The only way to assess correctly the model error is to estimate the predictive error $RMSE_{pred}$, which can be done in practice by using the "leave-one-out" procedure.

11.3 Model Identification and Trade-Off Between Model Parameters: Explanations and Consequences

To illustrate the issue of trade-off and its consequences, we will use the example given in Chap. 2. The reader must refer to this chapter to have all the details regarding the investigated structure and the various investigation strategies. The following discussion only focusses on a few investigation strategies that correspond to those of experts E3, E7 and E3A (Table 11.3). Expert E3 has an EQL2 and a medium TRP (UPV test results), and he has 12 cores taken using predefined coring. Expert E7 has an EQL3 and a high TRP, and he has 8 cores taken using conditional coring. Expert E3A must be considered apart because, although it has similar objectives and means than expert E3, it wrongly assumes to have a high TRP, and takes only 5 cores, which is not enough.

The conversion model identification stage leads to a pair of model parameters (a, b) whose values are influenced by the test result uncertainty. Each time the same process is repeated with the same characteristics (TRP, number of cores…), the values of (a, b) change. These values are in fact the outputs of a random process governed by test result uncertainties (since the input (test results) are never strictly identical). Table 11.4 gives the first two statistical moments for the two model parameters of the linear conversion model.

As a consequence, there is a trade-off between model parameters that was first identified by Breysse [1] from literature results, before its mathematical bases were analyzed by Al Wash et al. [2] who have shown that each identified (a, b) set corresponds to the local minimum of some objective function. A well-known consequence

Table 11.3 Experts, their objectives and means

Expert	E3	E3A	E7
EQL	EQL2	EQL2	EQL3
TRP	Medium	Medium	High
Number of cores	12	5 (*)	8
PC / CC	PC	PC	CC

*Number based on a wrong assumption

Table 11.4 Mean and standard deviation (sd) of the two model parameters (a = intercept, b = slope) for the three experts

Expert	E3	E3A	E7
a (mean/sd)	-84.5 / 26.2	-92.1 / 52.3	-95.8 / 19.1
b (mean/sd)	0.027 / 0.006	0.029 / 0.012	0.030 / 0.004

of this trade-off is that, when the two model parameters are plotted in the same graph, they exhibit a strong linear correlation (Fig. 11.1).

As can be seen in Table 11.4, the mean values of the two model parameters are quite identical for the three experts, but their standard deviation and coefficient of variations are highly sensitive to the number of cores and to the TRP levels: the range of fluctuation of model parameters is larger for expert E3 than for expert E7 and it is larger for expert E3A than for expert E3. The coefficients of variation for a and b are equal to 20% and 15% (expert E7), 31% and 23% (expert E3) and 57% and 42% (expert E3A), respectively. The practical consequence of this trade-off is that the model equation $f_{c\,est} = a + bV$ is also changing.

From each simulation, one can compute, as explained previously in §1.5.6 and in [3, 4] both the fitting error $RMSE_{fit}$ and the predictive error $RMSE_{pred}$. These two variables can be plotted against the b model parameter value (it works similarly with parameter a) as shown in Figs. 11.2 and 11.3. These figures show important differences. If we focus on expert E3 (EQL2, medium TRP, 12 cores), Fig. 11.2 (left) shows a point cloud with no apparent correlation between the parameter value and the fitting error. The fitting error varies between 1.5 MPa and 5.1 MPa, its value depending on the fact that all data pairs are (by the effect of chance) more or less along a straight line in the (UPV, strength) diagram. Figure 11.2 (right) reveals a different pattern, since the predictive error has a smaller range of variation, between 2.3 MPa and 4.3 MPa.

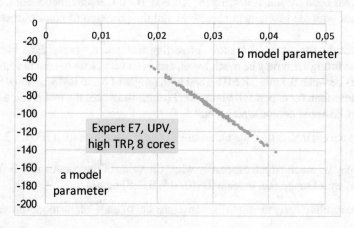

Fig. 11.1 Trade-off between the two model parameters from 200 simulations of Expert E7 (b = slope, a = intercept).

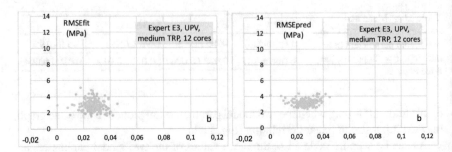

Fig. 11.2 Fitting error RMSE$_{fit}$ (left) and prediction error RMSE$_{pred}$ (right) as a function of the value of the b model parameter (200 simulations), for expert E3

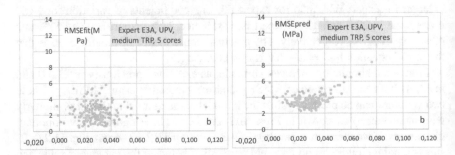

Fig. 11.3 Fitting error RMSE$_{fit}$ (left) and prediction error RMSE$_{pred}$ (right) as a function of the value of the b model parameter (200 simulations), for expert E3A

This predictive error is the real measure of the model uncertainty, and it appears to be statistically much more stable (this is a consequence of the fact that the number of cores has been chosen, following the RILEM TC-ISC Recommendations, in good agreement with the assessment objectives and the precision of non-destructive test results.

The results for expert E3A show interesting features. As expert E3A has only 5 cores, the fitting error may be much lower, and even very close to zero in some cases (it ranges between 0.5 and 5.8 MPa), and both ranges of b and RMSE$_{fit}$ show broader fluctuations than for expert E3. Figure 11.3 (right) illustrates the real predictive error of expert E3A and demonstrates that:

(a) It is never smaller than the minimum values reached by expert E3 (i.e. 2.3 MPa) and

(b) It may be much larger, especially when the value of the parameter b is far from the central range (i.e. outside the [0.015, 0.035] range).

These results clearly illustrate the consequences of trade-off on model uncertainty:

– Thanks to the small number of cores, the identified model can, by the effect of chance, fit well the dataset (therefore giving a low value of RMSE$_{fit}$) but this does not mean that the model is relevant, as shown by RMSE$_{pred}$ values,

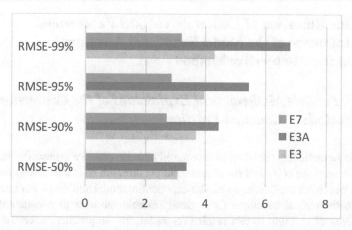

Fig. 11.4 Values (in MPa) of the prediction error RMSE$_{pred}$, for four different percentiles (from 50 to 99%) for the three experts

- Because of the large statistical fluctuations (resulting from both medium TRP test results and a small number of cores), the identified model parameter can be far from its (unknown) true value. In such cases, the RMSE$_{pred}$ value can be very large. This is for instance the case when the identification stage leads to negative values of slope b model parameter.

Figure 11.4 provides additional results, which are given for the three experts in terms of different percentiles of the predictive errors, from RMSE$_{pred}$-50% to RMSE$_{pred}$-99%. Each of these percentiles can be seen as the complementary risk of exceedance (from 50 to 1%, respectively). One must keep in mind that the simulation corresponded to a structure in which the overall variability of the concrete strength was equal to 5.3 MPa.

Figure 11.4 shows that expert E7 (in green, corresponding to high TRP) has the best results, with only 1% risk of a prediction error larger than 3.15 MPa (which corresponds to 60% of the overall strength variability). Expert E3 with medium TRP has logically poorer results. Regarding expert E3A, who has not enough cores, there is a significant risk to have a prediction error larger than the concrete variability.

All these considerations confirm what has been explained at § 11.2.2 about the **relevance of RMSE$_{pred}$ as being the only means of correctly assessing the model uncertainty.**

11.4 Identification of Conversion Model Parameters: Influence of the Model Shape and Advantage of the Bi-objective Method

11.4.1 Possible Mathematical Expressions of the Conversion Model—Common Practice

Three mathematical expressions are commonly used in practice to describe the dependence between the NDT test result and concrete strength with univariate conversion models, which are the linear model, the exponential model and the power law model. These mathematical functions theoretically enable the user to consider different sensitivities of strength to test result (Tr) variations. In practice, once the dataset of (core strength test results, NDT test results) is obtained, the investigator compares the three conversion models and chooses the model that provides the best fit (i.e. the one with the highest coefficient of determination or the lowest $RMSE_{fit}$ value). The common way for identifying the model parameters is through a linear regression that is applied slightly differently according to the model expression.

In the case of a linear conversion model M_L:

$$f_c = a + b\,Tr \qquad (11.8)$$

the linear regression is straightforward.

In the case of an exponential conversion model M_E:

$$f_c = a\,e^{b\,Tr} \qquad (11.9)$$

By taking the logarithm of both terms, the following linear equation is obtained:

$$\ln(f_c) = \ln a + b\,Tr \qquad (11.10)$$

This equation is linear and looks similar to Eq. 11.8 once $Y = f_c$ is replaced by $\ln Y = \ln f_c$.

In the case of a power law conversion model M_P:

$$f_c = a\,Tr^b \qquad (11.11)$$

By taking the logarithm of both terms, the following linear equation is obtained:

$$\ln(f_c) = \ln a + b\ln(Tr) \qquad (11.12)$$

This equation is linear and looks similar to Eq. 11.8 once $Y = f_c$ is replaced by $\ln Y = \ln f_c$ and $X = Tr$ is replaced by $\ln(Tr)$.

11.4.2   Possible Mathematical Expressions of the Conversion Model—Bi-objective Method

The bi-objective approach addresses two targets simultaneously by fitting both the experimental mean strength $f_{c, mean}$ and the strength standard deviation $sd(f_c)$. When using univariate conversion models, this requires solving a problem with two equations and two unknowns, whose solution is straightforward. The detailed process for identifying the model parameters must however be adapted depending on the mathematical shape of the conversion model.

In the case of a linear conversion model M_L:

The two conditions can be written as:

$$f_{c\,est\,mean} = a + b\,Tr_{mean} = f_{c\,mean} \tag{11.13}$$

and

$$sd(f_{c\,est}) = b\,sd(Tr) = sd(f_c) \tag{11.14}$$

where $f_{c\,mean}$ and Tr_{mean} are the mean values of measured core strength and NDT test results, respectively, and where $sd(f_c)$ and $sd(Tr)$ are the standard deviations of the measured core strength and NDT test results, respectively. Consequently, the values of the two model parameters can be calculated by:

$$b = sd(f_c)/sd(Tr) \tag{11.15}$$

and

$$a = f_{c\,mean} - b\,Tr_{mean} \tag{11.16}$$

In the case of an exponential conversion model M_E as defined by Eq. 11.9, after linearization, the two conditions can be written as:

$$(\ln(f_{c\,est}))_{mean} = \ln(a) + b\,Tr_{mean} = (\ln(f_c))_{mean} \tag{11.17}$$

and

$$sd(\ln(f_{c\,est})) = b\,sd(Tr) = sd(\ln(f_c)) \tag{11.18}$$

Consequently, the identification process applied for the linear conversion model can also be applied in this case and the values of a and b can be calculated as follows:

$$b = sd(\ln(f_c))\,/\,sd(Tr) \tag{11.19}$$

and

$$\ln a = (\ln f_c)_{mean} - b\ Tr_{mean} \qquad (11.20)$$

In the case of a power law conversion model M_P as defined by Eq. 11.11, after linearization, the two conditions can be written as:

$$(\ln (f_{c\ est}))_{mean} = \ln (a) + b\ (\ln (Tr))_{mean} = (\ln (f_c))_{mean} \qquad (11.21)$$

and

$$sd(\ln (f_{c\ est})) = b\ sd(\ln(Tr)) = sd(\ln (f_c)) \qquad (11.22)$$

Consequently, the same identification process for the linear conversion model can also be applied in this case and the values of a and b can be calculated as follows:

$$b = sd(\ln (f_c)) / sd(\ln(Tr)) \qquad (11.23)$$

and

$$\ln a = (\ln f_c)_{mean} - b\ (\ln(Tr))_{mean} \qquad (11.24)$$

11.4.3 Comparison of the Performance of Different Possible Univariate Conversion Models Identified Using Linear Regression

To illustrate the issue of trade-off and its consequences, we will use the example given in Chap. 2. The reader must refer to this chapter to have all the details regarding the investigated structure and the various investigation strategies. The following discussion only focusses on one investigation strategy that corresponds to that of expert E7 who has an EQL3 and a high TRP, with 8 cores and conditional coring.

Once the dataset of 8 pair results (core strength test results, UPV test results), expert E7 can still have several options regarding:

(a) The mathematical expression of the conversion model,
(b) The way to identify the conversion model parameters, either using a linear regression approach (see §11.4.1) or using a bi-objective approach (see §11.4.2).

By combining all these options, there are 6 (= 3 × 2) possibilities, we analyze first what is the performance of the different possible conversion models, using a linear regression identification approach. Table 11.5 compares the performances, for various percentiles of the statistical distribution regarding the three common

Table 11.5 Performances of the three possible conversion models (expert E7), using a linear regression approach

	percentile	linear	exponential	power
r^2	50%	0.90	0.90	0.91
	10%	0.78	0.78	0.78
	5%	0.76	0.76	0.76
	1%	0.68	0.67	0.66
$RMSE_{fit}$ (MPa)	50%	2.18	2.25	2.19
	90%	3.06	3.04	3.02
	95%	3.33	3.30	3.28
	99%	3.61	3.83	3.71
$RMSE_{pred}$ (MPa)	50%	**2.28**	2.31	2.31
	90%	**2.68**	2.77	2.74
	95%	2.89	2.87	**2.82**
	99%	**3.32**	3.58	3.59

measures: the coefficient of determination r^2, the fitting local error $RMSE_{fit}$ and the prediction error on the local strength $RMSE_{pred}$. The results of Table 11.5 are the outputs of a series of 200 simulations for each model. It can be noted that, regarding the linear model, the results in Table 11.5 are not exactly similar to those given in the previous section, Table 11.6. This is explained by the fact that a new series of simulations was carried out and is a sign of a small sensitivity to statistical fluctuations that do not alter previous conclusions.

The results in Table 11.5 show that the three possible models have very similar performances. If one focuses on the prediction error $RMSE_{pred}$ that is the only adequate way to evaluate the model prediction error, the linear model appears to have a performance that is slightly better than that of the other two options. It can be added that, if for each of the 200 simulations, the best model was identified, the linear model would have the best performance 101 times, against 55 times for the exponential model and 44 times for the power law model.

11.4.4 Comparison of the Performance Between the Identification of Parameters Using Linear Regression and Using the Bi-objective Method

We keep here the linear model option and compare the two approaches for the identification of the conversion model parameters, i.e. the linear regression and the bi-objective method using Eqs. 11.15 and 11.16. Since the difference between the different possible univariate conversion models is negligible, only the linear conversion model is kept for this comparison.

	Percentile	Linear regression	Bi-objective
Table 11.6 Performance of the two possible identification approaches (expert E7, linear model) — $RMSE_{pred}$ (MPa)	50%	2.28	2.28
	90%	2.68	2.70
	95%	2.89	2.83
	99%	3.32	3.30
Error on $f_{c\,mean}$ (MPa)	1%	− 2.55	− 2.48
	5%	− 1.43	− 1.42
	10%	− 1.09	− 1.08
	50%	+ 0.26	+ 0.25
	90%	+ 1.44	+ 1.46
	95%	+ 1.78	+ 1.78
	99%	+ 2.41	+ 2.40
Error on $sd(f_c)$ (MPa)	1%	− 1.87	− 1.54
	5%	− 1.56	− 1.27
	10%	− 1.36	− 1.04
	50%	− 0.58	− 0.27
	90%	+ 0.26	+ 0.55
	95%	+ 0.41	+ 0.71
	99%	+ 0.83	+ 1.16

Table 11.6 synthesizes the performance of the two identification approaches regarding three parameters, namely the mean strength assessment, the estimation of the concrete strength standard deviation and the local error. Since for the first two parameters, the deviation from the true value can be either positive or negative, lower and higher percentiles are compared, whereas only the higher percentiles (corresponding to largest errors) are considered for $RMSE_{pred}$.

Results for the mean strength assessment and local error show that both approaches lead to very similar performances. Therefore, although it does not provide the "best regression" between core strength and non-destructive test results, the bi-objective approach leads to a performance on these two targets that is as good as the one obtained with regression.

However, the conclusion is different when analyzing the assessment of concrete variability $sd(f_c)$. The bi-objective approach was proposed to improve this assessment. By looking at the percentiles, it appears that the two distributions are close, with a simple translation (of about + 0.3 MPa) of all values: whereas the linear regression approach tends to underestimate the concrete variability, this tendency is significantly reduced when using the bi-objective approach (the 50% percentile is divided by a factor 2). With the bi-objective approach, the center value of the estimation error of the concrete variability can be seen to be much closer to zero.

Short summary

The results of simulations analyzed in this Appendix provide important results that justify the fact that the RILEM TC 249-ISC Recommendations devote only a limited effort to the choice of the mathematical shape of the conversion model.

A literature review would easily prove that this issue has been commonly addressed by investigators and practitioners who have tried to identify "the best conversion model". It was shown here that **this issue is secondary**, and that the choice of the conversion model has only negligible effects on the final uncertainty of the final strength (it is reminded that $RMSE_{pred}$ is the only way to address it correctly). In most cases, **there is no argument for using a model other than the linear one, which is the simplest.**[4]

It was also shown why the **bi-objective approach must be privileged**, as it provides, without any additional cost, a better estimation of concrete variability, without reducing the performance with respect to other assessment targets (i.e. mean strength assessment, local strength assessment).

Acknowledgements The 6th author would like to acknowledge the financial support by Base Funding—UIDB/04,708/2020 of CONSTRUCT—Instituto de I&D em Estruturas e Construções, funded by national funds through FCT/MCTES (PIDDAC).

References

1. Breysse, D., Fernández-Martínez, J.L.: Assessing concrete strength with rebound hammer: review of key issues and ideas for more reliable conclusions. Mater. Struct. **47**, 1589–1604 (2014)
2. Fernandez-Martinez, J.L., Fernandez-Muniz, Z., Breysse, D.: Uncertainty analysis in linear and nonlinear regression revisited: application to concrete strength estimation. Inverse. Probl. Sci. Eng. **27**(12), 1740–1764 (2019)
3. Alwash, M., Breysse, D., Sbartaï, Z.M.: Using Monte-Carlo simulations to evaluate the efficiency of different strategies for nondestructive assessment of concrete strength. Mater. Struct. **50**, 90 (2017). https://doi.org/10.1617/s11527-016-0962-x
4. Breysse, D., Villain, G., Sbartaï, Z.M., Garnier, V.: Construction of conversion models of observables into indicators. In: J.-P. Balayssac, V. Garnier (eds.), Non-Destructive Testing and Evaluation of Civil Engineering Structures (ISTE Press, Elsevier, 2018), Chapter 7, pp. 231–257

---

[4]This conclusion applies to most cases. Using a non-linear model (exponential or power) may become interesting only when the pair datasets exhibit a visibly strong non-linear tendency. This requires the range of variation of test results to be very wide (i.e.in a highly contrasted dataset that mixes results from test locations with very bad properties with others that have good properties).

Chapter 12
For Those Who Want to Go Further

Xavier Romão, Denys Breysse, Jean-Paul Balayssac, and David Corbett

Abstract Further analyses and results are presented and discussed in order to provide additional information and justifications regarding the efficiency of the investigation strategy proposed by the Recommendations and these guidelines. These additional results are obtained from an extensive simulation study based on the case study described in Chap. 8 that accounts for the randomness of the assessment process and of the concrete properties. The presented results address the assessment of the Test Result Precision (TRP) level, the assessment of the mean strength and strength standard deviation of concrete, and look at the effect of uncertainty on the values of the conversion model parameters and the resulting prediction error. Furthermore, other analyses address the differences resulting from using different TRPs, from selecting different Estimation Quality Levels (EQL), from using different methods to select the location of cores to be extracted (i.e. predefined coring or conditional coring), from considering the bi-objective approach to determine the variability of concrete, and from combining two NDT techniques.

This chapter is based on the same case-study that was developed in Chap. 8 for illustrating the investigation methodology, based on a synthetic case study. The purpose is to provide the interested reader with additional information and explanations about the efficiency of the investigation strategy and to justify, through this example, why some choices have been made by the RILEM committee that led to the Recommendations.

X. Romão (✉)
CONSTRUCT-LESE, Faculty of Engineering, University of Porto, Porto, Portugal
e-mail: xnr@fe.up.pt

D. Breysse
University Bordeaux, I2M-UMR CNRS 5295, Talence, France

J.-P. Balayssac
LMDC, Université de Toulouse, INSA/UPS Génie Civil, Toulouse, France

D. Corbett
Proceq SA, Zurich, Switzerland

© RILEM 2021   359
D. Breysse and J.-P. Balayssac (eds.), *Non-Destructive In Situ Strength Assessment of Concrete*, RILEM State-of-the-Art Reports 32,
https://doi.org/10.1007/978-3-030-64900-5_12

The data used in this chapter are exactly identical to those described in Chap. 8 and the reader must refer to this chapter to get all the information about the synthetic building and the different investigation strategies that are analysed and compared.

12.1 Objectives—The Advantage of Repeating the Simulation

Our purpose is to provide useful information regarding the efficiency of the strategy recommended by the RILEM guidelines. The results analyzed herein were obtained for the same synthetic case study described in Chap. 8. The only difference is that, instead of simulating several experts working on a unique building, each investigation program is repeated many times, which is easy with synthetic simulations (and of course impossible in real practice, because of time and costs restrictions). Therefore, each expert, instead of having ONE assessment corresponding to ONE building, can deliver a series of N assessments corresponding to N different buildings, each building resulting from the same synthetic rules (i.e. synthetic concrete) but having slightly different properties because of the randomness of the simulations (when simulating both the concrete properties and the measurements).

The interest of carrying out many simulations is that it provides a more representative picture of the process, i.e. of the performance of each investigation strategy. The final performances can be analyzed in terms of risk of success or failure regarding their initial objectives. This is in full agreement with the revised paradigm of risk that underlines the RILEM Recommendations (see §6.1.2 *Target tolerance interval, number of cores and risk of wrong assessment*). The real level of risk (i.e. that the assessed value falls outside the tolerance interval) can be computed and can be compared with target requirements defined in terms of acceptable risk in the Recommendations (see Table 1 in §1.1 *Scope*). A similar approach was used within the RILEM TC 249-ISC when TC members were involved in a benchmark exercise (see Chap. 6), which enabled to compare the performance of the different strategies used in practice.

In this chapter, the results of experts E1 to E8 (defined in Chap. 8, Table 8.1), whose investigation program was repeated 200 times, are analyzed. The results of the 200 simulations are processed in order to derive their average behavior, scatter and risk of failure regarding each assessment target.

First, we will illustrate some results regarding the test result precision (TRP) assessment (§2) and the concrete strength assessment (§3). Then we will look at the effect of uncertainty on the values of the conversion model parameters and the resulting prediction error (§4). Finally, we will analyze more carefully the differences between some experts, namely those resulting from using different TRP (§5), having different estimation quality levels (EQL) (§6), having different ways of taking cores (i.e. predefined coring (PC) vs conditional coring (CC), §7), using the bi-objective approach (§8) or the combination of two NDT techniques (§9).

12.2 Assessment of the TRP Level

Assessing the TRP is a key step (T3) of the recommended investigation strategy, since it is a crucial input to define the relevant number of cores for complying with the EQL targets. We have shown in Chap.10.2 how to assess the TRP level by simply repeating the NDT test a few times at a few test locations. In this example, like for a real on site investigation, the process delivers a single within-test-repeatability (WTR) which corresponds to a TRP level (poor, medium or high precision), according to the thresholds defined in the Recommendations. Because of the intrinsic randomness of the process, the estimated TRP fluctuates when the process is repeated. The several simulations (N = 200) make it possible to derive the cumulative distribution of the assessed TRP (Figs. 12.1 and 12.2).

When ONE investigator performs ONE investigation program on ONE building, it is the same as if he was randomly sampling ONE value from these distributions.

- If one looks at expert E5, most WTR values are below the standard deviation threshold ($sd_{rep} \leq 50$ m/s) that defines high precision TRP. There is only a very low probability that the estimated WTR would exceed this threshold, which would lead him to assume that he uses a medium precision TRP.

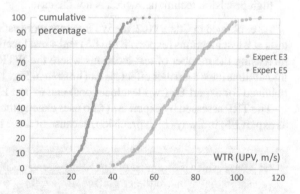

Fig. 12.1 Cumulative distribution of the assessed WTR for experts E3 and E5, using UPV test results

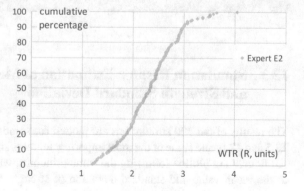

Fig. 12.2 Cumulative distribution of the assessed WTR for expert E2, using rebound test results

Fig. 12.3 Cumulative distribution of the assessed WTR for experts E5 and E7, using UPV test results

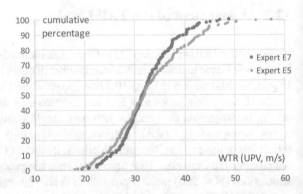

- The same considerations apply to expert E2 with respect to the rebound test. In most cases, the precision is assessed as medium ($1 < sd_{rep} \leq 3$), but there is a very low probability that the estimated WTR would be above 3, which would lead him to assume that he uses a poor precision TRP.
- Expert E3 may have the opposite problem: in most cases, the WTR values fall in the range 50 m/s $< sd_{rep} \leq$ 125 m/s which leads to a medium TRP. However, in about 1 case out of 10, the investigator could erroneously consider that he uses a high precision technique, which is not the case.

As explained in Chap. 10.2, the WTR values were estimated by repeating the NDT tests at 2 test locations (experts E2 to E6) and 4 test locations (experts E7 and E8). By increasing the number of test locations where the WTR is estimated, the statistical fluctuations have a smaller effect, as can be seen in Fig. 12.3 that compares the results obtained for expert E5 (two test locations) and expert E7 (4 test locations).

The TRP assessed by expert E7 (who uses exactly the same type of measurements as expert E5) is always high precision, because of these lower statistical fluctuations.

Of course, since the number of cores can be reduced if the investigator uses high precision, one can see the advantage of devoting some resources to carefully assess the TRP level.

12.3 Simulation Results: Estimation of Mean Strength and Strength Standard Deviation

The results of the 200 simulations are plotted for some of the experts in Figs. 12.4, 12.5 and 12.6, and some of the most important features are summarized in Table 12.1. All figures are plotted following the same principles where each point corresponds to the mean value and standard deviation of strength for one simulation. The red

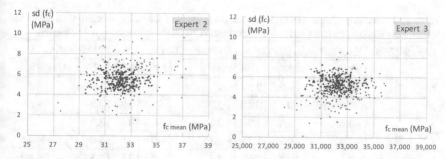

Fig. 12.4 Distribution of assessed mean strength and strength standard deviation, expert E2 (left) and expert E3 (right)

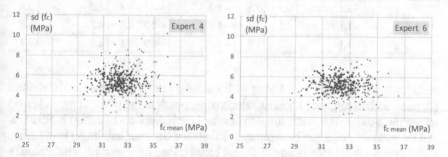

Fig. 12.5 Distribution of assessed mean strength and strength standard deviation, expert E4 (left) and expert E5 (right)

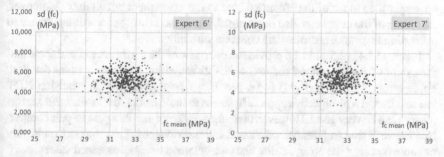

Fig. 12.6 Distribution of assessed mean strength and strength standard deviation using the bi-objective method, expert E6′ (left) and expert E7′ (right)

cloud corresponds to the exact synthetic values, those of the building to be assessed, and this cloud is identical in all graphs. The blue cloud corresponds to the expert's results and changes according to the expert.

The extent of the red cloud is a result of the simulation rules (see Chap. 8) that define the statistical properties of concrete strength for beams and columns, where each simulation leads to a different strength pattern across the components. The

Table 12.1 Expert performances regarding the assessment of the mean strength of concrete, the standard deviation of concrete strength and the local error on strength (n.a. = not addressed)

Experts	EQL	Error on $f_{c\ mean}$ (MPa)			Error on $sd(f_c)$ (MPa)			RMSE (MPa)	
		5%	50%	95%	5%	50%	95%	50%	95%
1	1	−3.2	−0.2	3.2	n.a			5.9	8.7
2	2	−2.4	0.1	2.4	−3.7	−1.1	2.2	3.9	5.9
3		−1.6	0.0	1.7	−2.2	−0.6	0.8	3.0	3.7
4		−1.2	0.2	1.5	−2.1	−0.9	0.3	3.0	3.6
5		−1.3	0.5	2.5	−2.2	−0.1	2.6	2.5	4.2
6		−1.4	0.4	2.1	−1.8	−0.5	0.7	2.4	3.1
6'		−1.8	0.1	2.1	−1.7	−0.3	1.0	2.4	3.2
7	3	−1.1	0.3	1.6	−1.7	−0.7	0.4	2.3	2.8
7'		**−1.1**	**0.3**	**1.6**	**−1.3**	**−0.3**	**0.8**	**2.3**	**2.8**
8		−1.1	0.3	1.6	−1.3	−0.3	1.1	1.9	2.7

mean concrete strength is 32.3 MPa with a standard deviation between simulations (or between buildings) of 1.0 MPa, which corresponds to the width of the red cloud (along the x-axis). The mean strength standard deviation (in a building) is 5.3 MPa, and its standard deviation between simulations (or between buildings) is 0.7 MPa, which corresponds to the height of the red cloud (along the y-axis). Each simulation delivers one pair of points, i.e. a red point for reference properties and a blue point for the estimated values. The distance between these two points measures the error on mean strength and standard deviation assessment. The extents of the blue clouds are larger than those of the red cloud, due to the assessment uncertainty. Moreover, the center of the cloud can also be changed, e.g. if the investigation strategy tends to overestimate or underestimate the concrete variability.

Figures 12.4, 12.5 and 12.6 provide an overview of the performance, and show that the extent of the blue cloud reduces as the investigation program is more detailed. Furthermore, Fig. 12.6 (right) for expert E7', who uses both high precision NDT test results and a bi-objective approach, seems to indicate that the assessment performances can be very good. However, these figures do not enable a pair-to-pair comparison between true (reference) values and estimated values. This comparison is only possible by using the error, i.e. the distance between true and assessed values.

Table 12.1 synthesizes the performances for all experts (cases with the bi-objective method have been added for experts E6' and E7', as this option provides a better assessment of the concrete strength standard deviation). The 5 and 95% percentiles correspond to a 5%-risk of under estimation and to a 5%-risk of over estimation of the property (mean value or standard deviation), respectively. The values of RMSE-50% and RMSE-95% correspond to a 50%-risk and a 5%-risk to have a mean local error larger than this value, respectively.

All performances must be analyzed keeping in mind the target tolerance intervals corresponding to each EQL (see Chap. 1, Table 1.1). For a concrete whose strength is about 30 MPa, the 10 and 15% requirements on the mean strength lead to 3 MPa (EQL2 and EQL3) and 4.5 MPa (EQL1), respectively. These target tolerance intervals are satisfied for all experts, as are those on the strength standard deviation and on the RMSE-95%.

The performances of experts E2 and E3, who both use medium precision TRP (the former with rebound, the latter with UPV, see Table 8.1), can be compared. The TRP levels were defined in Table 1.4 in such a way that the same precision level (whatever the NDT type) would lead to an equivalent accuracy regarding the concrete strength assessment. In Fig. 12.4, it appears that this is the case with a slightly larger error for rebound. However, it must be noted that UPV test results obtained on site usually correspond to a medium precision (or even better), whereas rebound tests must be carried out carefully to obtain a similar medium precision label.

It can be seen that the most efficient expert is expert E7' whose performances are significantly more efficient than those required for the EQL3 level. On the other extreme, expert E1 is the least efficient. As can be seen from the grey cells, the RMSE-95% is larger than the strength standard deviation in the building (in other words, the predictive model predicts nothing). This means that this expert cannot address this issue with the resources he has devoted to the investigation. Luckily, at EQL1, only the concrete mean strength is addressed, and expert E1 performs adequately regarding this limited target.

12.4 Uncertainty and Its Effects: Trade-Off Between Model Parameters and Model Error

The consequences of the uncertainties at all stages of the measurement and investigation process have been discussed in Chap. 2. It was shown in recent research that these uncertainties lead to what has been named a "trade-off" between the parameters defining the equation of the conversion model.[1] This trade-off exists whatever the mathematical shape of the conversion model (linear in our examples, but could also be a power law or an exponential law). The conversion model parameters can be seen as the statistical outputs of a random process, and it has been proven that, because of the mathematical roots of the problem, there is a strong (explicit) relationship between the parameter values (here a and b). As a result, repeating the investigation many times is similar to sampling the (a, b) pair among their population. The (a, b) pairs identified during the 200 repetitions can thus be plotted so as to visualize this relationship. Without going further into the theory (the interested reader can refer to

---

[1] D. Breysse, J.L. Fernández-Martínez, Assessing concrete strength with rebound hammer: review of key issues and ideas for more reliable conclusions. Mater Struct 47, 1589–1604 (2014).

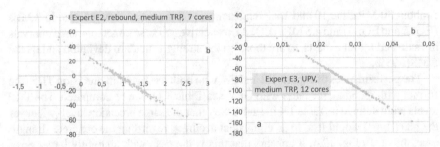

Fig. 12.7 Trade-off between model parameters a and b for expert E2 (with rebound) and expert E3 (with UPV)

additional reading material for more detailed explanations[2]) we can plot these pairs for experts E2 and E3 (Fig. 12.7) who identify respectively:

– a linear model with rebound $f_c = a + b \times R$, for expert E2, and
– a linear model with ultrasonic pulse velocity $f_c = a + b \times UPV$, for expert E3.

Regarding expert E3, the reader can check that the parameters that were previously identified in one specific case (Chap. 8, Table 8.12), with the equation $f_c = 0.0292$ UPV-90.3, are just one particular output among many others. The range of fluctuations of parameters a and b can be wide. Still, this fluctuation decreases when the effect of uncertainties is reduced, for instance with a higher TRP precision, by taking more cores, or by using conditional coring. Figure 12.7 (left) also shows that, with a small number of cores (here 7 cores, and medium precision rebound measurements), parameter b may in few cases become negative… which would mean a smaller strength for a higher rebound index. Of course, such a result contradicts physics and is meaningless. The number of cores (here 7 cores) is such that the frequency of such situations remains small. This problem can of course arise in real investigations.

> If the investigator is faced with this problem when identifying the conversion model, he must (a) question his test results, (b) reject the model, (c) probably take additional measurements (i.e. additional core strength test results).

For the set of 200 simulations, Fig. 12.8 (left and right) plot how the local error RMSE varies with the value of the model parameter b (a similar plot could be drawn for parameter a). Both figures show a typical boomerang-shaped cloud, whose mathematical analysis was carried out in the previously cited reference. On one hand, smaller RMSE values are obtained when the model error is the lowest, corresponding here to b values of about 1.0 for rebound and 0.025 for UPV. This minimum RMSE is not zero since there are always some measurement variabilities for both NDT

---
[2]J.L. Fernandez-Martinez, Z. Fernandez-Muniz, D. Breysse, Uncertainty analysis in linear and nonlinear regression revisited: application to concrete strength estimation. Inverse Probl Sci Eng 27(12), 1740–1764 (2019).

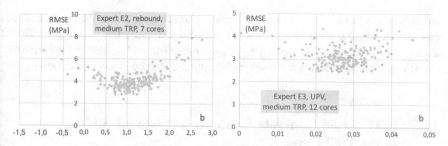

Fig. 12.8 Effect of the model uncertainty on the model error RMSE for expert E2 (with rebound) and expert E3 (with UPV)

test results and for destructive strength measurements on cores. On the other hand, the resulting RMSE can increase a lot if parameter b (therefore also parameter a) is far from the "real" value. The RMSE-50% and RMSE-95% values of Table 12.1 (i.e. 3.1 MPa and 5.9 MPa for expert E2 and 3.0 MPa and 3.7 MPa for expert E3, respectively) correspond to values along the y-axis of these two figures such that only 50% and 5% of the points, respectively, are above them.

It is thus clear that decreasing the prediction model error would imply controlling and checking the consequences of the trade-off between model parameters. One must note that the leave-one-out method for assessing the prediction error (Task 9 in the RILEM TC 249-ISC Recommendations) is an adequate way to do that.

The last two figures (Fig. 12.9) show how the error of the standard deviation of concrete strength varies as a function of the model parameter b. The clouds show a strong correlation, but no symmetry. This can be easily understood if one considers that a stronger b value (i.e. a steeper slope in the strength/NDT test result diagram)

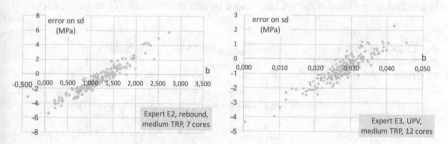

Fig. 12.9 Effect of the model uncertainty on the error of the standard deviation of concrete strength for expert E2 (with rebound) and expert E3 (with UPV)

tends to overestimate the concrete variability whereas a smoother slope will underestimate it. Finally, these two clouds are not centered along the y-axis, thus illustrating the common tendency of the regression models to underestimate the concrete variability. This issue is why the bi-objective approach has been proposed (see §12.8).

12.5 NDT Based Concrete Strength Assessment: Effect of the TRP Level

Figure 12.10 plots the cumulative distributions of the local error RMSE for experts E3 and E5. The differences between these two experts is that the former uses medium precision UPV test results, whereas the latter uses high precision test results (orange and green colors are used in this figure and the following ones to distinguish between medium and high precision test results).

As a direct consequence, expert E5 needs only 5 cores whereas expert E3 had to take 12 cores. Two main features can be seen from these results:

- the green curve (expert E5) is shifted towards the left side, which corresponds to a decrease for the RMSE-50% value of 0.5 MPa (see Table 12.1). This decrease occurs for 80% of all cases.
- however, in a limited number of cases (about 10%), the local error for expert E5 is larger than that of expert E3 (in fact, the simulations even lead in three cases out of 200 to RMSE values larger than 7 MPa, which have not been plotted in Fig. 12.10 for more clarity). This comes directly from the trade-off discussed before and the conversion model uncertainty, and is an effect of the much smaller number of cores. This phenomenon would progressively disappear as the number of cores would increase.

Of course, the situations where the RMSE value is large, due to the random uncertainties on test results and to the resulting trade-off between the conversion model parameters, may be encountered in real investigations. A number of cores that is too low increases the risk of carrying a wrong assessment of concrete properties,

Fig. 12.10 Cumulative distribution of the local error RMSE for experts E3 and E5

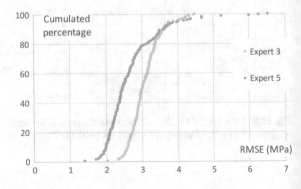

thus justifying how the required number of cores has been defined in the RILEM TC 249-ISC Recommendations. The recommended number of cores has been calculated and checked by RILEM-TC experts such that the target requirements defined for each EQL are satisfied, with an accepted level of risk (see Chap. 2). It is therefore very important to know and to check the precision of the test results.

This can be illustrated through the example of an expert (let us call him expert E3a) who would not check the TRP of test results (which is in fact medium) but who nevertheless would (wrongly) assume they are high precision test results. As a consequence, having assumed that the concrete mean strength is no more than 40 MPa and that the coefficient of variation is no more than 20% (see Chap. 8, Task 5), he would take only 5 cores (Table 12.2) instead of 12, as logically derived from Table 12.3.

Figures 12.11 and 12.12 compare the final performances obtained by expert 3a to those that would be obtained if the correct methodology had been followed with the same techniques (expert E3). Both figures show that the performances are worse as a consequence of the overly optimistic assumption of the TRP and of the resulting

Table 12.2 Minimum number of cores for EQL2 (UPV measurements, absolute/absolute requirements), with high precision (TRP1)

fc mean	High precision (TRP1)				
	cov (%)				
	10	15	20	25	30
10	5	5	5	5	5
15	5	5	5	5	5
20	5	5	5	5	5
25	5	5	5	5	5
30	5	5	5	5	5
35	5	5	5	5	5
40	5	5	5	5	5
45	5	5	5	5	6
50	5	5	5	6	8

Table 12.3 Minimum number of cores for EQL2 (UPV measurements, absolute/absolute requirements), with medium precision (TRP2)

fc mean	Medium precision (TRP2)				
	cov (%)				
	10	15	20	25	30
10	6	6	6	6	7
15	6	6	6	6	6
20	6	6	6	6	6
25	6	6	6	6	6
30	6	6	6	7	9
35	6	6	8	11	13
40	6	8	12	15	18
45	7	11	16	20	25
50	9	15	21	27	33

Fig. 12.11 Cumulative distribution of the local error RMSE for experts E3 and E3a (two occurrences for expert 3A lead to a RMSE value exceeding 10 MPa)

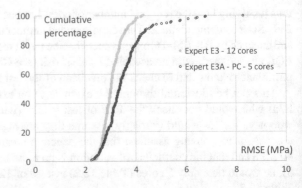

Fig. 12.12 Cumulative distribution of the error on the mean strength estimate for experts E3 and E3a

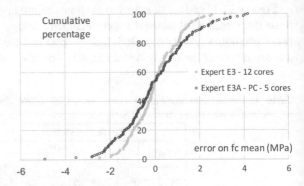

selected number of cores that is too small. The RMSE value is often larger than the standard deviation of concrete strength over the whole building (5.3 MPa). The error on the mean strength estimate is about 50% larger and can frequently be more than 3 MPa.

The reader (and the investigator) must be aware of the statistical nature of the non-destructive concrete strength assessment problem. More precisely, he **must always check that the real NDT precision is really as he has assumed** (see Task T3). When the randomness of the process eventually leads him to **conversion models that are contrary to the basis of physics (typically with a negative slope), he must reject them.**

12.6   NDT-Based Concrete Strength Assessment: Effect of the EQL Requirement, from EQL1 to EQL3

Figure 12.13 plots the cumulative distribution of the local error RMSE for experts E1, E2, E6 and E7 whose characteristics are summarized in Table 12.4. The differences between these four experts is that expert E1 has an EQL1 requirement, whereas experts E2 and E6 have an EQL2 requirement and that expert E7 has a EQL3 requirement. In agreement with the color code, experts E1 and E2 use medium precision TRP and experts E6 and E7 use high precision TRP. For each expert, the number of cores resulted from a double constraint, i.e. on means (TRP) and on objectives (i.e. EQL), as explained in Chap. 2 and illustrated in Chap. 8. As can be checked from Table 12.4, the required number of cores increases with the EQL and, for a given EQL, decreases if the TRP improves.

The curves clearly show a decrease of the error when the requirement is more ambitious. The RMSE-50% values fall from 5.9 MPa for expert E1 and 3.9 MPa for expert E2 to 2.4 MPa and 2.3 MPa for experts E6 and E7, respectively. The RMSE-95% values follow the same logical tendency. The minimum local error converges towards about 1.8 MPa, which appears as an irreducible threshold (due to high precision NDT results and to the measurement errors on core strengths). The reader should be reminded that all of these errors satisfy the requirements defined for the corresponding EQLs (RMSE is not addressed at EQL1).

Fig. 12.13 Cumulative distribution of the local error RMSE for experts E1, E2, E6 and E7

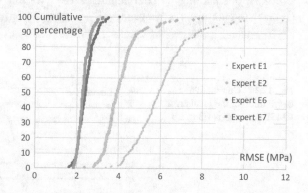

Table 12.4 Characteristics of the four experts and number of cores

Expert	EQL	Test result precision	N_c
E1	EQL 1	TRP 2 (medium)	7
E2	EQL 2	TRP 2 (medium)	12
E6	EQL 2	TRP 1 (high)	5
E7	EQL 3	TRP 1 (high)	8

12.7 NDT Based Concrete Strength Assessment: The Advantage of Conditional Coring

Figure 12.14 plots the respective cumulative distribution of the local error RMSE for experts E3 to E6 who all have an EQL2 requirement. In agreement with the color code, experts E3 and E4 use medium precision TRP while experts E5 and E6 use high precision TRP. The other difference is that experts E4 and E6 use conditional coring.

The effect of conditional coring is particularly visible for experts E5 and E6 who both use five cores. The possible dangers of having such a limited number of cores was discussed at §5 when looking at the highest percentiles of the curve of expert E5. By prioritizing conditional coring instead of having a predefined core location, this problem is mostly solved and the dark green curve of expert E6 shows that the risk of a "wrong" conversion model is eliminated.

When the two curves of experts E3 and E4 are compared, the effect of conditional coring is less apparent, for two reasons: on one hand, these experts have 12 cores and the risk of "bad" sampling is statistically speaking much lower, and on the other hand, because medium TRP non-destructive test results are involved, the added value of conditional coring is more limited.

> Conditional coring improves the performance of the investigation, particularly for a limited number of cores and when the selection of cores is based on high precision non-destructive test results.

Fig. 12.14 Cumulative distribution of the local error RMSE for experts E3 to E6

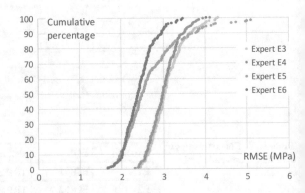

12.8 NDT Based Concrete Strength Assessment: The Advantage of the Bi-objective Method for Assessing Concrete Variability

The assessment of concrete strength variability is one of the objectives of these guidelines. The bi-objective method is recommended as it is more suited to capture this variability (see the RILEM TC 249-ISC Recommendations, §7.3.1). Figure 12.15 compares the cumulative distributions of the error of the mean strength estimate for experts E6′, E7 and E7′, where the ' denotes the use of the bi-objective approach instead of the best regression model (see also Chap. 8 for information regarding the model identification). Figure 12.16 compares the performances using the cumulative distributions of the error for the concrete strength standard deviation assessment whereas Table 12.5 synthesizes the performances of the same experts regarding the concrete strength standard deviation assessment.

The curves of Fig. 12.15 can be compared to values in Table 12.1. They show that the estimation is logically better for EQL3 (experts E7 and E7′, with 8 cores) than for EQL2 (expert E6, with 5 cores), with the same type of high precision test results.

Fig. 12.15 Cumulative distributions of the error of the mean strength estimate for experts 6′, 7 and 7′ (' denotes the use of the bi-objective approach)

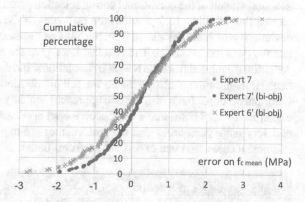

Fig. 12.16 Cumulative distributions of the error of the concrete strength standard deviation estimate for experts 6′, 7 and 7′ (' denotes the use of the bi-objective approach)

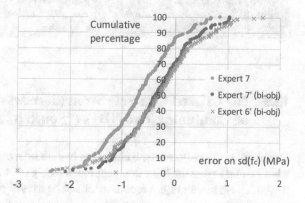

Table 12.5 Concrete strength standard deviation estimated by experts (all values in MPa)

	expert E6'	expert E7	expert E7'
	CC + bi-objective	CC	CC + bi-objective
From cores strength test results (mean ± st.dev.)	8.1 ± 1.6 (from 5 cores)	7.0 ± 1.3 (from 8 cores)	
sd(f_c) (mean ± st. dev.)	5.1 ± 1.1	4.7 ± 1.0	5.1 ± 1.0
mean error on sd(f_c)	−0.3	−0.7	−0.3
error on sd(f_c) [5%, 95%]	[−1.7, + 1.0]	[−1.7, + 0.4]	[−1.3, + 0.8]

The curves of experts E7 and E7' cannot be distinguished and are quite identical, meaning that the bi-objective approach does not reduce the performances regarding the assessment of mean strength (which is confirmed by values in Table 12.1).

The real standard deviation of concrete strength in the building (see §3 in this chapter) is 5.3 ± 0.7 MPa, where ± 0.7 corresponds to the variability between simulations. The first line in Table 12.5 gives the standard deviation as it can be estimated by the core datasets. These 3 experts logically overestimate the variability, as a direct result of the conditional coring selection process, as explained before. The second line in Table 12.5 provides the estimates of the concrete strength standard deviations which are calculated from the NDT assessment, by using NDT test results and the conversion model. The two last lines give the mean error on the standard deviation calculated from the $N = 200$ simulations, and the two-sided 10% risk confidence interval. Figure 12.16 presents the full distributions and shows that expert E6' (with only 5 cores but a bi-objective approach) globally reaches smaller errors than expert E7 (with 8 cores and a regression model approach). It is only for very small and very large percentiles that the performances of expert E6' worsen, because of insuficient cores, and that expert E7' performs better.

All these results confirm that the experts are able to capture the concrete variability and that the combination of high precision test results, conditional coring and the bi-objective approach (i.e. expert E7') provides the best results.

12.9 NDT Based Concrete Strength Assessment: Interest of Combining Two NDTs ("SonReb" Approach)

Expert E8 is the only one who combines two non-destructive techniques, following what can be called the "SonReb" approach (see §1.5.4.2 Single *vs* combined NDT methods). He has an EQL3 requirement and he uses high precision UPV test results and medium precision rebound test results (in real practice, it is difficult to reach high

Fig. 12.17 Cumulative distributions of the local error RMSE (EQL3, Experts 7′ and 8)

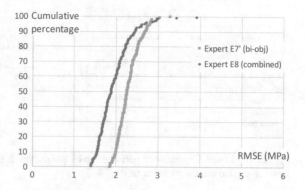

precision for the rebound test, i.e. a standard deviation of test results lower than ONE rebound unit). In addition, he uses conditional coring, the location of cores being defined on the basis of the UPV test results. Figure 12.17 compares the cumulative distribution of local error RMSE obtained by expert E8 with that of expert E7′ whose estimates are based on UPV test results only and uses a bi-objective approach. This comparison is relevant since, for each synthetic simulation, both experts use the same UPV test results and the same core strength results. The only difference is that expert E8 also uses the rebound test results obtained at the same core location in order to identify a bivariate model:

$$f_c = a + b \, UPV + c \, R$$

Figure 12.17 shows that expert E8 obtains in many cases better results than expert E7′ with the bi-objective approach (who is the best performing of all experts, see previous sections). However, in some cases (here in about 5% of cases), the performance deteriorates and the right tail of the expert E8 cumulative curve shows that the combination of test results may provide less adequate concrete strength estimates.

This controversial effect is well-known and explains why some authors are reluctant to combine NDT methods. This effect can be better understood by looking at what happens in the conversion model identification stage. The trade-off effect between model parameters, which was discussed before (see §4) still exists, but its effects are exacerbated since the bivariate model has 3 parameters (a, b, c) compared with 2 parameters for univariate models. Figure 12.18 illustrates the (b, c) trade-off and its effect on the local error RMSE.

The typical trade-off discussed at §4 is confirmed, but the correlation between the two slope parameters (b measures the sensitivity to UPV, c measures the sensitivity to R) is poorer than with a univariate model, as the trade-off is now a 3-dimensional phenomenon. Thus, Fig. 12.18 (left) is the projection of a 3D-cloud in two dimensions. Similarly, Fig. 12.18 (right) also shows a typical symmetrical U-shaped cloud with larger RMSE values corresponding to the values of the model parameters that

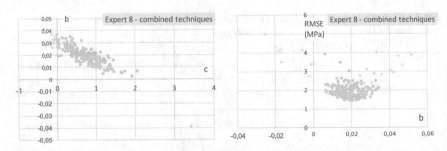

Fig. 12.18 Expert E8, trade-off between model parameters b (right) and c (left), and relation between c value and the model error RMSE

are the farthest from the center. The performances of all experts that are summarized in Table 12.1 show that expert E8 globally obtains a very good performance, which is the best of all experts regarding mean strength assessment and very close to experts E6′ and E7′ regarding the assessment of mean strength and strength standard deviation.

The performances of expert E8 can however be improved easily, just by considering that conversion models that have a negative slope for either UPV or R are not physically possible. Therefore, when such a model is identified (at stage T8 of the investigation and assessment process), it must be rejected and replaced with a univariate model using only the high precision test results (e.g. UPV test results here). In the synthetic simulations, this occurred 11 times (among 200 simulations), which corresponds to 10 cases (i.e. 5%) where the c parameter (slope against R) was negative and 1 case (i.e. 0.5%) where the b parameter (slope against UPV) was negative. These cases are identified in Fig. 12.18 with a circle, which also confirms that they mostly correspond to the cases leading to the highest RMSE values.

By just checking the physical consistency of the bivariate conversion model, and identifying an alternative univariate model, the performance on local errors is easily and significantly improved. More specifically, the 5%-risk RMSE value, which corresponds to the 95th percentile of the cumulative distribution decreases from 2.7 MPa to 2.6 MPa (the 1%-risk RMSE value is decreased from 3.4 MPa to 3.0 MPa). The risk of large errors is mostly controlled.

This confirms that combining two non-destructive techniques can bring some added-value. The conditions are that this combination must be based on high precision test results and that the physical consistency of the conversion model must be checked. However, it must be remembered that this option also requires more resources, since the number of NDT test results doubles.

Acknowledgements The first author would like to acknowledge the financial support by Base Funding—UIDB/04708/2020 of CONSTRUCT—Instituto de I&D em Estruturas e Construções, funded by national funds through FCT/MCTES (PIDDAC).

Index

A

Accuracy, 335
Assessment performances
  detection of defective areas, 262
  standard deviation estimation, 213, 261, 296, 322, 342, 362
  strength estimation, 213, 261, 296, 323, 342, 362
Assessment strategy
  efficiency, 213, 265, 323

B

Benchmark, 184, 220
  rules, 187, 222
Bi-objective approach, 44, 71, 294, 296, 348, 373

C

Carbonation
  influence, 19, 228, 256
Combination of ND methods, 40, 70, 103, 111, 313, 374
Confidence level, 335
Conversion model
  and test regions, 140, 304
  calibration, 46, 51, 207, 215, 313
  charts, 106, 211
  error, 140, 365
  identification, 43, 59, 66, 294, 348
  mathematical shape, 41, 69, 313, 348
  model parameters, 270, 293, 313, 341
  multivariate, 40, 103, 268, 309, 313
  parameters, 41
  prior conversion model, 205

  regression of a specific model, 107, 211, 256, 294, 348
  trade-off (identification), 268, 342, 365
  type, 39, 313, 348
  uncertainty, 48, 59
Core
  location of, 25, 51, 252, 289, 313
  number of, 20, 26, 323
  preparation, 28
  testing, 30
Coring
  conditional, 25, 51, 67, 240, 289, 372
  predefined, 25, 289

D

Data fusion, 110

E

Error
  assessment of, 51, 108, 296, 342
  fitting ($RMSE_{fit}$), 48, 310, 313, 342
  on local strength, 90, 140
  prediction ($RMSE_{pred}$), 48, 296, 313, 323, 325, 342, 342
Estimation Quality Level (EQL), 6, 86, 283, 313

G

Guidelines, 3

I

Investigation domain, 21, 23

© RILEM 2021

D. Breysse and J.-P. Balayssac (eds.), *Non-Destructive In Situ Strength Assessment of Concrete*, RILEM State-of-the-Art Reports 32,
https://doi.org/10.1007/978-3-030-64900-5

Investigation program
  flowchart, 51, 267, 282, 313
  key tasks, 51
Investigation strategy
  examples, 204, 240, 285

K
Knowledge level, 196, 236

M
Moisture
  influence, 28, 103
Multi-objective strength assessment, 84, 348

N
Neural networks, 109
Non-destructive methods, 15
  electrical capacity, 103
  electrical resistivity, 103
  ground penetrating radar (GPR), 103
  micro-core testing, 15
  penetration resistance, 15
  pull-out test, 15, 103
  rebound hammer, 15, 103
  ultrasonic pulse velocity, 15, 103
Number of cores
  influence (of), 49, 62, 202, 335
  minimum, 84, 304
  recommended, 52, 53, 64, 81, 313
  recommended (tables), 86, 90, 288, 313, 368

O
Outliers, 37
  definition, 162
  identification, 163
  processing of, 174

P
Precision, 335

R
Real case study, 125, 141, 304
Reinforcement
  influence, 18
Repeatability, 331
Reproducibility, 331
Risk

of wrong assessment, 59
Risk curves
  definition, 63
  modelling, 73
  use, 72
Risk model, 75

S
Simulation
  principles of synthetic simulations, 65, 360
  of investigation strategies, 196, 231, 280
  of synthetic structures, 118, 191, 226, 280
SonReb, 106, 304, 374
Statistical methods
  analysis of distribution, 266
  Anova, 125
  cross validation, 141, 296
  robust statistics, 166
Strength
  local strength, 48
  mean, 48, 90
  reference strength, 15
  standard deviation, 48, 90
Strength assessment
  precision, 90
  process, 6
  risk, 52
  requirements, 51, 84, 90, 371
Strength estimation process, 59

T
Target tolerance interval, 6, 52, 90
Temperature
  influence, 103
Test location, 21
Test reading, 21
Test region, 21, 51
  characterization, 255, 262, 313
  identification, 118, 234, 262, 287, 304, 309
Test result, 21
Test result precision, 33, 51, 360
  assessment, 35, 53, 274, 285, 313, 360
  for pull-out, 35
  for rebound, 35, 285
  for ultrasonic pulse velocity, 35, 285
  influence of, 62, 69, 86, 202, 342, 368

U
Uncertainty
    of the conversion model, 59, 215, 343

on estimated strength, 59, 296, 343
on NDT test results, 59, 215

Printed in the United States
by Baker & Taylor Publisher Services